Mass Spectrometry/ Mass Spectrometry: Techniques and Applications of Tandem Mass Spectrometry

Mass Spectrometry/ Mass Spectrometry: Techniques and Applications of Tandem Mass Spectrometry

Kenneth L. Busch

Department of Chemistry, Indiana University, Bloomington, IN 47405

Gary L. Glish and Scott A. McLuckey

Analytical Chemistry Division, Oak Ridge National Laboratory, Oak Ridge, TN 37831

VCH

Kenneth L. Busch
Department of Chemistry
Indiana University
Bloomington, IN 47405

Gary L. Glish and Scott A. McLuckey
Analytical Chemistry Division
Oak Ridge National Laboratory
Oak Ridge, TN 37831

Library of Congress Cataloging-in-Publication Data

Busch, Kenneth L.
  Mass spectrometry/mass spectrometry.

  Bibliography: p.
  Includes index.
  1. Mass spectrometry. I. Glish, Gary L. II. McLuckey,
Scott A. III. Title.
QD96.M3B87 1988   543'.0873   88-33755
ISBN 0-89573-280-7

Printed in the United States of America.

ISBN 0-89573-275-0 VCH Publishers
ISBN 3-527-26785-9 VCH Verlagsgesellschaft

Distributed in North America by:

VCH Publishers, Inc.
220 East 23rd Street, Suite 909
New York, New York 10010

Distributed Worldwide by:

VCH Verlagsgeselleschaft mbH
P.O. Box 1260/1280
D-6940 Weinheim
Federal Republic of Germany

Acknowledgments

One emphasis of this book is the diversity apparent in the techniques and application of MS/MS. A similar diversity characterizes the individuals who have helped the authors in the completion of this book. Foremost among these individuals is Prof. R. Graham Cooks of Purdue University, in whose lab all the authors spent informative and enjoyable years. We also acknowledge Dr. David H. Smith of Oak Ridge National Laboratory and Prof. Thomas A. Lehman of Bethel College, each of whom provided detailed reviews of several chapters. The contributions of the students of KLB at Indiana University and the colleagues of GLG and SAM at Oak Ridge are also acknowledged.

SAM thanks his wife for her considerable understanding during preparation of the manuscript, which proceeded in parallel with their courtship. KLB and GLG each thank their wife and daughter for putting up with us through the term of this project. We each also acknowledge a new son. In the last months, the completion schedules of the boys and that of the book were identical. Both boys arrived on time; the text followed by a few sleepless weeks. Finally, we would like to thank Ed Immergut and Eleanor Riemer at VCH for their help and patience.

Preface

Mass spectrometry/mass spectrometry is entering its second decade of application in the area of complex mixture analysis and its third decade of use in basic studies of ion chemistry in the gas phase. As many applications become routine, even to the point of "push-button" MS/MS experiments, others continue to demand the most sophisticated instrumentation, the most innovative experiments, and the most dedicated and skillful scientists.

MS/MS is notable for the diversity of its forms. In keeping with the theme of diversity, this text is intended to be useful to the novice desiring to learn about the technique of MS/MS, as well as to current practitioners of the technique. For those in the latter category, it is hoped this text will stimulate new ideas for experiments. For the authors, these new ideas were diversions from the completion of this project. For readers already familiar with MS/MS, the various chapters can be read in any order. However, for novices who desire to develop a solid foundation in MS/MS, it is suggested that the text be read in the order of presentation.

Chapter 1 provides a brief historical view of MS/MS, followed by an overview of the basic concept of the technique. This chapter will show that, while MS/MS experiments can appear to be complex, the basic tenets are simple, and that all MS/MS experiments are derivations from elementary concepts.

Chapter 2 discusses the instrumentation used in MS/MS with an emphasis on the various analyzers used to acquire MS/MS data. We feel that this comprehension is necessary to understand the options available with today's complex instruments and to take advantage of the full capabilities of MS/MS.

In Chapter 3, the different types of reactions that are used in MS/MS are discussed. This chapter covers the chemical and physical bases, to the degree that they are currently understood, for the various reactions used in MS/MS. Reading this chapter may not be necessary for someone interested in a specific problem. However, an understanding of the underlying physics and chemistry governing reactions in an MS/MS experiment is clearly important in evaluation of the data obtained.

Chapter 4 presents examples of the use of MS/MS in fundamental studies of chemical systems. This chapter is meant to show the wide range of MS/MS applications in organic and physical chemistry research.

Chapters 5 and 6 cover analytical aspects of MS/MS. Chapter 5 considers sampling and ionization methodology in MS/MS and analytical concerns in the interpretation of

MS/MS data. Sampling considerations, in particular, can be of critical importance, since improper sampling can lead to the failure of what should be a successful experiment. The same is true for choices to be made among the ionization methods, although the care that must be exercised in this selection has been more widely recognized. Chapter 6 provides a wide variety of examples of analytical applications of MS/MS in different areas.

Chapter 7 summarizes likely areas of emphasis in future MS/MS research. The number of specific problems to which MS/MS can yet be applied is immense. The applications are reported at such a rate that this book could profitably be updated every few years. In assessing the future of MS/MS, certainly we will have missed anticipating some important upcoming developments, but we hope that this book may play a role in sparking the ideas for some of them.

Table of Contents

Mass Spectrometry/ Mass Spectrometry: Techniques and Applications of Tandem Mass Spectrometry

Introduction to Mass Spectrometry/ Mass Spectrometry

1.1. History of Mass Spectrometry/Mass Spectrometry

"I feel sure that there are many problems in Chemistry which could be solved with far greater ease by this than by any other method. The method is surprisingly sensitive—more so even than that of spectrum analysis—requires an infinitesimal amount of material and does not require this to be specially purified: ...."[1] While this statement was made by Thomson about the technique of mass spectrometry, it may be even more apt in describing mass spectrometry/mass spectrometry (MS/MS). If MS/MS is defined as the detection of ions that, after their initial formation in the source, have undergone a change in mass and/or charge during the course of their analysis with a mass spectrometer, it can be argued that Thomson, considered by many to be the father of mass spectrometry, was also the forefather of MS/MS.

In Thomson's day, the pressures within the mass spectrometer were relatively high by today's standards, and numerous ion/molecule interactions took place as a result. A mass spectrum of that era contained not only the singly and multiply charged ions generated in the "ion source," but also many products due to these ion/molecule reactions. Such ions included products of the dissociations of metastable (Section 3.2) and collisionally activated ions (Section 3.3), fast neutrals from charge exchange processes (Section 3.5.1.1), as well as ionic products of charge inversion (Section 3.5.2.1), charge stripping (Section 3.5.2.2), and collisional ionization (Section 3.5.2.3). The nature of each of these reactions, with the exception of the metastable and collision-induced dissociation processes, was identified by Thomson. To demonstrate experimentally the processes of neutralization and collisional ionization, Thomson built the first MS/MS instrument, consisting of two magnets in series, with the field of one magnet oriented perpendicularly to the other.[2]

For several decades following Thomson's pioneering work in mass spectrometry, much effort went into the attainment of a better vacuum to prevent the "problem" that produced the extra peaks in Thomson's mass spectra. During this time, Aston and Dempster were among the leading practitioners of mass spectrometry, using the technique to determine the exact masses and relative abundances of the isotopes of the elements.[3,4] During this period, there was no exploration of the reactions now grouped under the term MS/MS. In addition, there was little progress in the application of mass spectrometry to organic analysis at all. In fact, there was some debate over what use, if any, mass spectrometry would have in chemical analysis once the masses and abundances of the isotopes had all been determined.

The next major development leading to the technique of MS/MS occurred in 1945 when Hipple and Condon observed and explained the presence of metastable ions in a mass spectrum.[5] After this, several investigators began the required fundamental studies on the use of metastable ions as a source of chemical and physical information. This work was the driving force that has led to MS/MS as it is practiced today.

In the early 1960s, the first experiments were performed in which instruments were used in unconventional modes to study the metastable decompositions of ions. These studies were made possible by developments in two different areas. The first development was the introduction of the accelerating-voltage scan on sector instruments by Barber and Elliott.[6] Concurrently, time-of-flight instruments were modified to separate metastable ions from the stable ions by the use of retarding grids and drift spaces.[7,8] The next major development was the discovery of the enhancement of magnitude and quantity of peaks in a "metastable ion" spectrum upon introduction of a collision gas into a localized region of the mass spectrometer.[9-11] In some ways, this brought mass spectrometry full circle in that a gas was purposefully introduced into a mass spectrometer to cause reactions that Thomson had observed because of the poor vacuum of his time. The fragmentation of a polyatomic ion following an energetic collision with a target gas is referred to as collision-induced dissociation (CID). At the time of the initial experiments using CID, the focus of the studies was very narrow, and directed primarily toward exploration of the physical aspects of the phenomenon. Not until a few years later, at the time of publication of the book *Metastable Ions*,[12] did MS/MS start to gather momentum. Publication of this book marked the beginning of the modern era of MS/MS, a period in which instruments would be designed expressly for MS/MS experiments. The availability of that instrumentation led to the rapid transformation and expansion of the technique from predominantly physical/organic investigations using metastable ions to a wide variety of new applications, especially those of analytical focus.

Demonstrations of the utility of MS/MS in analytical applications provided continuing impetus for instrument development, including the incorporation of computers into more automated MS/MS instruments, and the exploration of a diverse mix of instrument types and configurations. The next chapter focuses in detail on these developments with respect to the configuration of the MS/MS instrument. It is worthwhile to note that many of the instruments that have been constructed during this time are similar in basic geometry to instruments previously assembled for fundamental physics research. However, this new generation of instruments was designed for fundamentally different experiments. While the application of MS/MS to analytical problems helped spur instrument development, the converse is also true. The relationship between instrumentation and applications is genuinely symbiotic. Many newer fundamental applications have evolved as MS/MS instrumentation has become more sophisticated and more accessible. As these new applications have developed, the division between these MS/MS experiments and experiments of a more fundamental nature using similar techniques has become blurred. It is unfortunate that MS/MS experiments that are conceptually similar are described by names that vary with the scientific constituency. Although the fundamental applications of metastable ion study and ion structure characterization provided the base from which MS/MS was to develop, the analytical applications are currently more responsible for its present widespread recognition and use. The wide variety of analytical applications of MS/MS is discussed in Chapter 6.

1.2. Concepts and Principles

This section begins with the most fundamental discussion of ion dissociation in MS/MS, redefining terms that have already been introduced in a more general context. While the instrumentation and applications of MS/MS are extremely varied, there is a single basic concept involved: the measurement of the mass-to-charge ratios (or a related quantity) of ions before and after reaction within the mass spectrometer. The most commonly measured process, by far, is a change in mass. This is generically depicted in equation (1-1) (the charges on the ions are arbitrarily represented as single positive charges, and the values of the masses are defined by $m_p = m_d + m_n$):

$$m_p^+ \rightarrow m_d^+ + m_n \tag{1-1}$$

Generally, m_p^+ is referred to as the parent ion. There is a difference between the term parent ion and molecular ion. The molecular ion is the ion of mass equivalent to the molecular weight of the compound introduced into the ion source. While the molecular ion can be selected as a parent ion in an MS/MS experiment, parent ions need not be molecular ions. Fragment ions formed in the ion source are often selected as parent ions in MS/MS experiments.

The species m_d^+ in equation (1-1) is usually termed the daughter ion to indicate a direct relationship between it and the parent ion. The last species in equation (1-1), m_n, is the neutral fragment formed in the dissociation reaction. Since mass spectrometers can only analyze charged particles, this species is not generally detected. However, since m_p^+ and m_d^+ are measured, the value of m_n is inferred. Also, while m_n is generally written as a single neutral species, in many cases there may well be two or more neutral fragments formed in a reaction, with the sum of their masses being equal to m_n.

The basic MS/MS experiment can be described as mass selection of a parent ion in the first stage of analysis within the instrument, and then analysis of the daughter ion(s), often formed in a CID process, in the second stage of analysis. Of the three species in equation (1-1), m_p^+, m_d^+ and m_n, usually only the daughter ion, m_d^+, is actually detected. However, the first mass analyzer is *set* to pass m_p^+, and the relationship of equation (1-1) determines m_n implicitly. An activation barrier must be surmounted before the general reaction depicted in equation (1-1) can occur. The energy to overcome this barrier, the critical energy, can come from one of two sources. The first of these is excess energy deposited in the parent ion during the ionization process. An ion with the appropriate dissociation rate versus internal energy content (see Section 3.2) may survive long enough to be extracted from the ion source before it fragments, but may then fragment prior to detection. Such ions are called *metastable ions*. Metastable ion dissociations can be detected when they occur in certain portions of the mass spectrometer called reaction regions.

The percentage of metastable ions relative to all ions leaving the ion source is generally very small, and the time the ions spend in the reaction regions is generally only a small percentage of the total transit time through the mass spectrometer. Since metastable ions are characterized by a relatively narrow range of internal energies (the more energetic range of ions will fragment in the source, and the less energetic will not fragment), there are few fragmentation routes that are available for their decomposition. To increase the percentage of dissociating ions and also the number of fragmentation routes accessible to a given ion, means of adding internal energy to the parent ion in a reaction region have been devised. This is the second means by which an energy barrier may be overcome. The most common method is to admit a collision gas to a reaction region to induce dissociations by collision of the parent ion with the neutral target gas. This process, known as collision-induced dissociation (CID), can be considered to be the heart of MS/MS [equation (1-2)].

$$m_p^+ \xrightarrow{N} m_d^+ + m_n \tag{1-2}$$

Here the only difference from equation (1-1) is the extra energy imparted to m_p^+, in this case by collision with a neutral target gas (N).

The CID process results from the conversion of translational energy of the parent ion into internal energy by collision with a neutral target gas (N) admitted to a reaction region. Thus, the population of ions sampled in the CID process is different than that involved in metastable decompositions. Those ions that undergo dissociation in a CID process are those that would normally be stable; they would not dissociate without collisional activation. Details of collisional activation (CA) and CID processes are discussed in Chapter 3.

While equations (1-1) and (1-2) depict the most common reaction in MS/MS experiments, charge permutation reactions (those that involve changes in charge) can also play an important role in MS/MS experiments. A common charge permutation reaction, charge stripping, is shown in equation (1-3).

$$m_p^+ \xrightarrow{N} m_p^{2+} + e^- \tag{1-3}$$

Equation (1-3) involves the loss of an electron from the parent ion rather than the loss of a neutral fragment as shown in equation (1-2). The loss of an electron from a positive ion usually requires substantially more energy than does a dissociative process. This higher activation energy requirement results in much lower cross sections for charge stripping than for CID; implications of this difference in MS/MS reactions for ion structural studies are discussed in Chapter 4. Often the reaction depicted in equation (1-3) will be accompanied by loss of hydrogen(s) from the doubly charged ions, a dissociation prompted by the high activation energy. These fragment ions usually provide additional ion structural information in charge-stripping spectra.

A combination of charge permutation reactions occurs in the technique known as neutralization–reionization.[13] Here the parent ion is first neutralized and subsequently reionized in a collision as depicted in equation (1-4).

$$m_p^+ \xrightarrow{N_x} m_p \xrightarrow{N_y} m_d^+ + m_n \tag{1-4}$$

In addition to formation of neutral parent ions, neutral daughter ions can also be formed and reionized; the measured spectrum is a convolution of the results of these processes. Neutralization–reionization has proven to be very useful in detailed ion structural studies, as discussed in more detail in Chapter 4.

Another common MS/MS reaction includes a combination of the change in mass of equation (1-2) and the change in charge of equation (1-3). The combined process is generally referred to as charge inversion, and the most common form of the reaction is shown generically in equation (1-5).

$$m_p^- \rightarrow m_d^+ + m_n + 2e^- \tag{1-5}$$

Note that when the parent ion is negatively charged (as shown), this reaction is also a form of the charge-stripping reaction. The reaction shown may also proceed via equation (1-4), and in some cases the positively charged parent ion is observed. In this reaction, positively charged daughter ions from a negatively charged parent ion are detected after collisional activation. Such reactions have proven to be very useful in certain analytical applications discussed in Chapter 6. The fundamentals of charge permutation reactions are discussed in Section 3.5.

The energetics of the reactions depicted in equations (1-2), (1-3), and (1-4) are different. Whether or not a particular reaction is observed depends upon the kinetic energy of the ion when it collides with the target gas N. The kinetic energy of the parent ion is usually determined by the type of instrument used to perform the MS/MS experiment. As discussed in Chapter 2, for instruments in which the ions have relatively low kinetic energies, the energy of the collisional activation process may not be sufficient to access charge permutation reactions. However, in these low-energy instruments, a variant of the reaction shown in equation (1-2) is possible. If the target gas is a chemically reactive species, then instead of the condition that $m_p = m_d + m_n$, a different algebraic relationship, $m_p + N = m_d + m_n$, can hold in an ion/molecule *association* reaction. In such an experiment, the target gas is a chemical participant in the reaction. An ion/molecule reaction of the parent ion with the target gas can lead to daughter ions of masses greater than the parent ion. Dynamical aspects of this reaction are covered in Section 3.4, and applications of such reactions can be both of fundamental interest (Chapter 4) and analytical application (Chapter 6).

Most MS/MS experiments fit into the category described by equation (1-2). Any one of three parameters, the mass-to-charge ratio of the parent ion, the mass-to-charge ratio of the daughter ion, or the mass of the neutral fragment(s), can be designated as the independent variable in an MS/MS experiment. It is this flexibility that makes MS/MS such a powerful analytical technique.

The most common mode of MS/MS operation is the *daughter ion scan*. In this experiment, the parent ion mass is fixed and the masses of all daughter ions formed from that parent ion are measured, i.e., m_{d1}^+, m_{d2}^+, or m_{d3}^+. The daughter ion scan is essentially the mass spectrum of an ion in a mass spectrum, as shown graphically in Figure 1-1. The first stage of mass analysis selects ions of a particular m/z (the parent ion) from all ions initially formed (which taken together make up the normal mass spectrum). This mass-selected parent ion is passed onto the reaction region. Daughter

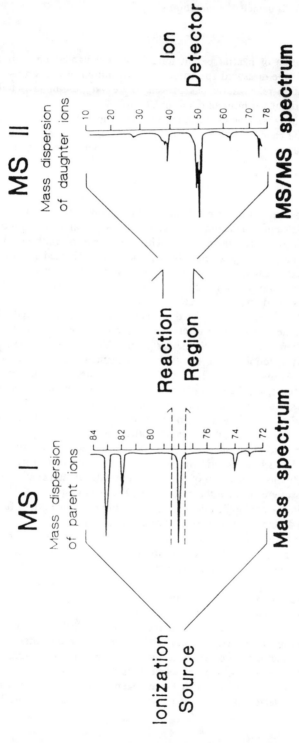

Figure 1-1 ■ A diagram of an MS/MS experiment depicting the two different stages of mass dispersion.

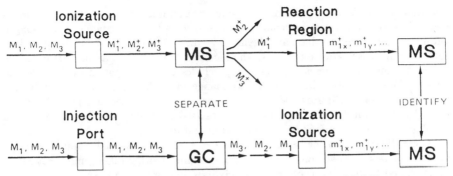

Figure 1-2 ■ A schematic comparison of an MS/MS experiment with a GC/MS experiment.

ions formed from dissociations of the parent ion in a reaction region are then analyzed by mass. These daughter ions are generally characteristic of the structure of the parent ion and thus provide a means of determining that structure. The daughter ion scan is used in almost all fundamental applications of MS/MS (Chapter 4) as well as in the majority of analytical applications (Chapter 6). Other scan modes are becoming more popular, however, and add to the advantages MS/MS has over other analytical techniques.

Early descriptions of the analytical characteristics of MS/MS contrasted the method with gas chromatography/mass spectrometry (GC/MS). Modern MS/MS includes many experiments that have no direct counterpart in GC/MS, but a summary of the comparison between daughter ion MS/MS and GC/MS will be included here since it highlights the function of parent ion selection in MS/MS as analogous to separation of compounds by gas chromatography, and the nature of the information provided in an MS/MS experiment. Figure 1-2 shows a schematic comparison of GC/MS with a daughter ion MS/MS experiment. In GC/MS, the first step of the analysis is the injection of the sample mixture onto the GC column. The analogous step in MS/MS is the introduction of the sample mixture into the ion source. The second step in both analytical schemes is the separation of one component of the mixture from the other components. In GC/MS, the separation of the components occurs in time as each sequentially elutes from the column. The retention time is a function of the chemical composition of the analyte, the carrier gas, and the stationary phase. Separation in time in GC/MS is contrasted to separation in space in MS/MS. Ions are selected on the basis of a physical property (mass-to-charge ratio); this property is independent of the type of mass spectrometer being used. These differences in the separation step are important to the relative capabilities and applications of GC/MS and MS/MS.

In GC/MS, once a specific analyte is separated from the rest of the mixture, it is transferred into the ion source of the mass spectrometer to be ionized. In MS/MS, the separated species is transferred into a reaction region in which it typically undergoes collision-induced dissociation. The last step, identification, is essentially the same in both techniques. This step involves analysis of the ions formed in the previous step (i.e., by ionization or CID) with a mass analyzer. A gas-chromatographic separation involves a specific type of column, stationary phase, and mobile phase; for complex mixtures, the exact order of retention times established in one GC separation may be difficult to duplicate with different instruments with similar but not identical experimental parameters. However, flow rates and temperatures used for a GC separation can be changed within a wide range of values, an advantage in the use of GC for difficult separations such as that of isomeric compounds. The fact that MS/MS "separation" is based on physical rather than chemical properties precludes direct separation of isomers, but balances this weakness with an ability to change the order of analysis for individual mixture components during the course of the experiment, and a higher degree of reproducibility on different instruments. Furthermore, in MS/MS, there is no "dead time" between the "elution" of one compound and the next.

These differences in the separation step make MS/MS especially appropriate for analysis of targeted compounds in complex mixtures. On the other hand, GC/MS is better suited to the analysis of unknown mixtures, due in part to the existence of large reference libraries of mass spectra to help in the identification. Such extensive libraries of MS/MS spectra do not exist at the present (see Chapter 5), and spectra must be interpreted from first principles or compared with spectra of standard compounds if these are available. Similarly, libraries of liquid chromatography/mass spectrometry (LC/MS) are not widely available, but the method is developing into a routine applications tool quite rapidly. However, until MS/MS libraries are compiled and standardized operating conditions are developed, GC/MS usually will be a more powerful, albeit slower, instrumental method for analysis of unknown compounds in mixtures.

The ionization method selected for an MS/MS experiment, as for a GC/MS or LC/MS analysis, offers the analyst different options that confer special sensitivity or selectivity. While electron ionization (EI) is the most common general ionization technique in mass spectrometry, chemical ionization (CI) is often preferred in MS/MS. These two ionization techniques have been used in the majority of MS/MS applications to date with electron ionization predominant in studies of ion structure, and chemical ionization widely used in mixture analyses. However, newer ionization techniques are being increasingly used, including, but not limited to, the desorption ionization techniques of fast atom bombardment (FAB), secondary ion mass spectrometry (SIMS), and laser desorption (LD), and nebulization ionization techniques such as thermospray, electrospray, and liquid ionization.

The practical requirements for an ionization method are discussed in Chapter 5. One major criterion must be fulfilled regardless of which ionization technique is used: for any compound in the source, the ion formed must be representative in structure of its neutral precursor. Ionization processes that induce an isomerization of a compound in the ion source should be avoided. While isomerization in the ion source does not preclude analysis by MS/MS, it can lead to erroneous compound identification. To date, such problems have been very rarely reported; however, this may be more a result of a failure to recognize such effects rather than their nonoccurrence, especially in chemically reactive environments such as chemical ionization sources.

A second desirable criterion for an ionization method in MS/MS, especially for mixture analysis, is that each compound produce as few (ideally only one) ions of different mass-to-charge ratio as possible upon ionization. The advantage is a simpler mass spectrum with fewer interferences in the selection of the desired parent ions. Ideally, each of the ions in the mass spectrum corresponds to a component in the mixture. For this reason, the "soft" ionization techniques, such as chemical ionization, which tend to give few fragment ions, are preferable to "harder" ionization techniques, such as electron ionization. Increased sensitivity and lower detection limits are also usually achieved if all the ion current for a given compound is concentrated in a single ion and not diluted in ions of several different m/z values, many of which may not be characteristic of the neutral molecule.

A unique analytical aspect of MS/MS is the ability to screen unknown mixtures rapidly for compound *classes*. Referring to the basic MS/MS reaction given in equation (1-2), any of the three species can be designated as the independent variable in an MS/MS experiment. The preceding comparison of GC/MS and MS/MS focused only on cases in which the parent ion, m_p^+, was the independent variable. If, however, the daughter ion, m_d^+, or the neutral fragment, m_n, is specified as the independent variable, new information is available in the MS/MS experiment that has no GC/MS analog. The functional group specificity of these parent ion and neutral loss MS/MS experiments are apparent in the applications outlined in Chapter 6.

When an ion dissociates, the stabilities of the daughter ion and neutral fragment both influence fragmentation routes. Many ions dissociate by loss of a small neutral molecule associated with a functional group or substructure of the parent ion. For example, nitroaromatic cations are well known for the loss of NO$\cdot$. (30 daltons) from the molecular ion $M^+\cdot$ or from the protonated molecule $(M + H)^+$. An MS/MS experiment can be designed to specify $m_n = 30$; this is termed a *neutral loss* scan. The data so obtained identify *all* parent ions formed from a complex mixture that fragment by loss of 30 daltons. The data are acquired in a single scan of the mass spectrometer. The ions detected are then identified as candidate nitroaromatic compounds. Confirmation of this functional group identity can then be obtained by recording daughter ion MS/MS spectra of the parent ions identified in the neutral loss scan. A significant advantage in speed of analysis is gained by avoiding the need to identify every component in a mixture if only nitroaromatic compounds are of interest.

An experiment in which m_d^+ is the independent variable is known as a *parent ion* scan. A compound class may include a particular substructure that forms a very stable ion. Parent ions derived from members of this class tend, therefore, to dissociate via processes that lead to the ion of this substructure. Thus, for cases in which there is a common daughter ion, the daughter ion can be fixed as the independent variable and a scan performed to detect all its parent ions.

A final aspect of MS/MS to be discussed is that of the sensitivity of the technique. As Section 5.1 details, sample preparation methods used in GC/MS or LC/MS often include a sample concentration step. By contrast, in MS/MS all species are continually "consumed" during the analysis of each individual component. However, MS/MS provides detection limits comparable to GC/MS when the analysis is targeted to a few specific compounds. MS/MS provides better detection limits than a single stage of mass spectrometry when several compounds are introduced into the ion source at the same time. Since ions are obviously lost in each stage of mass analysis, it would seem that the addition of the second stage of mass analysis should result in a decrease in sensitivity and poorer detection limits. However, in most instances the sensitivity and detection limits of a mass spectrometer are not limited by detector noise but by "chemical" noise.

Chemical noise is defined as that signal measured in the mass spectrum due to generation of ions from other components in a sample, and "background" contamination of the sample, the sample introduction system, and the ion source of the mass spectrometer. At high sensitivities, an ion signal at each mass usually can be seen even

Mass Spectrum

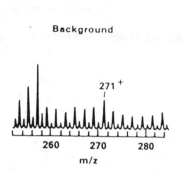

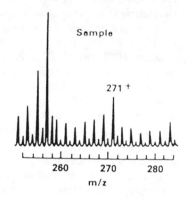

MS/MS Spectrum

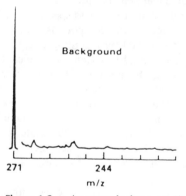

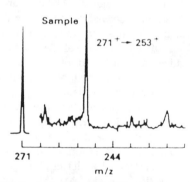

Figure 1-3 ■ An example demonstrating the high signal-to-noise ratio in an MS/MS spectrum vs. a mass spectrum due to the elimination of the "chemical noise" in the MS/MS spectrum.

when no sample is explicitly present. Thus, while the detection electronics of a mass spectrometer are sufficiently sensitive to operate with signal fluxes of less than one detected ion per second, the signal from background contaminants—chemical noise—is usually much higher than this level. Improved sensitivity and detection limits for MS/MS arise because the first stage of mass analysis "filters" out much of this chemical noise. Figure 1-3 gives an example of this, comparing background mass spectra and background MS/MS spectra with spectra obtained under identical conditions when a small amount of sample is introduced into the ion source. The improved signal-to-noise ratios for the signals observed in the MS/MS experiment are apparent, and this improvement is the source of the lower detection limits attainable with MS/MS.

To achieve the lowest possible detection limits for a targeted compound in daughter ion MS/MS, an experiment known as single-reaction monitoring (SRM) is performed. In this experiment, all the variables in equation (1-2) are fixed. For the example given in Figure 1-3, a single-reaction monitoring experiment would result when the first mass analyzer was set to pass the parent ion m/z 271 and the second mass analyzer was set to pass the daughter ion m/z 253. This single-reaction monitoring experiment is analogous to single-ion monitoring (SIM) used in GC/MS. While it is apparent that single-reaction monitoring offers less specificity than a complete MS/MS spectrum, it nonetheless provides increased specificity over a single stage of mass spectrometric analysis. Generation of a signal at the detector in a single-reaction monitoring MS/MS experiment requires that *two* criteria be met: the mass-to-charge ratio of the parent ion and the mass-to-charge ratio of the daughter ion must satisfy the selected values.

With some instruments, specificity can be enhanced by taking the analysis one step further, as in an MS/MS/MS experiment. Here the selected daughter ion fragments in a second reaction region, and a granddaughter ion is selected by a third stage of mass analysis. In this type of experiment, three criteria, all related, must be met to produce a signal at the detector.

Theoretically, additional stages of mass analysis can be added to the experiment ad infinitum. A recent morphological analysis of all possible combinations of MS/MS scans shows that the number of unique and useful scans increases exponentially (2^n where n is the number of stages of mass analysis).[14] In the course of this analysis, the scan mode where $m_d^+ = f(m_p^+)$ has been identified as a generalization of the (fixed) neutral loss scan.[15] Thus, the complexity of the experiment increases as additional analysis stages are added. Such experiments potentially offer much more information than the usual MS/MS experiments, and will increasingly be used in studies of complex multistep ion chemistry. However, since signal is decreased by transmission losses through each stage of the experiment, the practical limit of MS/MS/MS... is likely to be three to four stages of mass analysis in beam instruments, and perhaps seven or eight individual steps in ion-trapping instruments.

The flexibility of MS/MS as a general analytical method is apparent in several aspects of the experiment. The first of these is the range of molecules to which MS/MS can be applied. This range is limited only by the ability to form a gas-phase ion characteristic of the neutral molecule and the capability to induce a subsequent reaction. The new desorption ionization methods have made the more significant limit the ability to induce fragmentation in very large ions with molecular weights of thousands of daltons.

The second area in which MS/MS offers numerous options is in the type of data obtained. In the basic MS/MS reaction [equation (1-2)], there are three species, any one of which can be designated as the independent variable in an MS/MS experiment. Table 1-1 shows the relationships between the various parameters for the different MS/MS experiments. Additional flexibility in the type of data acquired results from the ability to study and use charge permutation reactions and ion/molecule association reactions. Another major area of flexibility in MS/MS is the instrument design itself. There is a wide variety of useful instrumental configurations used for MS/MS, each offering certain advantages. Given a specified problem to solve, an MS/MS instrument can be designed to optimize the capabilities appropriate to solution of that problem. This aspect will be discussed in detail in the next chapter.

Table 1-1 ■ Fixed and Variable Parameters for Various MS/MS Experiments

Scan	m_p^+	m_d^+	m_n
Daughter ion	Fix	Vary	Vary
Parent ion	Vary	Fix	Vary
Neutral loss	Vary	Vary	Fix
Single-reaction monitoring	Fix	Fix	Fix

1.3. Nomenclature

Mass spectrometry in general, and MS/MS in particular, is burdened with acronyms and multiple shorthand designations for experiments and procedures. Considerable debate continues over the use of acronyms and specialized nomenclature; if used in moderation and explicitly defined, acronyms and technical nomenclature are undeniably convenient. It is appropriate to define at this time some of the nomenclature and acronyms used throughout this book and to equate these with other terms that are encountered in the MS/MS literature. There is no standard nomenclature for MS/MS; the terms used in this text are those that seem most prevalent or most descriptive.

We begin with the acronym for the technique itself, *MS/MS*. Several other terms appear in the literature that describe the identical experiment. A commonly used alternative is *tandem mass spectrometry*. The acronym MS/MS is consistent with the general trend in abbreviating the so-called "hyphenated" techniques, such as GC/MS, GC/IR, and LC/MS. MS/MS is also likely to remain unique in the fields of chemistry and physics, unlike the acronym for tandem mass spectrometry, TMS, which is in common use as the abbreviation of trimethylsilyl, and has also been recently used as an acronym for thermospray ionization mass spectrometry. Several other acronyms have been used to describe particular MS/MS experiments. These are: MIKES, for mass-analyzed ion kinetic energy spectrometry; CAMS, for collisional activation mass spectrometry; and DADI, for direct analysis of daughter ions. These terms were coined in the early days of MS/MS and have generally been used to describe only daughter ion MS/MS, usually on a sector instrument consisting of a magnetic sector followed by an electric sector.

The process in which an ion collides with a target gas and a portion of the ion translational energy is converted into internal energy of the ion is referred to as *collisional activation* (CA). In cases in which there is subsequent dissociation of that ion in the same reaction region, the process is termed *collision-induced dissociation* (CID). Many authors used the alternative collision-activated dissocation (CAD). Note that collisional activation has a different, discrete meaning as used in this text.

A *reaction region* is a region in the mass spectrometer, usually between analyzers in a sector instrument, in which an ion can undergo a change in mass and/or charge. In sector-based mass spectrometers, this reaction region is often referred to as a "field-free region." Other MS/MS instruments have reaction regions in which electric fields are used to constrain the fragment ions, and therefore reaction region is a more general term than field-free region.

Parent ion and *daughter ion* were defined in the previous section. They are sometimes referred to as the precursor ion and fragment ion, respectively, but, parent ion and daughter ion are the more commonly used terms. Daughter ion is a more accurate term than is fragment ion, which can be any ion in a mass spectrum of lower mass than the molecular ion. A daughter ion is the result of a dissociation of a specified parent ion.

Many different types of charged-particle analyzers are used in mass spectrometry, and most have been used in MS/MS. A shorthand method to describe instrument configuration has been generally adopted in the literature in which B represents a magnetic sector, E an electric sector, and Q a quadrupole mass filter. A time-of-flight analyzer is designated as TOF, and a Wien filter, although rarely used in MS/MS so far, is designated by a W. The order of the symbols represents the order of the analyzers traversed by the ion beam. Using this terminology, a conventional double-focusing mass spectrometer, constructed as an electric sector followed by a magnetic sector, is designated as an EB instrument.

Since most MS/MS instruments have been developed recently, there are few alternative designations in the literature for them. An exception is instruments of BE geometry, often referred to as "reverse" geometry mass spectrometers, since the order of the sectors is reversed from that used conventionally in mass spectrometry. The BE geometry instrument has also been termed a MIKE spectrometer, but this term is less frequently used in current literature.

Nomenclature that describes instruments based on multiple quadrupole mass filters can also be ambiguous. In instruments with multiple quadrupoles, one quadrupole is usually used in a nonmass-filtering mode as a reaction region to focus daughter ions. Thus, in a triple quadrupole instrument (QQQ), only the first and third quadrupoles are actually used as mass filters. The general literature does not differentiate the two different uses of the quadrupoles in the acronym, although occassionally a lower case "q" is used to denote the collision quadrupole. The usual "QQQ" convention will be followed in this text, making sure to note explicitly quadrupoles not used as mass filters.

Some of the analyzers used in mass spectrometry (B, E, TOF, and W) generally operate on ions with keV energies while others (Q and W) operate on ions with eV energies. An MS/MS instrument that includes analyzers from more than one category is generally referred to as a hybrid instrument. Other abbreviations and acronyms will be defined as they appear.

2

MS/MS Instrumentation

2.1. Introduction

Mass analysis in mass spectrometry is usually accomplished with one of a few basic types of instrument—the quadrupole mass filter or the single magnetic sector for low-resolution work, and Nier–Johnson and Mattauch–Herzog geometry double-focusing instruments for exact mass analysis. The latter double-focusing instruments use electric sectors followed by magnetic sectors to provide direction and velocity focusing, providing the capability for high-resolution mass measurements. A factor in the predominance of these instruments, in comparison to the many other types of mass analyzers that might be constructed, is that basic mass spectrometry is conceptually simple—a measurement of ion intensity versus mass-to-charge ratio. These instruments readily provide this information, and their designs are well understood and highly optimized. There has been little need to use other types of instruments to obtain the same type of data. The popularity of these instruments vary with time. For instance, the time-of-flight mass spectrometer has enjoyed a period in which it was very popular, followed by a period of near extinction, and now is undergoing a renaissance as new areas of mass spectrometry are developed that make use of some of the unique features of time-of-flight analysis.

In contrast, although MS/MS is relatively new, there is already a wide variety of instrumentation in use. This greater diversity of MS/MS instruments is a result of the broader range of types of data that can be obtained in an MS/MS experiment, and the fact that there are also several approaches to each experiment, depending upon the exact form of the data desired. There are more instrumental parameters to consider in MS/MS instrumentation (such as collision energy, parent ion resolution, and daughter ion resolution) than with conventional mass spectrometry instrumentation; certain types of analyzers enhance certain characteristics of the MS/MS spectrum. As will be seen throughout this chapter, MS/MS instruments have been specifically designed to take advantage of certain characteristics of various analyzers for different types of experiments.

The focus of this chapter is the principles and concepts of operation of MS/MS instruments. Data systems will not be discussed, although it is important to make brief mention of them here and to recognize that they are essential in the successful implementation of many MS/MS experiments. While instrumental development has seen vigorous activity, data systems for MS/MS instruments seem to have lagged in development. In comparison to conventional mass spectrometry, software for control of the MS/MS experiments must be much more extensive, especially if the entire array of experiments is to be accessible to the operator. For a few MS/MS experiments, data systems comparable to those used on conventional mass spectrometers can be used to acquire data as long as the operator manually controls other parameters of the experiment. In other cases, complex scans may require three or four different electric and magnetic fields to be changed simultaneously and as complicated functions of each other. Realistically, this can only be accomplished under computer control. Modern networked computers introduced during the past few years may now provide the distributed control and processing speed necessary for realization of the full MS/MS potential of present-day instruments.

2.2. Principles of Charged-Particle Analysis

In an MS/MS experiment, the dissociation of a parent ion generates daughter ions with different kinetic energies and momenta than the parent ion, as dictated by the change in mass. Mass spectrometers are devices that provide a measurement of mass-to-charge ratios, but in MS/MS, consideration of the analyzer parameters of merit in terms of energy and momentum is also necessary to understand fully the advantages and disadvantages of the various configurations of MS/MS instruments. The discussion of the principles involved in the analysis of charged particles will be limited to that necessary to understand the actual physical property upon which instrument operation is based and to understand the effect of the change in mass in the MS/MS experiment on ion motion through the different types of analyzers.

The first step in most mass spectrometric experiments, following ionization, is acceleration of the ions out of the ion source into the mass analyzer. The kinetic energy of an ion after it has been fully accelerated from the ion source, assuming that the ion was formed with negligible kinetic energy, is:

$$mv^2/2 = zV \tag{2-1}$$

where m is the mass of a *single* ion (in kilograms), v is the velocity of the ion in (meters/second), z is the charge of the ion (in Coulombs) (i.e., $[z'e]$, where z' is the integral number of units of charge and e is the fundamental unit of charge, 1.6×10^{-19} C), and V is the potential drop through which the ion is accelerated (in volts). Typically, the mass of the ion and the accelerating voltage are known and equation (2-1) is used to calculate the ion velocity, rearranged as shown in equation (2-2):

$$v = (2dz'V/m')^{1/2} \tag{2-2}$$

In this equation, given the units described above, d is a constant, 9.64×10^7 ($1.6 \times 10^{-19} \times 6.023 \times 10^{26}$).

2.2.1. Electric Sectors

A cylindrical electric sector disperses ions according to their kinetic energy-to-charge ratio and thus, for ions subjected to the same acceleration potential, provides no mass separation in conventional operation. It is used as a direction-focusing element. The force on the ion acts in the direction of the electric field. The equation of motion of an ion in an electric field is:

$$mv^2/r = zE \qquad (2\text{-}3)$$

In this equation m, v, and z are as defined previously, r is the radius of deflection (in meters), and E is the electric field strength (in volts/meter). Equation 2-3 can be rearranged to give:

$$mv^2/z = Er \qquad (2\text{-}4)$$

In conventional operation at a constant electric field ($E = E_p$) and with a constant accelerating voltage $V = V_p$, all the quantities in equation (2-4) are fixed and independent of mass since, from equation (2-1), $mv^2 = 2zV$. The electric field, E_p, is set such that $r = r_c$, where r_c is the central axis of the electric sector, and ions formed in the ion source travel in a path that follows this radius of curvature.

However, if an ion fragments after it has been accelerated, the newly formed ion no longer has the same kinetic energy-to-charge ratio as the ion accelerated out of the source: the mass has changed but the velocity has remained the same (assuming fragmentation occurs in a nonaccelerating field). Thus, if the electric field strength remains fixed at E_p, the daughter ion will experience a different radius of deflection, r_x, and will not pass along the central axis, r_c, of the electric sector with the undissociated or "main" beam of ions (the parent ions, m_p^+, in an MS/MS experiment). To pass the daughter ions, m_d^+, along the central axis, the electric field must be set according to equation (2-5):

$$E_p m_d z_p / m_p z_d = E_d \qquad (2\text{-}5)$$

where z_p and z_d are the charges of the parent ion and daughter ion, respectively. This relationship follows directly from equation (2-4) and from the fact that $v_p = v_d$. There is a linear relationship between the requisite electric field strength E for passage through the analyzer and the daughter ion mass. However, there is no unique solution to equation (2-4) for the daughter ion mass *unless* the parent ion mass is known. The implications of this will be discussed in Section 2.3.1.

2.2.2. Magnetic Sectors

A magnetic sector, like an electric sector, acts as a direction-focusing element for charged particles. An ion in a magnetic field (B) travels in a circular path in a plane normal to the direction of the magnetic field. The motion of the ion is described by:

$$mv^2/r = zvB \qquad (2\text{-}6)$$

where B is in units of tesla and the other parameters are as previously defined. If this equation is rearranged, it can be seen that the magnetic sector disperses ions according to their momentum-to-charge ratios [equation (2-7)]:

$$mv/z = Br \qquad (2\text{-}7)$$

Thus, ions of different mass but the same kinetic energy, emanating from the same point (e.g., the source slit), follow different trajectories through a fixed magnetic field. At a given magnetic field strength, ions of only one mass-to-charge ratio will follow a trajectory along the central radius, r_c, of the magnetic sector. The mass-to-charge ratio of ions that follow this trajectory is determined by equation (2-8):

$$m/z = B^2 r^2 / 2V \qquad (2-8)$$

where $r = r_c$. This equation is obtained by substituting the expression for the ion velocity from equation (2-1) into equation (2-7). The mass-to-charge ratio of the ion focused along this trajectory is thus a function of the magnetic field strength and accelerating voltage. Either parameter can be varied to change the mass of the ion that follows this trajectory.

If an ion fragments after acceleration from the ion source, but prior to entering the magnetic field, equation (2-8) no longer applies. The velocity of the daughter ion, m_d^+, formed in the dissociation is the same as the parent ion, m_p^+, and is different from that of stable ions of mass-to-charge ratio m_d^+ formed in the ion source. The velocity of the daughter ion is determined from equation (2-1) and is given by equation (2-9):

$$v = \left(2zV/m_p \right)^{1/2} \qquad (2-9)$$

Substituting velocity from equation (2-9) into equation (2-7) and rearranging gives:

$$\left(m_d^2/m_p \right)/z = B^2 r^2 / 2V \qquad (2-10)$$

Thus, the daughter ion is passed through the analyzer and appears in a normal mass spectrum at an apparent mass of m^* where:

$$m^* = m_d^2/m_p \qquad (2-11)$$

An equation analogous to equation (2-5) for the electric sector may be written for the magnetic sector:

$$B_p m_d z_p / m_p z_d = B_d \qquad (2-12)$$

In contrast to the electric sector situation, there are few nonunique solutions given the condition that the charges, z_d and z_p, must be integers, and the masses, m_d and m_p, are nominally integers. This is a result of the fact that B_p is different for every different m_p; in the case of the electric sector, E_p is independent of m_p.

2.2.3. Quadrupole Mass Filters

Whereas sector instruments typically operate with acceleration voltages in the keV range, quadrupoles operate with ion kinetic energies in the range of a few to tens of electron volts. In operation, a dc voltage, V_1, and an rf voltage, $V_0 \cos t$ (t is time), are applied to the rods, with adjacent rods having opposite polarities. Ideally, the rods of the quadrupole are hyperbolic in shape, in which case the equations of force are:

$$F_y = ma_y = z(V_1 + V_0 \cos t)\left(2y/r_0^2\right) \qquad (2-13)$$

$$F_z = ma_z = -z(V_1 + V_0 \cos t)\left(2z/r_0^2\right) \qquad (2-14)$$

$$F_x = ma_x = 0 \qquad (2-15)$$

For these equations, the x direction is along the axis of the quadrupole and r_0 is the radius of the largest circle that can be inscribed within the quadrupole in the yz plane. Detailed analysis of the trajectories of ions in a quadrupole is beyond the scope of this book, but it can be seen from equations (2-13) and (2-14) that only ion mass and charge are important in describing the trajectories. A daughter ion formed in a dissociation process will be passed through the quadrupole mass filter at the same ratio of the rf and dc voltages as an ion of the same mass formed in the ion source. For quadrupoles, as opposed to magnetic and electric sectors, the force responsible for the mass-filtering action is nominally independent of velocity. Quadrupole performance is, however, dependent upon the number of rf cycles that the ion spends in the field, which is dependent upon the ion velocity. For sectors, velocity is a parameter upon which separation of ions by mass is explicitly dependent. This difference has implications that favor sectors for certain types of MS/MS experiments and quadrupoles for others, as will be seen later.

In some applications, quadrupoles are operated without a dc voltage in what is known as the "rf-only" or "total-ion" mode. In this mode, a quadrupole acts as a strong focusing lens that is analogous to a high pass filter. The amplitude of the rf voltage determines the low mass cutoff, which is slightly lower than the mass that would be transmitted through the quadrupole if the dc voltage were applied. Theoretically, all ions of mass-to-charge ratio greater than the low mass cutoff value are passed through the rf-only quadrupole. In practice, however, there is some discrimination in transmission that is dependent upon velocity[16] and upon mass, with ions of higher mass subject to a lower transmission.[17] This is discussed in more detail in Section 2.7.2.

2.2.4. Time-of-Flight Analysis

Time-of-flight (TOF) analysis is the simplest type of charged-particle analysis. The only field involved in time-of-flight analysis is the acceleration field. From equation (2-1), using a constant accelerating voltage, every ion of different mass has a different velocity. Therefore, each ion traverses a fixed distance in a different amount of time. The flight time, t (in seconds), of an ion over a distance, L (in meters), is given by:

$$t = L/v \qquad (2-16)$$

Substituting for velocity using equation (2-1) gives:

$$t = Lm^{1/2}/(2zV)^{1/2} \qquad (2-17)$$

This is the principle of time-of-flight analysis. For the case in which an ion fragments after acceleration, its velocity is unchanged; therefore in time-of-flight analysis where no other accelerating or decelerating fields are present, daughter ions formed after complete acceleration are detected with the same flight time as the parent ion. The charge of the ion can also change after acceleration without changing the ion velocity. Thus, differently charged (notably neutral) species are still detected at the time corresponding to the parent ion in a simple linear TOF instrument.

2.2.5. The Wien Filter

The Wien filter is another type of velocity analyzer (sometimes called a velocity filter). It utilizes crossed electric and magnetic fields. Since the force of an electric field is in the direction of the field and the force of a magnetic field is perpendicular to the field, the crossed fields exert opposing forces in the same plane upon ions traveling through such an arrangement. For a given mass, the fields can be adjusted so that they balance each other and there is no net force on the ion. This is described by equating the relationships (2-3) and (2-6):

$$zE = zvB \tag{2-18}$$

Equation (2-18) can be rearranged to give:

$$E/B = v \tag{2-19}$$

Thus, the ratio of the electric and magnetic fields determines the velocity and therefore, for ions of fixed kinetic energy-to-charge ratio, the mass of the ion that passes through a Wien filter without deflection. Ions of lower velocities (larger masses) experience increasing deflections in the direction of the electric field. The opposite is true for lower mass ions.

The Wien filter operates by varying one of the fields such that a null condition exists for ions of varying mass. Ions of the appropriate mass (velocity) pass along the ion axis and through the resolving slit. Like the time-of-flight analyzer, since there is no change in velocity if an ion fragments after acceleration, the resulting daughter ions are passed through the Wien filter at the field strengths appropriate to pass the parent ion.

2.2.6. Fourier Transform–Ion Cyclotron Resonance (FT–ICR)

All the previously discussed techniques for analysis of charged particles are beam techniques. The ions are continually formed in one region, extracted from that region, accelerated, and then travel through the analyzer(s), subsequently reaching a detector. In FT–ICR, analysis and detection (and often ionization) occur as discrete events in the same region, separated in time rather than space.

The ICR experiment is carried out in a cell located in a magnetic field. After the ions are formed in the ICR cell, usually by a pulse of electrons, they travel according to equation (2-6). The time, t, that it takes for an ion to make a complete revolution within the ICR cell is:

$$t = 2r/v \tag{2-20}$$

Substituting for the radius from equation (2-6) into equation (2-20) gives:

$$t = 2(m/zB) \tag{2-21}$$

Since time is inversely related to frequency, the cyclotron frequency of an ion is:

$$f_c = 1/t = Bz/2m \tag{2-22}$$

This can be rearranged to give:

$$m/z = B/2f_c \tag{2-23}$$

Thus, if the magnetic field is known and the cyclotron frequency measured, the mass-to-charge ratio of an ion is determined.

At this point, it is pertinent to mention several other aspects of the ICR experiment. At a given magnetic field strength, ions of each mass have a characteristic cyclotron frequency, f_c. If an rf voltage is applied to opposite plates of an ICR cell (those that are parallel with the magnetic field) with a frequency f_c, ions with that cyclotron frequency absorb power, increasing their velocity and radius. Early ICR experiments used this principle to determine mass-to-charge ratios by measurement of power absorption as the frequency or magnetic field was slowly scanned. With addition of enough energy to the ion, the radius can be increased to the point at which the ion collides with a cell wall. This procedure is termed ion ejection.

The FT–ICR experiment improves upon the early ICR experiment by taking advantage of the mass/frequency relationship. Conceptually, the FT–ICR experiment involves: exciting all ions with a broadband rf excitation pulse; detecting an image current (i.e., the current induced in a plate due to proximity of charged particles) of the excited ions with respect to time; transforming this varying time-domain signal into the frequency domain (via a fast Fourier transform); and then determining the masses from the relationship in equation (2-23). The image current is that current induced in the receiver plates by the cyclical motion of the ions in the cell. As a positive ion approaches a plate, for example, electrons travel in an external circuit toward that plate, and as the ions recede, the electrons travel in the other direction. The oscillating current caused by the movement of the electrons is the image current.

2.2.7. Ion Trap Mass Spectrometers

The ion trap mass spectrometer (ITMS) could be classified in the same category of mass filter as a quadrupole. It is, in fact, a three-dimensional quadrupole, and similar equations for ion motion apply, taking into account the three-dimensional characteristics of the ITMS. However, due to its newness as an analytical mass spectrometer, and especially as an MS/MS instrument, it seems appropriate to discuss the ITMS in more detail. In addition, its operation as an MS/MS instrument is quite different from the operation of instruments based on conventional quadrupoles.

While there have been several modes of operation of an ion trap, one method has recently been offered in a commercial instrument. This method, termed mass-selective instability,[18] will be discussed here. The ion trap consists of a hyperbolic cross-section center-ring electrode (a doughnut) and two hyperbolic cross-section end-cap electrodes. Figure 2-1 shows a cross section of an ITMS. To obtain a mass spectrum, an rf voltage of variable amplitude and fixed frequency is applied to the center electrode and the end caps are grounded. Since no dc voltages are used, the device can be thought of as a three-dimensional rf-only quadrupole. All ions above a certain m/z, determined by the rf amplitude, have trajectories that keep them trapped within the electrodes. As the rf amplitude is increased, ions of increasing mass sequentially become unstable in the z direction and are ejected from the trap through the end caps.

The operating sequence of the ITMS begins with the injection of a pulse of electrons into the ion trap to ionize a gaseous sample. The ring electrode is held at a low rf amplitude to trap all of the ions formed during this ionization pulse. The rf amplitude is then scanned upward, and ions of increasing mass are sequentially ejected from the ion trap and detected with an electron multiplier. A key parameter in the operation of the ion trap is the background gas pressure. Typically, helium is added as a bath gas at a pressure of about 10^{-3} Torr. Collisions of the ions with the bath gas damp the trajectories of the ions toward the center of the ion trap, providing better resolution and sensitivity. As will be seen later (Section 2.6.2), the bath gas used to improve performance of the ITMS in its normal mode of operation is also necessary for

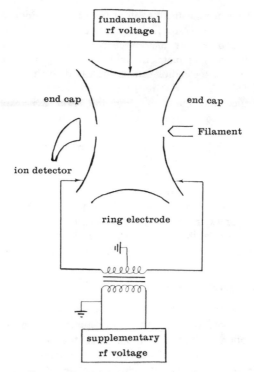

Figure 2-1 ■ Cross-sectional schematic of an ion trap mass spectrometer.

performance of MS/MS experiments, the complete operation of which is also described in Section 2.6.2.

2.3. Sector-Based MS/MS Instruments

The principles of mass analysis for the measurement of mass spectra have been summarized in the previous section. For MS/MS, new means of scanning single mass analyzers and the use of multi-analyzer instruments or experiments are required. The discussion in this section deals with the use of sector instruments for MS/MS experiments.

2.3.1. One-Sector Instruments

Sector mass spectrometers have been the mainstay of mass spectrometry since the beginning of the technique. They were also the instruments used in the early stages of MS/MS work and continue to play an important role in MS/MS. The simplest sector MS/MS instrument is a single-stage magnetic or electric sector. Daughter ions formed in the reaction region prior to the magnetic sector appear in the normal mass spectrum at the position determined by equation (2-11). Such an arrangement has a serious drawback: there is no parent ion selection. Any of the ions that exit the source can fragment and the daughter ions formed will be detected in the mass spectrum. This can result in a very complicated spectrum that requires substantial effort for interpretation.

If a single electric sector is used instead of a magnetic sector, the ratio of the daughter ion mass to the parent ion mass can be determined from equation (2-5), but neither mass is known. This technique was first used by Beynon et al.[19] and has been termed ion kinetic energy spectrometry (IKES). While the IKES experiment provides useful information for simple systems in which the parent ion can be assigned from knowledge of the sample, complex samples provide IKE spectra far too complicated to be of practical use. However, the focus of these early experiments was kinetic energy spectroscopy, and the IKES experiments were well suited for these studies.

2.3.2. Two-Sector MS/MS Instruments

The most common type of sector instrument used for MS/MS is the two-sector instrument composed of an electric sector and a magnetic sector. Experiments of the type now designated as MS/MS were introduced by Barber and Elliot, who used a two-sector mass spectrometer operated in a nonconventional mode to obtain MS/MS information.[6] This experiment entailed scanning the accelerating voltage with the electric and magnetic sectors held at constant values in an instrument of EB geometry. This type of experiment is now termed an accelerating- (or high-) voltage scan and falls into the general classification of a parent ion scan. The mass spectrometer is set up so that the desired daughter ion is transmitted to the detector. By fixing the electric and magnetic sector fields, the value $m_d v$ for ions to be transmitted through both analyzers to the detector is a constant [from equations (2-3) and (2-6)]. From equation (2-1), if v is taken to be a constant and if V_d is increased to V_p, then m_d must increase to m_p. If m_p fragments to m_d in the reaction region prior to the first sector, the product $m_d v$ possesses the value of momentum necessary for transmission to the detector. Thus, the mass of the parent ion is determined from the following equation:

$$m_d V_p / V_d = m_p \qquad (2\text{-}24)$$

Accelerating-voltage scans have been used extensively to obtain kinetic energy release data (see Chapter 3).

The major breakthrough in instrumentation that paved the way to modern MS/MS seems simple in retrospect. It occurred when the order of the magnetic and electric sectors in a conventional EB mass spectrometer was reversed to give an instrument of BE or "reverse" geometry. This was first described by Beynon et al.,[20] and then carried out by them and others.[21-24] These were not the first reversed geometry mass spectrometers to be constructed, however. For instance, White and co-workers had used the BE portion of a three-stage instrument, built primarily for high abundance sensitivity isotope ratio measurements, for several such experiments.[25]

Early BE instruments were mainly designed for studies to elucidate reaction mechanisms and ion structures via observation of metastable and CID reactions. In particular, these instruments were designed to measure accurately the kinetic energy release associated with the dissociation of mass-selected ions. Kinetic energy release is discussed in detail in Section 3.2. The result of kinetic energy release (which occurs in *every* dissociation) is a spread of velocities of the daughter ions of each mass. Since an electric sector separates charged particles as a function of their velocity, the width of

**Parent
Ion**

**Daughter
Ion**

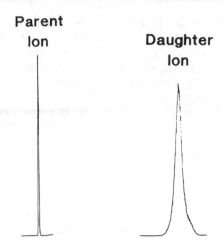

Figure 2-2 ■ Comparison of width of the parent ion of protonated acetophenylhydrazone with the width of the daughter ion formed by loss of ammonia. The data were obtained with a BE instrument by scanning the electric sector.

daughter ion peaks is substantially greater than that of the parent ion. An example of this effect is shown in Figure 2-2.

The advantage of the BE instruments over the EB geometries and single-sector instruments is the ability to mass-select the parent ion prior to dissociation. A schematic of a BE instrument is shown in Figure 2-3. In the typical mode of operation, ions are formed in the ion source, extracted, and accelerated in the normal manner. As the ions travel through the magnetic sector, dispersion according to ion mass occurs (see Section 2.2.2). The magnetic field is set such that ions of the selected mass-to-charge ratio travel along the path of radius r_c through the magnetic sector to reach the mass resolving slit. Ions of other mass-to-charge ratios describe trajectories with other radii and are lost in collisions with the walls of the mass spectrometer. Those mass-selected ions that fragment in the reaction region after the magnetic sector but prior to the electric sector (reaction region 2) can be passed to the detector when the electric sector is scanned (an E scan) to the appropriate field strength [equation (2-5)]. Since m_p is

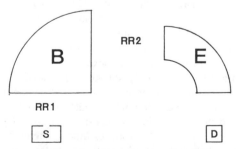

Figure 2-3 ■ Schematic representation of a BE instrument. S is the ion source, D is the detector, and RR is the reaction region.

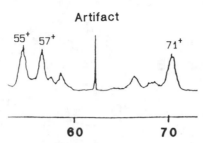

Figure 2-4 ■ Portion of the daughter ion MS/MS spectrum obtained by an E scan for dissociations occurring in reaction region 2. The artifact peak results from a dissociation in reaction region 1 that results in the formation of an ion with the appropriate momentum to be passed through the magnet at the same field strength as the parent ion.

now known, m_d can be unambiguously determined. This E scan provides a daughter ion MS/MS spectrum.

Artifact peaks occasionally occur in this type of experiment. They result from higher mass ions that fragment prior to the magnet to form daughter ions that have the appropriate momentum [equation (2-7)] to be passed through the magnet at the setting appropriate for the selected parent ion.[26-28] This dissociation can occur either in the accelerating region or in the field-free reaction region between the acceleration region and magnetic field (reaction region one). These particular artifact peaks are readily identified in an MS/MS spectrum because the peaks are narrow relative to the true daughter ion peaks (see Figure 2-4). This is due to the fact that the static magnetic field passes only a narrow range of velocities; thus, while the kinetic energy release still occurs in the dissociation of this ion that leads to an artifact peak, the majority of the velocity spread is filtered out by the magnetic sector. However, should this artifact ion dissociate again, after passing through the magnetic sector (reaction region two), the ion formed would be indistinguishable from true daughter ions.

Following the demonstration of the utility of daughter ion MS/MS scans with BE instruments, linked scans, in which two of the fields are scanned simultaneously, were developed.[29-37] This allowed MS/MS experiments to be performed using sector instruments of conventional geometry, i.e., EB instruments with the dissociation occurring in reaction region 1, prior to the electric sector. These linked scans can also be performed on BE geometry instruments. The basis of the linked scan for daughter ions is that, when an ion fragments in a reaction region and experiences no further acceleration or deceleration, the velocity of the daughter ion is the same as that of the parent ion, as discussed in Sections 2.2.1 and 2.2.2 (disregarding the kinetic energy release). The trajectory of an ion through a magnetic field and an electric field is a function of both mass and velocity. By maintaining the ratio of the fields such that the ion velocity necessary to pass through both sectors is constant, a daughter ion scan is performed.[29-31] Derivation of the condition in which the ion velocity is constant is straightforward. Division of equation (2-7) by equation (2-4) gives:

$$B/E = 1/v \qquad (2\text{-}25)$$

Thus, if the ratio of the magnetic and electric fields is kept constant, only ions of a

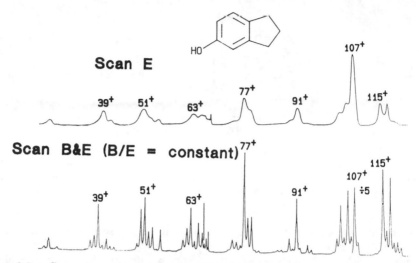

Figure 2-5 ■ Comparison of the daughter ion MS/MS spectra of protonated 5-indanol (m/z 135) obtained by different methods on a BE instrument. The top spectrum was obtained by scanning the electric sector for dissociations occurring in reaction region 2 while the bottom spectrum was obtained by a linked scan at constant B/E of the magnetic and electric sectors for dissociations occurring in reaction region 1.

given velocity can pass through both sectors to the detector. By fixing the velocity in this manner the mass of the parent ion is fixed, since each parent ion has a different velocity.

A linked scan at constant B/E provides spectra with much better mass resolution than do E scans in a BE instrument. A comparison of the two types of spectra is shown in Figure 2-5. The peaks in the linked scan are narrow, for the same reason that artifact peaks are narrow in E scans, because only a small range of velocities can pass through both sectors. In fact, both types of peaks, linked scan and E-scan artifact, are a result of the same process—a dissociation in the first reaction region. Since the linked scan at constant B/E filters out most of the velocity spread, the kinetic energy release information is lost in this scan. However, the kinetic energy release still plays an important part in the overall appearance of the spectra obtained when a linked scan at constant B/E is performed. The kinetic energy release associated with dissociation of ions with slightly different velocities from that of the desired parent ion [e.g., an ion peak 1 dalton lower than the parent ion, such as $(M - H)^+$] can produce daughter ions that have the correct velocity to be passed through both sectors. Since these ions are a result of the same process as the true daughter ions, there is no simple way to distinguish peaks for the two processes, although some rather complicated methods have been proposed.[38]

To execute a parent ion scan for dissociations occurring in the reaction region preceding the first sector on two-sector instruments of either geometry (reaction region 1), another type of linked scan is necessary.[32] Whereas in the daughter ion scan, the constant quantity that relates all the ions of interest is velocity, in a parent ion scan, the constant parameter is the daughter ion mass. From equations (2-7) and (2-4), it can be seen that division of the square of equation (2-7) by equation (2-4) removes the velocity term and only a mass term is left:

$$B^2/E = m \qquad (2\text{-}26)$$

Thus, by keeping the ratio of the electric field to the square of the magnetic field constant, all ions that dissociate to a given daughter ion in the reaction region prior to the first sector are detected. A linked scan at a constant B^2/E is an alternative to the accelerating-voltage scan previously discussed. An advantage of this linked scan is that the accelerating voltage is held constant. Therefore, ion source defocusing, which occurs during an accelerating-voltage scan, is not a problem.

For the linked scan at constant B^2/E, in contrast to the linked scan at constant B/E, kinetic energy release information is not lost. Since both fields are being scanned proportional to the velocity squared, the magnetic sector does not filter out part of the velocity spread. A consequence of this is that the mass resolution of the spectrum is degraded as in an E-scan daughter ion spectrum on a BE geometry instrument. In this case, however, it is the parent ion resolution that degrades.

The two linked scans discussed previously are by far the most popular and useful. However, there are numerous other linked scans that can be performed, although not all with both EB and BE instruments. In considering the possibilities, one has to remember that the accelerating voltage, V, is also a parameter that can be varied, although it generally provides data inferior to that obtained by scanning the analyzers due to source defocusing. Since many of these linked scans are of limited usefulness, some will be mentioned at this time but will not be discussed in detail. A list of all the linked scans can be found in Appendix A.

Perhaps the most useful of the remaining linked scans is that for a neutral loss in the reaction region preceding the first sector.[33] This is more complex than the preceding two, with the necessary ratio:

$$B^2(1 - E)/E^2 = \text{constant} \qquad (2\text{-}27)$$

Another linked scan, applicable only to dissociation in reaction region 2 (between the magnetic and electric sectors) of a BE geometry instrument, provides a parent ion spectrum.[34] In this linked scan, the product B^2E must be kept constant. In contrast to a linked scan at constant B^2/E, a linked scan at constant B^2E provides improved mass resolution, but no kinetic energy release information. A linked scan at constant V/E^2 is analogous to the linked scan at constant B/E.[35,36] This scan is inferior to the linked scan at constant B/E due to ion source defocusing as V is changed, as discussed previously.

As mass spectrometers have become more fully computerized, complex scans of the fields can readily be performed without requiring a rigid operational relationship between the various fields. For instance, the magnetic field strength and electric sector voltage can be measured and the computer can then calculate the appropriate setting of each sector independently [according to equations (2-5) and (2-12)]. A neutral loss scan for dissociations occurring between the two sectors of a BE geometry instrument has been implemented in this manner (although an explicit relationship between the fields can also be used). The mass of the parent ion was determined from the magnetic field strength and then the appropriate electric sector voltage was calculated and applied to the electric sector.[37]

In a comparison of the BE and EB geometries, the former is found to be preferable overall for MS/MS experiments. The BE instrument can perform most of the MS/MS experiments that the EB can (i.e., linked scans for first reaction region dissociations), and, in addition, can perform these experiments in the second reaction region, between the B and E. The reaction region between the E and B in an EB instrument is of little utility for MS/MS experiments, but can be used for MS/MS/MS experiments.

Another two-sector instrument that has been used for MS/MS experiments is the BB geometry. BB instruments have been used in isotope ratio measurements[39] and in photodissociation studies.[40,41] The first BB geometry instrument constructed expressly for MS/MS work was reported by Louter et al. in 1980.[42] An advantage of the BB instrument for MS/MS is somewhat better daughter ion resolution when daughter ion MS/MS spectra are obtained by scanning the second analyzer (B_2 versus E improves resolution by a factor of 4). For high-energy CID, this follows from the fact that the magnetic field analyzes as a function of the velocity [equation (2-7)], while the electric field analyzes as a function of the velocity squared. Thus, the effect of the velocity spread from the kinetic energy release is reduced when scanning B_2 on the BB geometry as compared to an E scan on a BE geometry instrument. Still better daughter ion resolution can be obtained by deceleration of the parent ions into an electrically floated collision cell, followed by low energy CID, and then reacceleration of the ions prior to analysis with B_2. This experiment allows ready access to not only the typical low and high collision energies, but also to the intermediate range of collision energies.[43] This feature makes the BB instrument useful for experiments in which a wide range of collision energies must be covered as in fundamental studies of collisional activation mechanisms.

There are several disadvantages of the BB geometry relative to the BE. Most of these disadvantages are of an operational nature relating to the difficulty in measurement and control of magnetic fields. Thus, it seems likely that the BB geometry instrument will have a limited demand for MS/MS experiments.

Two MS/MS instruments have been developed by groups at Michigan State University that do not fall neatly into any of the categories of this chapter, but seem most appropriately discussed in this section . The first of these instruments uses momentum analysis with a magnetic sector in conjunction with time-of-flight measurement (BTOF) to obtain MS/MS data.[44] This instrument performs the time-of-flight measurement simultaneously with the momentum dispersion and thus offers the possibility of very rapid data acquisition. Combination of equation (2-16) with equation (2-7), yields:

$$m/z = Btr/L \qquad (2\text{-}28)$$

From equation (2-28) it can be seen that the mass of the ions arriving at the detector can unambiguously be determined regardless of the energy spread of the beam. The ions arriving at the detector will include both stable ions that form the normal mass spectrum and daughter ions.

Since the daughter ions nominally retain the velocity of the parent ion, the time of flight of the ion can be used to assign the parent ion mass. Thus, the experiment is based upon ion velocity as is the linked scan at constant B/E. Therefore, as discussed previously for the linked scan, parent ion resolution is poor in the BTOF instrument due to the kinetic energy release associated with dissociation.

While data acquisition can be very rapid with the BTOF instrument, there is no means to obtain a daughter ion MS/MS spectrum, parent ion MS/MS spectrum, or neutral loss MS/MS spectrum without scanning the magnet. The complete MS/MS data matrix must be acquired, i.e., a TOF spectrum at each setting of the B, and then the appropriate data extracted from that data matrix. Thus, the main advantages of the BTOF instrument are its relative simplicity and its ability to obtain rapidly the complete MS/MS data matrix, advantages that suggest the possibility of some form of GC/MS/MS.

The other instrument developed at Michigan State University is analogous to the BTOF, but instead of using a magnetic sector, it uses an electric sector, giving an ETOF geometry, and the analysis is done sequentially instead of simultaneously.[45] (An instrument in which the analysis is performed simultaneously could also be constructed and it would differ little in MS/MS performance from this ETOF instrument.) The operation of the ETOF instrument is analogous to the BTOF with the exception that energy analysis is done instead of momentum analysis. Equation (2-29) is obtained from the combination of equation (2-4) with equation (2-16):

$$m/z = Er(t/L)^2 \qquad (2-29)$$

Again, the mass of the ions reaching the detector is determined unambiguously by the electric field strength and the flight time. Otherwise, the ETOF is essentially the same as the BTOF. The ETOF will have poorer parent ion resolution due to the kinetic energy release since E is proportional to v^2 while B is proportional to v. An advantage of the ETOF, as mentioned previously, is the fact that an electric sector is easier to control and monitor than a magnetic sector.

Several other instruments incorporating time-of-flight analyzers have also been used for MS/MS experiments. The simplest of these is a time-of-flight mass spectrometer with a 90° reflectron.[46] In this instrument, both the ions and neutrals are detected in coincidence, the ions after a deflection of 90° from the initial flight direction with the reflectron, and the neutrals at a 0° detector since neutral molecules are not deflected. A major limitation to this type of MS/MS system, however, is that the ion flux must be very low to make an unambiguous coincidence measurement. Typical values are one ion per time-of-flight pulse. This arrangement also has a limited resolution due to the kinetic energy release.

Another MS/MS instrument using time of flight has a Wien filter preceding a time-of-flight analyzer (WTOF).[47] The WTOF instrument has very limited parent ion resolution, with a resolution of between 20 and 30 with ion energies of 1.5 keV. While such a poor parent ion resolution limits the utility of this geometry, it is useful for the specific application for which it was designed, i.e., the MS/MS analysis of metal cluster ions. In this application, resolution requirements are minimal and the simplicity and high transmission of the Wien filter are advantageous.

Another MS/MS instrument incorporating time of flight was built to study surface-induced dissociations[48] (see Sections 2.7.2 and 3.3.4 for discussions of surface-induced dissociation). This instrument consists of a time-of-flight mass spectrometer for the first stage of analysis and a second time-of-flight mass spectrometer for the second stage of analysis (TOFTOF). These two time-of-flight instruments are oriented 90° with respect to each other and a solid metal surface is at the vertex of the ion flight paths, oriented at a 45° angle with respect to each time-of-flight system. The first TOF is somewhat unconventional in that it includes a set of deflection plates along the drift path of the ions. These plates are used to mass-select the parent ion. The plates normally deflect the ion beam out of the drift path to the reaction region. They are turned off briefly, at the appropriate time after the ionization pulse, to allow the desired parent ion to pass undeflected to the reaction region. The TOFTOF has somewhat better parent ion resolution than the WTOF, but it is still quite limited. The daughter ion resolution is also quite limited. This particular instrument was built to increase easily the accessible energy range for the surface-induced dissociation experiment. This is another example in which an instrument was built for a specific experimental purpose, and a common parameter that is usually considered very important in a mass spectrometer, resolution, was of secondary concern.

2.3.3. Three- and Four-Sector MS/MS Instruments

The early studies in MS/MS with BE and EB geometry instruments laid the groundwork for the technique. As MS/MS gained popularity, new instruments were developed. These instrument developments took several different paths. One approach involved the use of quadrupoles (Section 2.4), a second the addition of more sectors to EB and BE instruments, producing instruments similar to those previously constructed for purposes other than MS/MS,[25,49,50] and a final approach toward the development of hybrid instruments.

Mathematically, there are eight possible combinations of magnetic and electric sectors in a three-sector instrument. Of these eight possible combinations, three have been tried in MS/MS applications. Among the untried geometries, three, EEE, BBB, and EEB, provide little or no advantage over the two-sector instruments. The other two untried geometries, EBB and BBE, are viable MS/MS geometries, but appear to provide no unique advantages over one or more of the geometries that have been constructed.

A logical step in the evolution of three-sector MS/MS instruments was to add a sector onto the end of a conventional instrument. For MS/MS purposes, an electric sector is the simplest addition from both operational and physical/mechanical aspects. Several groups added electric sectors to their EB instruments to obtain EBE geometries (Figure 2-6).[51-54] There are several advantages in this configuration. First, the additional electric sector obviates the need to link-scan to obtain daughter ion MS/MS spectra on the EB geometry. Daughter ion spectra are obtained by selecting the parent ion with the EB portion and scanning the second E to observe dissociations occurring in reaction region 3. If so designed, the EB portion of the instrument can be used for high-resolution parent ion measurement and separation. To date, all but the first EBE MS/MS instrument[51] have consisted of a double-focusing configuration for the first two sectors, although few examples in which high resolution is needed in the first stage of an MS/MS experiment have been demonstrated. Parent ion and neutral loss scans are implemented by linked scanning in the same manner as with two-sector instruments, using either reaction region 1, 2, or 3. (For reaction regions 1 and 2, the linked scans for EB or BE instruments using their reaction region 1 apply; for reaction region 3 in the EBE, the BE linked scans using reaction region 2 apply.) The success of these instruments led quickly to the introduction of a commercial EBE instrument.[55]

Similar in concept to the EBE geometry is the BEE geometry, which can be obtained by adding an electric sector to a BE instrument. This was first done several

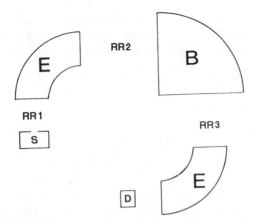

Figure 2-6 ■ Schematic representation of an EBE instrument. In general, only reaction region 3 is used for the basic MS/MS experiments and only the dissociation of high kinetic energy ions can be observed.

years after the first EBE instruments were described.[56] The BEE geometry instrument is quite similar in capabilities to the EBE geometry instrument. The first two sectors can be used in a double-focusing mode to provide high-resolution parent ion measurement and separation, and similar MS/MS/MS experiments can be performed on both.

An alternative three-sector arrangement is the BEB geometry (Figure 2-7). Such an instrument is now available commercially.[57] It has been argued that this is the most useful of any MS/MS instrument consisting of three analyzers.[58] This argument is quite convincing if the reaction regions in such an instrument contain collision cells

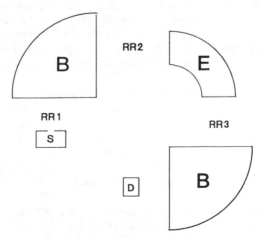

Figure 2-7 ■ Schematic representation of a BEB instrument. Reaction regions 2 and 3 can be used for the basic MS/MS experiments and the dissociations observed can be of either low or high kinetic energy ions in both reaction regions.

that can be floated at an independent electrical potential. This would allow the basic MS/MS experiments to be performed using reaction regions 2 and 3, at both high and low collision energies. (The latter experiments require an additional stage of deceleration and acceleration in the analysis.) This is in contrast to the EBE instrument geometry in which, for the basic MS/MS experiments, only reaction region 3 is useful and then only with high-energy ions.

There are several features of the BEB instrument that makes it superior to other three-sector analyzer geometries. The first two sectors can be used in a high-resolution mode in the same manner as the EBE and BEE instruments. The daughter ion MS/MS spectrum is then obtained from dissociations in reaction region 3 by scanning the second magnet, with the same daughter ion resolution considerations as the BB geometry. However, the real advantage of the BEB is in the other possible modes of operation. In a variation of the preceding mode of operation, the first magnet can be used to select the parent ion while the remaining EB portion is operated in a high-resolution mode and link-scanned at a constant B/E, with the dissociations occurring in reaction region 2. This provides high resolution of the daughter ions. It should be noted that in this mode of operation the problem of artifact peaks associated with linked scans on two-sector instruments is alleviated with the mass selection of parent ions by the first magnet.

The ability to float the collision cell in the reaction region, described here for the BEB, but of course possible in all multisector MS/MS instruments, provides several other modes of operation that have yet to be demonstrated. Low-energy CID and ion/molecule reactions are possible. If these reactions are carried out in the reaction region between the first magnet and electric sector (i.e., reaction region 2), the product ions might be analyzed at high resolution by scanning the second magnet in the normal manner [i.e., equation (2-8) is applicable] after the ions have been reaccelerated. Transmission losses in such an experiment would be quite high. Alternatively, if the collision cell in the reaction region between the electric sector and second magnet is floated, parent ions can be selected at high resolution, and the product ions analyzed by the second magnet, again by a normal scan after reacceleration. Daughter ion resolution in this mode of operation is similar to that of a single-sector mass spectrometer. This is a result of the dissociation occurring at low ion translational energies with the subsequent acceleration of the daughter ions. The daughter ions, therefore, have only a relatively small translational energy distribution (typically a few tenths of a percent).

One of the few experiments that cannot be performed on an instrument of BEB geometry is one in which the parent ions are separated at high resolution and the daughter ions are measured at high resolution. Such an experiment with a sector instrument requires two double-focusing mass spectrometers on either side of a reaction region. The three-sector instrument built in McLafferty's laboratory was an intermediate step in their construction of a four-sector MS/MS instrument of EBEB geometry.[52] Several commercial instrument manufacturers now provide these four-sector instruments.

While four-sector instruments are theoretically capable of performing high-resolution parent ion separation and high-resolution daughter ion analysis in a single experiment, this has yet to be reported. Only in the best of cases will such an experiment be possible because of the large losses in ion transmission associated with operation of a mass spectrometer at high resolution. While the first stage of high resolution is exactly the same as any conventional double-focusing mass spectrometer, the daughter ion MS/MS spectrum for high-energy CID is obtained by a linked scan at constant B/E of the second double-focusing mass spectrometer (as is done for high daughter ion resolution with a BEB). The ion losses in the second stage of this MS/MS experiment are much larger than the already substantial losses occurring in a conventional high-resolution experiment. This is due to the much larger energy spread of the daughter ions relative to the parent ions (from the kinetic energy release). Due to these increased ion losses in the daughter ion high-resolution scan, the best resolution obtainable on the daughter ions will never match that obtainable on parent ions. A resolution of 5500 has been reported for daughter ion analysis, without high-resolution parent ion selection.[59] The resolution will undoubtedly increase with improved instrumentation and with the improved four-sector MS/MS instruments now becoming available.

An alternative mode of operation with low-energy CID is available if the collision cells in the reaction regions can be electrically floated. This is directly analogous to the case of the BEB, with the difference in the theoretical capability for high-resolution parent ion selection and high-resolution daughter ion measurement with the four-sector instruments, rather than one or the other with the BEB. Again, the daughter ion losses will be substantially greater in this mode of operation. In part this is due to the larger energy distribution of the daughter ions. Although this energy distribution will be much smaller than for the higher-energy CID experiment, it will still be much larger than that of the parent ions. Additional ion transmission losses occur in the deceleration necessary for low-energy CID experiments in a sector MS/MS instrument. In the low-energy CID mode of operation, the best parent ion resolution will again always be greater than the best daughter ion resolution. Due to the different factors leading to ion losses in the low- and high-energy CID experiments, it is not apparent from first principles which method should be more sensitive or which should provide the higher limit for daughter ion resolution.

There are four possible geometries of tandem double-focusing MS/MS instruments, of which two, EBEB and BEEB, have been constructed. The other two, EBBE and BEBE, have yet to be assembled. For basic MS/MS experiments that involve the use of the central reaction region (reaction region 3), there is essentially no difference in capabilities between any of the four possible geometries. It is only when different reaction regions or additional stages of mass separation (i.e., MS/MS/MS) are utilized that differences between potential capabilities of the various geometries occur. In the latter case, these differences occur for experiments in which at least one stage of the experiment uses low-energy CID. In such experiments, the earlier in the analysis scheme that a magnetic sector is present, the greater the number of possible experiments. Thus, the most flexible geometry is that of the BEBE instrument.

A unique experiment that can theoretically be performed on an appropriately designed BEBE instrument is parent ion selection with the first magnet, low-energy CID (reaction region 2), high-resolution daughter ion selection with the following EB, high-energy CID (reaction region 4), and granddaughter ion analysis by scanning the second electric sector. Another possible experiment that could be performed on the BEBE and EBBE instruments is high-resolution parent ion selection, low-energy CID (reaction region 3), daughter ion selection with the second magnet, high-energy CID (reaction region 4), and granddaughter ion analysis by scanning the second electric sector. Occasionally, such experiments would provide information that could not be obtained by other methods; however, in the vast majority of applications any of the four four-sector geometries would be equally useful.

There is one other four-sector instrument that has been built.[60] It is distinct from the four-sector instruments described previously in that is not composed of two conventional double-focusing instruments assembled in tandem. The first stage of this instrument is of double-focusing EB geometry. However, the second stage consists of an electric sector and a crossed electric and magnetic field (Wien filter). The daughter ion mass resolution of this instrument is only moderate, although it is better than that of two-sector and most three-sector instruments. The advantage of this configuration is that daughter ion MS/MS spectra are obtained by a simple and rapid scan of the electric fields, with the magnetic field remaining fixed. However, it seems very unlikely that this one advantage will make this instrument competitive with other four-sector instruments.

2.4. Quadrupole Instruments

Quadrupole mass spectrometers are a mainstay of gas chromatography/mass spectrometry (GC/MS) and liquid chromatography/mass spectrometry (LC/MS). Some of the features that have made quadrupoles so popular are: rapid scanning, relatively small size, simplicity of operation, and relatively low cost. These same features have made multiquadrupole systems attractive for MS/MS.

The first multiquadrupole instruments were built to study photodissociation.[61,62] In 1978, the first multiple-quadrupole system developed expressly for general MS/MS applications was reported by Yost and Enke.[63] This instrument consisted of three quadrupoles (QQQ) in sequence. A schematic of a QQQ system is shown in Figure 2-8. The first and third quadrupoles are operated in the normal manner with a combination of rf and dc voltages. The center quadrupole is operated in the rf-only or total-ion mode (see Section 2.2.3) and is used as the reaction region.[64] The use of an rf-only quadrupole will be discussed in more detail in Section 2.7.

After Yost and Enke developed their QQQ system, a second QQQ instrument was built in Hunt's laboratory[65] and a QQ system was built in Cooks' laboratory.[66] The latter instrument used a conventional reaction region instead of an rf-only quadrupole; otherwise, the basic operating procedures are identical for the QQQ and QQ. Subsequently, the first commercial triple quadrupole was offered, and now there are several vendors that offer such systems.

A major motivation for the development of the QQQ system was the desire to improve the daughter ion resolution available with the standard MS/MS instrument at that time, the BE instrument. For analytical applications, the unit mass resolution of

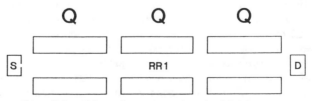

Figure 2-8 ■ Schematic representation of a QQQ instrument.

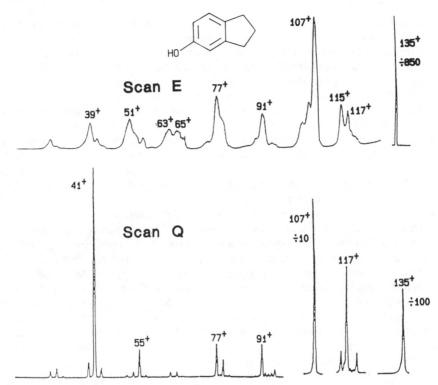

Figure 2-9 ■ Comparison of the daughter ion MS/MS spectra of 5-indanol obtained by scanning the electric sector for reaction region 2 dissociations (high kinetic energy ions) in a BE instrument (top) and by scanning the last quadrupole for reaction region 1 dissociations (low kinetic energy ions) in a multiquadrupole system (bottom).

daughter ions provided by the quadrupole is certainly desirable compared to the relatively poor daughter ion resolution of the BE instrument. Since the quadrupole is not dependent upon ion velocity for its mass-filtering action, the kinetic energy release does not degrade the daughter ion resolution in QQQ instruments as it does in the BE instruments. Figure 2-9 compares daughter ion MS/MS spectra obtained by scanning the last quadrupole in a multiquadrupole instrument and an *E* scan of a BE instrument, illustrating the difference in mass resolution.

Another important feature of the QQQ is simplicity of operation. Overall physical/mechanical operation is one aspect of this simplicity. Since quadrupole mass filters act upon ions with translational energies in the eV range, there is no need to float the ion source at kV potentials as with sector instruments. Thus, the QQQ instruments do not have problems with high-voltage arcing that will occur, at least occasionally, with sector instruments. Although the rf voltage applied to the quadrupole rods is typically in the kV range, the design of a quadrupole is such that this is a potential problem only if there is a loss of vacuum.

The other major aspect of the operational simplicity of the QQQ is the straightforward scanning procedures used for acquisition of the various types of MS/MS data.

With sectors, the relationship between ionic mass and field strength depends upon the history of the ion (i.e., whether it is a parent ion or daughter ion). In contrast, quadrupoles pass both parent and daughter ions under the same conditions. Thus, a daughter ion scan is obtained by simply setting the first quadrupole to pass the desired parent ion and scanning the last quadrupole in the normal manner.

Parent ion scans are performed by reversing the operating procedure used for the daughter ion scan. The last quadrupole is set to pass the desired daughter ion and the first quadrupole is scanned in the normal manner. Neutral loss (or gain) spectra are obtained by offsetting the mass of the first and last quadrupoles by the mass of the desired neutral loss (or gain) and then scanning them in concert at the same rate, a relatively simple experiment to implement.[67]

MS/MS scans are much simpler to perform on quadrupole instruments, either manually or under computer control, than they are on sector instruments. In fact, the length of this section describing quadrupole MS/MS instruments versus that describing the sector MS/MS instruments is a reflection of the simplicity of the quadrupole instruments and not of the popularity and importance of sectors versus quadrupoles. However, as with most techniques, this simplicity of operation is balanced against poorer parent ion transmission and less flexibility in the type of data which can be obtained (e.g., high-energy CID experiments are usually not practical). While the triple-quadrupole system cannot perform MS/MS/MS experiments, additional quadrupoles can be added to a system to provide such a capability.[68,69] It remains to be seen whether the number of problems for which such an experiment would be needed justifies the added complexity. However, certain fundamental applications can benefit from the additional stage(s) of analysis.

The use of quadrupole MS/MS instruments for the study of ion/molecule association reactions is likely to be increasingly widespread. The chemical nature of the mass-selected parent ion is not probed through its dissociation chemistry, but rather through its chemical associations with a reactive gas. Very complex chemistry can be studied in MS/MS experiments, with the reactants and products each characterized by mass spectrometry.

2.5. Hybrid Instruments

As discussed in Chapter 1, there are two energy ranges used for collision-induced dissociation. The sector instruments, which operate with keV ion energies, generally use keV CID, although by deceleration and subsequent acceleration (discussed in Section 2.3) the low-energy CID range can also be accessed. However, quadrupoles operate with ion energies in the few to tens of eV range, and the multiquadrupole instruments have been used exclusively in the low-collision-energy range. Details of the differences between the low- and high-energy CA processes are discussed in the next chapter. It needs to be mentioned that, while the MS/MS spectra from the two energy ranges often show similarities, there are differences due to the different internal energy distributions deposited into the ion by the collision. There are also differences in the type of data that can be obtained: high-energy collisions offer access to charge permutation reactions, whereas very low-energy collisions allow ion/molecule association reactions to be studied.

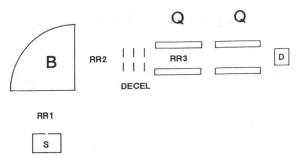

Figure 2-10 ■ Schematic representation of a BQQ instrument. Reaction regions 1 and 2 are for dissociations of ions with high kinetic energy and reaction region 3 is for dissociations or association reactions of ions with low kinetic energies. In the "decel" region of the instrument, the kinetic energy of the ions is changed from the keV range to the eV range.

The desire to access both CA energy ranges and to take advantage of the best performance characteristics of sectors and quadrupoles, without the complexity and cost of the four-sector instruments, led to the development of hybrid instruments consisting of both sectors and quadrupoles. Chemical physicists constructed hybrid instruments that predate those developed for MS/MS.[70-77] The first hybrid instrument built for general MS/MS applications was constructed by Cooks' group.[78,79] This BQQ instrument (shown schematically in Figure 2-10) consists of a magnetic sector followed by an rf-only quadrupole, used as a reaction region, and an analyzing quadrupole. It has the capability of moderate parent ion resolution (5000) with good transmission, followed by unit resolution of daughter ions. The usual mode of operation of this BQQ instrument is acceleration out of the ion source to an energy in the keV range, mass analysis with the magnet, deceleration to the eV energy range (typically 1–100 eV), low-energy CID in the first quadrupole (reaction region 3), and daughter ion analysis with the second quadrupole. Parent ion scans are performed by setting the last quadrupole to the desired daughter ion mass and scanning the magnetic field strength, with CID occurring in reaction region 3. Neutral loss spectra are obtained by linked scanning of the magnet and analyzing quadrupole such that the ratio of B^2/Q is constant, again using reaction region 3.

BQQ data are, as expected, very similar to those from QQQ systems. Some advantages of the BQQ over the QQQ are better parent ion resolution and transmission (including consideration of losses in the deceleration lens) and less background noise due to the 90° bend in the ion flight path through the magnet. The deceleration lens also acts as a filter against ions that dissociate after leaving the source but before deceleration. Most ions formed in this region of the instrument are not transmitted through the lens due to their lower kinetic energy. The problem of artifact peaks observed in E scans of BE instruments[26,27] is thereby alleviated.

The actual operation of this BQQ instrument is somewhat nonconventional. Since the sector portion of the instrument operates on ions with keV energies and the quadrupole portion on ions with eV energies, one or the other must be electrically floated with respect to the other. The BQQ built in Cooks' laboratory operated with the magnetic sector flight tube electrically floated and the quadrupoles at ground potential

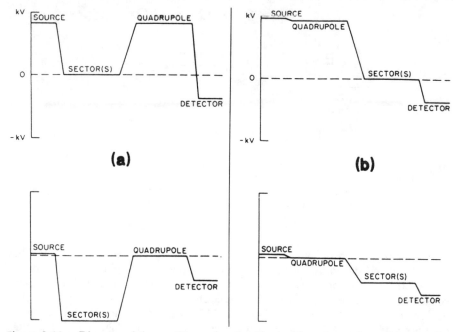

Figure 2-11 ■ Diagrams of the possible acceleration/deceleration schemes for positive ions that can be used with hybrid instruments of (*a*) sector/quadrupole geometry and (*b*) quadrupole/ sector geometry.

with the source operated near ground. This is shown schematically in Figure 2-11*b* for the analysis of positive ions, along with the other possible option for this instrument geometry, a QQB (Figure 2-11*b*).

This particular BQQ instrument was designed such that the quadrupoles could also be floated at potentials other than the acceleration/decleration potential, providing still other modes of operation. High-energy CID could then be performed in the reaction region prior to the magnet (reaction region 1) or the reaction region deceleration lens (reaction region 2), although there are problems with this latter mode unless some type of energy filtering is performed prior to analysis by the quadrupole. This will be discussed in more detail shortly. This and numerous other modes of operation have been demonstrated in Cooks' laboratory on a second generation hybrid instrument, a BEQQ.[80-82] Utilization of a double-focusing sector portion of a hybrid sector/quadrupole instrument was the obvious next step after the BQQ instrument. Such instruments were commercially offered shortly after the BQQ was reported.[80,83,84] These instruments, like triple-sector instruments, provide high-resolution separation of parent ions in the first stage of an MS/MS experiment. The two variations of this type of instrument, EBQQ and BEQQ, are shown schematically in Figure 2-12, with all of the reaction regions that can possibly be used for MS/MS experiments. It should be noted that the reaction region between the *E* and *B* in the EBQQ, reaction region 2, is not useful for MS/MS experiments but can be used as a reaction region in MS/MS/MS experiments. While these configurations have many reaction regions, only one instrument to date has actually demonstrated the use of all of them.[80] The EBQQ and BEQQ

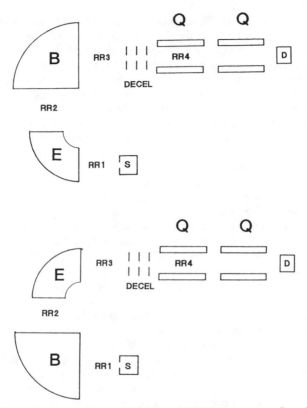

Figure 2-12 ■ Schematic representations of EBQQ and BEQQ instruments. Reaction region 4 is for dissociations of ions with low kinetic energies and the rest of the reaction regions are for dissociations of ions with high kinetic energy. Reaction region 2 is useful in the BEQQ geometry for basic MS/MS experiments, but this reaction region is not used in the EBQQ for the basic MS/MS experiments. All other reaction regions are similar between the two instruments. Additional energy filtering is needed to use effectively reaction region 3 in both instruments.

geometries offer no difference in performance for the basic MS/MS experiments in which the dissociations occur in the quadrupole used as a reaction region (reaction region 4). Daughter ion MS/MS spectra are obtained by mass selection of the parent ion with the sector portion of the instrument, deceleration of the ions, with CID in the rf-only quadrupole, and then a scan of the last quadrupole to detect the daughter ions formed. Parent ion scans involve setting the analyzing quadrupole to pass the desired daughter ion and scanning the magnetic sector, with deceleration and CID occurring as in the daughter ion scan. Neutral loss scans are performed by scanning the magnetic field and analyzing quadrupole such that $(B^2/Q) = $ constant, where Q is proportional to the rf voltage level of the quadrupole. In these modes of operation, the deceleration voltage is held constant, whereas in high-energy CID, the deceleration voltage varies as a function of the daughter ion mass.

While the EBQQ and BEQQ give identical results in the experiments described previously, there is a difference in performance due to geometry for high-energy CID experiments that favors the BEQQ geometry. On the EBQQ instrument, high-energy CID experiments involve CA in the first reaction region with linked scanning of the sectors as described in Section 2.3.2. The weakness of the linked scan is the poor parent ion resolution for daughter ion scans; the addition of a quadrupole mass analyzer subsequent to the sectors therefore provides no improvement in this resolution. In essence, complexity is added to the experiment without any gain in information and with a decrease in sensitivity. Therefore, an intermediate detector, after the magnet, is usually used and the ions never pass into the quadrupoles. While the BEQQ instrument can also be operated in the same mode, with high-energy CID in the first reaction region, there is also the option for a similar experiment with high-energy CID between the magnetic and electric sectors. For a two-sector instrument, the weakness in this mode of operation is poor daughter ion resolution. However, with an additional quadrupole mass analyzer, unit mass resolution of the daughter ions is achieved. An example of this is shown in Figure 2-13. Thus, the BEQQ can provide equivalent daughter ion resolution to the EBQQ for high-energy CID experiments, but with substantially better parent ion resolution. Conceivably, similar data could be obtained by CID between the last sector and the deceleration lens (reaction region 3). While this is true to a certain extent, the variation of kinetic energy of the ions after dissociation means that the magnitude of the deceleration voltage chosen must be varied based on daughter ion mass. When low mass ions (low kinetic energy) are analyzed, higher mass ions are not sufficiently decelerated to be rejected by the quadrupole mass analyzer; a large continuous background signal due to the parent ion is present that can totally obscure daughter ions. To alleviate this problem, another stage of energy analysis is necessary, although the resolution required can be minimal. In practice, one must be aware of this possibility even when this reaction region is not explicitly being used. A conventional parent ion scan of dissociations occurring in reaction region 4 has been found to contain artifacts from reactions occurring in reaction region 3.[85] Thus, it is especially important to minimize the amount of the target gas from the rf-only quadrupole that diffuses into the region preceding the quadrupoles.

Another high-energy CID experiment that can better be performed on the BEQQ rather than the EBQQ is the deconvolution of kinetic energy releases from dissociations resulting in unresolved daughter ions of adjacent masses. To obtain mass-resolved kinetic energy release data with a BEQQ, the quadrupole is set to pass the desired daughter ion while the electric sector is scanned over the appropriate range to observe dissociations occurring in reaction region 2. For the EBQQ geometry, linked scanning (B^2/E) for dissociations occurring in the first reaction region is necessary to obtain kinetic energy release data. However, this scan is subject to poor parent ion resolution, and, just as in the case of the linked scan at a constant B/E on the EBQQ, the final quadrupole mass analyzer does not provide a solution to this problem.

While the EBQQ instrument does not allow deconvolution of the kinetic energy release distribution in signals due to ions of adjacent masses, a hybrid instrument without a magnet can perform such an experiment. This has been demonstrated with an EQ hybrid instrument built in Beynon's laboratory.[86] A schematic of this instrument is shown in Figure 2-14. Although this instrument has only one reaction region and only high-energy CID experiments can be completed, all of the basic MS/MS experiments are possible. Performance characteristics are comparable to those for linked scans of dissociations occurring in reaction region 1 in two-sector instruments, with the added advantage of the measurement of mass-resolved kinetic energy release data. This instrument operates by electrically floating the source and the quadrupole (top diagram

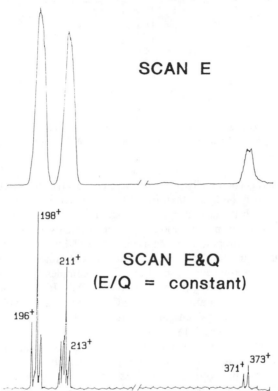

Figure 2-13 ■ Comparison of a portion of the MS/MS spectra of m/z 408 from hexachlorophene obtained by scanning the electric sector for dissociations occurring in reaction region 2 in a BE configuration (top) and by linked scanning of the electric sector and quadrupole for dissociations occurring in reaction region 2 of a BEQQ instrument (bottom). (Adapted from reference 82.)

in Figure 2-11a) and then scanning the voltage at which the quadrupole is floated in conjunction with the mass scan of the quadrupole such that all the ions passing through the quadrupole have the same kinetic energy.

The first MS/MS hybrid (the BQQ) was designed such that the source and detector could be interchanged to give an instrument of QQB geometry.[78,79] However, experiments using this configuration have never been formally reported. More recently, however, several quadrupole/sector hybrid MS/MS instruments have been reported. In 1983, a QB geometry instrument for MS/MS studies with field ionization was constructed.[87] There has been no subsequent report of this instrument in the literature. One reason for selection of a QB geometry over a BQ geometry is that the collision energy can be varied over a wide range without compromising the transmission efficiency through the second stage of analysis (the magnetic sector). Furthermore,

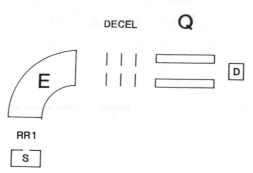

DECEL   Q

E

RR1

S

Figure 2-14 ■ Schematic representation of an EQ instrument. Only dissociations of high kinetic energy ions can be observed but with unit mass resolution of the daughter ions.

since the ion beam is continually being accelerated through the instrument, it is easier to maintain a focused ion beam than in a BQ instrument that requires a stage of ion deceleration, although the advantages of the latter in removing interferences are lost.

The next development in quadrupole/sector instruments was the construction of an instrument with a quadrupole mass analyzer followed by a time-of-flight mass spectrometer (QTOF), built at Oak Ridge National Laboratory.[88,89] This instrument represents a departure from the trend of increasing complexity and cost in that it was developed as a simple, low-cost instrument for MS/MS. It does not have the performance capabilities of the multisector and hybrid instruments with double-focusing capabilities, but it has several unique features. A schematic of the simplest QTOF instrument is shown in Figure 2-15.

Besides the relatively low cost, the other prominent advantage of the QTOF is a rapid data acquisition rate, similar to the BTOF and ETOF. This advantage is a direct result of the use of a time-of-flight mass spectrometer to analyze the daughter ions. Since a time-of-flight analysis provides all the masses in the mass spectrum (or daughter ion MS/MS spectrum) within a few hundred microseconds, spectra can be obtained much faster than with scanning techniques. This ability is particularly helpful for situations in which the parent ion has a very low intensity. In these instances, the QTOF will detect any daughter ion passed into the TOF, whereas MS/MS instruments based on a scanning method will reject the majority of ions formed at any given time in the scan cycle. While the low duty cycle of time-of-flight limits the magnitude of this advantage, the combination of this feature with the high transmission of a TOF makes the QTOF at least an order of magnitude more sensitive than scanning instruments and especially advantageous for pulsed-ionization experiments such as those using lasers.

Q   ACCEL   TOF

S

RR1

D

Figure 2-15 ■ Schematic representation of a QTOF instrument. With a normal linear TOF, only dissociations of ions with low kinetic energies can be observed. "Accel" represents the region of the instrument where the ions are accelerated from eV kinetic energies to keV kinetic energies.

A notable difference between the QTOF and the sector-based BTOF or ETOF instruments is that the former instrument can provide a daughter ion MS/MS spectrum without scanning the quadrupole. The BTOF and ETOF instruments described in Section 2.3.2 require acquisition of the TOF spectra at all settings of B or E, and then reconstruction of a daughter ion MS/MS spectrum from this complete data matrix. For applications in which only a few selected daughter ion MS/MS spectra are desired, the QTOF has an advantage in data acquisition time over the BTOF and ETOF. With the QTOF, parent ion and neutral loss MS/MS spectra can be reconstructed from the MS/MS data matrix, much like the BTOF and ETOF. Alternatively, the QTOF can operate in essentially a scanning mode for the time-of-flight analysis, using a boxcar integrator to obtain the time-of-flight data. For a parent ion scan, the boxcar integrator is set at a fixed time corresponding to the arrival time of the selected daughter ion at the detector, and the quadrupole is scanned. For a neutral loss spectrum, the quadrupole and boxcar integrator time window would have to be scanned simultaneously at a fixed ratio of $Q/TOF^{1/2}$. Use of these scan modes will provide some decrease in data acquisition time but with loss of the remainder of the MS/MS data.

A disadvantage of the QTOF is decreased daughter ion resolution. Due to the kinetic energy release of the dissociation, daughter ions have a spread of translational energies that is much greater than that of ions formed in the ion source. This spread leads to decreased resolution in a conventional time-of-flight mass spectrometer. Some of the newer energy-focusing methods such as the electrostatic mirror[90,91] or velocity compaction[92] should improve daughter ion resolution, but at a cost of greater complexity of the instrument.

Another approach recently implemented to improve daughter ion resolution in the QTOF instrument is the addition of an electric sector as part of the time-of-flight drift tube.[93] While this does not offer the improvement in mass resolution that is possible with the energy-focusing methods mentioned previously, the use of the electric sector does give the QETOF several other modes of operation. There is a second reaction region in this arrangement between the acceleration lens and the electric sector. In a basic high-collision-energy MS/MS experiment, the instrument can be considered to have a QE geometry. The experiment to observe daughter ions from high-energy CID is analogous to that of a BE instrument in which the quadrupole selects the parent ion and daughter ions are analyzed by scanning the electric sector.

Parent ion and neutral loss scans using high-energy CID can also be performed on this instrument by linked scanning of the quadrupole and electric sector. For parent ion scans, the product QE must be kept constant, while for neutral loss scans the product $Q(E - E_p)$ is a constant. This version of the QETOF can also provide MS/MS/MS experiments with CID in both reaction regions and operation of the latter portion of the instrument as an ETOF.[45]

The most recent quadrupole/sector hybrid instrument was developed at Oak Ridge National Laboratory by Glish et al.[94] It is on the opposite end of the cost/complexity scale from the QTOF. This instrument consists of a quadrupole mass analyzer preceding a double-focusing EB mass spectrometer to give an overall geometry of QEB (Figure 2-16). In this configuration, reaction region 1, while usually operated as a low-collision-energy reaction region, can be operated as a high-collision-energy reaction region by accelerating the ions immediately after they exit the quadrupole rather than between reaction regions 1 and 2.

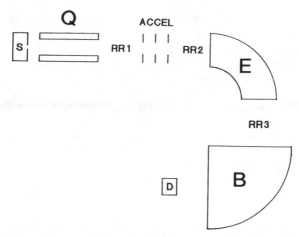

Figure 2-16 ■ Schematic representation of a QEB instrument. Reaction region 1 is used for the dissociation of low kinetic energy ions while reaction regions 2 and 3 are for the dissociations of ions with high kinetic energies.

Low-collision-energy MS/MS experiments are carried out in reaction region 1. Daughter ion MS/MS spectra from this region are obtained by scanning the magnet in the normal manner. Parent ion spectra are obtained by scanning the quadrupole, while obtaining neutral loss spectra requires linked scanning of the quadrupole and magnet. High-collision-energy MS/MS experiments can be performed in reaction region 2 by linked scanning of the sectors, and in some cases, also the quadrupole. The various scans can be found in Appendix A. Placing the quadrupole prior to the sectors alleviates the artifact problems associated with linked scans, and, thus, gives improved data in some experiments over the sole use of the sectors for such scans.

The most important feature of the QEB geometry may be the capability for high-resolution mass measurements on daughter ions and parent ions. This contrasts with sector/quadrupole geometries in which high-resolution mass measurement and separation of the parent ions are possible, but which provide only unit mass resolution of the daughter ions. High-resolution measurements of the daughter ion with the QEB can be done for both high- or low-collision-energy experiments. After low-energy CID in reaction region 1, the resulting daughter ions are accelerated and the EB portion of the instrument is operated in its normal manner, i.e., fixing E and scanning B. Conversely, high-energy CID daughter ion MS/MS spectra can be obtained by operation in a manner analogous to BEB or four-sector instruments in which the collisions occur in reaction region 2, and the EB portion of the instrument is link-scanned at a constant B/E.

Another feature that the QEB geometry offers is the ability to study ion/molecule association reactions, followed by MS/MS analysis of the product ions. These association reactions occur in reaction region 1, and MS/MS spectra of the resulting product ions are obtained via linked scanning at constant B/E for CID in reaction region 2.

The QEB can also be used for the experiments discussed previously for the BEQQ and EQ geometry for cases in which the kinetic energy release associated with dissociation causes overlap in the energies of daughter ions. This is done in one case by mass selecting the parent ion with the quadrupole mass analyzer and then link scanning the EB portion of the instrument at a constant B^2/E for dissociations occurring in reaction region 2. As discussed previously, the B^2/E linked scan retains the kinetic energy release information, and the quadrupole provides the parent ion mass resolution that can otherwise be a problem with this type of linked scan. The complementary operational mode, i.e., mass-resolving daughter ions (formed by high-energy CID) that are unresolved due to kinetic energy release, is the linked scan at constant B/E on a parent ion mass selected by the quadrupole, as discussed in relation to high-resolution mass measurement of daughter ions.

Overall, instruments of QEB geometry are comparable to those of BEQQ geometry in the types of data that can be readily obtained, although there are some differences. Mass resolution differences have been discussed previously. Other differences in MS/MS/MS experiments result from the order of the low- and high-collision energy reaction regions. As discussed previously, the QEB can provide daughter ion MS/MS spectra of selected ion/molecule reaction products. Conversely, the BEQQ (and EBQQ) can be used to study ion/molecule reactions of daughter ions formed in a preceding reaction region. Another MS/MS/MS experiment possible with the QEB is the charge permutation reactions of daughter ions formed in reaction region 1.

While an instrument of QBE geometry has yet to be built, it should be a slightly better geometry for MS/MS than the QEB. This is again due to the mass-selection capability of the magnet early in the experimental sequence of sectors (see Section 2.3.3). Overall, however, the differences between the QBE and QEB are less than those of the BEQQ and EBQQ because the quadrupole has already provided a stage of mass selection.

2.6. MS/MS with Ion-Trapping Techniques

A common feature of the previously discussed MS/MS instruments is that the ions move through these instruments as a "beam," and the different stages of analysis are separated in space. However, two recent developments in MS/MS experiments involve instruments in which the ions are trapped in magnetic and/or electric fields. In these instruments, the different stages of the MS/MS experiment are separated in time. These experiments are carried out with Fourier transform–ion cyclotron resonance spectrometers (FT–ICR) and three-dimensional quadrupoles known as ion trap mass spectrometers (ITMS).

2.6.1. MS/MS with an FT–ICR

As discussed in Section 2.2.6, ions are trapped by a combination of electric and magnetic fields in an FT–ICR. Ions of each mass-to-charge ratio have characteristic cyclotron frequencies. By irradiating an ion with energy at its cyclotron frequency, it can be excited to larger orbits within the FT–ICR cell. This leads to energetic collisions between the ion and the target gas and the ability to perform MS/MS in an FT–ICR.

The MS/MS experiment consists of several steps, in addition to the basic FT–ICR experiment of sequential ionization, excitation, detection, and data transformation. In the MS/MS experiment, all ions except the desired parent ion are excited (accelerated) and ejected from the cell after ionization. Then the parent ion is excited but not to an

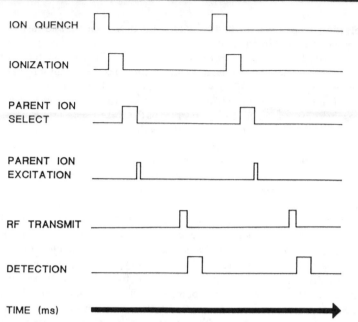

Figure 2-17 ■ Diagram of the pulse sequence used to manipulate the ions in an FT–ICR to obtain a daughter ion MS/MS spectrum.

orbit large enough to cause ejection from the cell. After excitation, the parent ion undergoes collisions with background gas or a collision gas introduced as a pulse into the FT–ICR cell. The daughter ions resulting from CID are then excited, detected, and the signal transformed to give the daughter ion MS/MS spectrum. Figure 2-17 shows the pulse sequences in an FT–ICR used to obtain a daughter ion MS/MS spectrum. The kinetic energy of the ion when it undergoes collisional activation is a function of the power and duration of the excitation rf pulse. In general, the CID reaction in an FT–ICR is in the low-collision-energy region. The maximum kinetic energy to which the ion can be accelerated is a function of the ion mass, magnetic field strength, and FT–ICR cell dimension. Cell dimension is critical since, if the ion is overly excited, the radius of its orbit will be such that it will strike the cell walls and be lost. From equation (2-7), it can be seen that the radius of ion orbit is directly proportional to mass and inversely proportional to magnetic field at a given velocity. Thus, the maximum kinetic energy to which an ion can be excited is inversely proportional to the ionic mass. The larger the cell and magnetic field, the higher this energy. Kiloelectron volt CA energies have been demonstrated in a large cell.[95]

The MS/MS experiment using FT–ICR was first suggested by McIver[96] and first performed in Freiser's laboratory.[97] Freiser's group then went on to demonstrate several capabilities of FT–ICR in MS/MS.[98,99] There are two types of MS/MS experiments especially suited to the FT–ICR type of instrument. The first of these is the multiple MS/MS experiment, i.e., MS/MS/MS, MS/MS/MS/MS, and so on.[98] However, the MS/MS/MS experiment in an FT–ICR is limited to just one of the eight possible MS/MS/MS experiments.[14] The parent ion is selected in the first stage and then reacted, followed by selection and reaction of a specific daughter ion and then detection of those reaction products (the granddaughter ions). Since the daughter ions formed in the MS/MS experiment can be trapped with good efficiency, additional ejection and excitation pulses can also be added to the pulse sequence to perform these multiple MS/MS experiments. In theory, if the daughter ions could be trapped with 100% efficiency, many stages of MS/MS could be completed, limited only by the reaction cross section. In actuality, Freiser has demonstrated MS^4 (MS/MS/MS/MS)[100] and the practical limit will probably be MS^5 or, in very favorable circumstances, MS^6. While this is a very impressive capability, the need for such an experiment to solve an analytical problem is probably rare. Even in the application of these experiments to fundamental studies of chemical phenomena, such capabilities are likely to be used infrequently.

The high-resolution capabilities of FT–ICR are the basis for the other MS/MS experiment to which this instrument is particularly well suited — the measurement of high-resolution MS/MS data. This was first demonstrated by Freiser's group for daughter ion MS/MS analysis.[99] Since ion velocity does not affect the mass measurement in FT–ICR, the kinetic energy release associated with the dissociation does not affect the resolution obtainable on daughter ions as it does with sector instruments. The difficulty of obtaining high mass resolution on daughter ions with FT–ICR is that the resolution is inversely related to the pressure in the system. A high background pressure of collision gas results in decreased mass resolution. This problem has been solved by three methods: introduction of a pulsed gas into the FT–ICR,[101] use of differentially pumped dual cells,[102] and injection of daughter ions formed outside of the FT–ICR.[103] Initial high-resolution experiments provided a daughter ion resolution of about 3000; the pulsed-valve method increased this by about an order of magnitude.[101] Injection of the daughter ions formed outside of the FT–ICR provided a daughter ion resolution of 140,000.[103] Using a differentially pumped cell, a daughter ion resolution of 211,000 has been shown.[104] These impressive resolution capabilities have yet to be applied to real-world analyses, but such applications must be in the foreseeable future.

For most MS/MS experiments using an FT–ICR, the resolution of parent ion selection was unity. However, recent work from Marshall's group with tailored excitation[105] has demonstrated the possibility to excite (or not excite) ions with much higher "resolution" than is possible with the standard excitation process. This may allow high-resolution mass separation of parent ions in an FT–ICR comparable to that obtained with three- and four-sector and sector/quadrupole hybrid instruments. At the same time, the FT–ICR experiment will be able to provide high mass resolution daughter ion measurements.

While FT–ICR offers several distinct advantages for MS/MS experiments, the method has certain disadvantages. A major disadvantage is the inability to perform parent ion and neutral loss scans. In the beam-type instruments, the different stages of MS/MS are performed by different analyzers. The requisite relationship between the two stages of analysis for a parent ion or neutral loss scan can be readily set and data recorded in a manner analogous to obtaining a regular mass spectrum. Conversely, with an FT–ICR in which the different stages of the experiment are separated in time, each parent ion has to be independently excited and the daughter ion MS/MS spectrum obtained. Once the complete MS/MS data matrix has been obtained, parent ion or neutral loss data can be extracted from it. With Marshall's tailored excitation method, specific daughter ions can be detected instead of the complete daughter ion

MS/MS spectrum. However, as in the QTOF instrument this strategy offers only a slight increase in data acquisition rate with a corresponding loss of much of the MS/MS data.

There is another possible problem with FT–ICR due to the fact that the entire MS/MS experiment occurs in the same region in space. In the single-cell instrument, the neutral sample is always present in the FT–ICR cell and may react with the parent and/or daughter ions. This could lead to erroneous conclusions about the reaction sequences. There has been little discussion of this problem in the literature. However, since one of the great strengths of FT–ICR and its predecessor, ICR, is the study of ion/molecule reactions, such reactions (and therefore such complications) would seem likely. Injection of externally formed ions into the FT–ICR or the dual-cell arrangement will greatly reduce the chances of such reactions. An alternate means of identification of such reactions would be to compare MS/MS spectra (and mass spectra) obtained with different time delays prior to detection and to search for peaks that show unusual changes with time.

While the majority of MS/MS experiments with any instrument involve the use of a collision gas, there are other methods of ion activation possible, as will be discussed in the next chapter. Two such methods, in particular, are well suited to the FT–ICR instrument. These are electron-induced dissociation and photodissociation. FT–ICR is especially suited to the latter since the ions are trapped in a well-defined position, and can be held there for an extended time. The photodissociation method has been used quite extensively in conventional ICR, especially by Dunbar's group[106] and by several researchers in FT–ICR.[107-109] Cody and Freiser have pioneered the use of electrons to excite ions trapped in an FT–ICR for MS/MS experiments.[110,111] Both of these excitation techniques have the advantage that the background pressure in the cell is kept as low as possible, and the activation method does not compromise the capability for high-resolution measurements.

2.6.2. MS/MS with an ITMS

There are several differences between an ITMS and an FT–ICR. Since the ITMS uses only electric fields, it is much smaller and simpler than an FT–ICR, especially an FT–ICR with a superconducting magnet. Recent advances in superconductivity give hope for smaller, simpler, and less costly superconducting magnets, but such advances are likely to be many years away. Also, while an FT–ICR is usually operated at the lowest pressures possible (10^{-8} Torr or lower), a bath gas is often added to an ITMS to allow operation at a pressure of about 10^{-3} Torr. A result of these differences is that the ITMS is, at least for the present, not capable of the high mass resolution that an FT–ICR can provide.

There is another major difference between an ITMS and an FT–ICR. The trapping plates of an FT–ICR are biased either positively or negatively, depending upon the polarity of ions to be analyzed; ions of the opposite polarity are not trapped. However, since an rf field is used in an ITMS, both positive and negative ions can be trapped simultaneously. They are also ejected simultaneously and thus can be analyzed simultaneously. In fact, if proper caution is not taken, this could cause problems in interpretation of spectra.

While there are instrumental differences between FT–ICR and ITMS, conceptually the performance of the MS/MS experiment is similar. As with FT–ICR, the sequence in time for an MS/MS experiment using an ITMS is as follows: ionization, ejection of all ions except the parent ion, excitation of the parent ion for CID, and detection of the daughter ions. As with FT–ICR, the various steps in which the ions are "manipulated" are performed by application of an rf voltage to an electrode. For the ITMS, however, generation of the appropriate rf signal is much easier since the manipulation depends mainly upon the rf amplitude instead of rf frequency. It is pertinent to note that the ITMS instrument can also be operated in an FT mode.[112]

MS/MS experiments with an ITMS were first demonstrated by Louris et al.[113] The ring electrode is always operated at a constant frequency and, as discussed in Section 2.2.7, the amplitude of this frequency determines the low mass cutoff for ions trapped in the ITMS. By increasing this amplitude, all ions of mass less than the desired parent ion are ejected from the trap. To eject ions of masses greater than the parent ion, it is necessary to add a supplementary rf voltage to the trap, applied to the end caps. The frequency and amplitude of this supplementary rf voltage are set such that ions of mass just greater than the parent ion are resonantly ejected from the trap prior to ramping the ring electrode rf amplitude to eject the lower mass ions. As the ring electrode rf amplitude is increased, the ions of mass higher than the parent ion will be successively brought into resonance with the supplementary rf voltage on the end caps and ejected. When the ring electrode rf amplitude ramp is stopped (at the point just short of ejection of the parent ion), if there are still higher masses present in the trap, they can be ejected by decreasing the frequency of the supplementary rf voltage on the end caps at the appropriate amplitude. An alternative method of mass selecting the parent ion is to apply a dc voltage (with the rf voltage) to the ring electrode at an appropriate level so that only ions of the desired mass-to-charge ratio have a stable trajectory in the ion trap, and all other masses are ejected.

At this point only the parent ion is left in the ion trap. It can be excited (accelerated) axially by application of a supplementary rf voltage of the appropriate frequency to the end caps, causing the parent ion to undergo more energetic collisions with the background bath gas. The daughter ions from CID are then detected by the

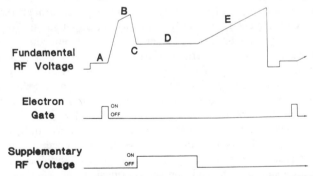

Figure 2-18 ■ Diagram of the rf voltage level on the ring electrode of an ITMS necessary to obtain a daughter ion MS/MS spectrum. A is the ionization, B is the increase in rf voltage to the level to select the parent ion (eject other ions from the ion trap), C is the decrease in rf voltage to the low mass cutoff level for daughter ions, D is the parent ion excitation (CID with supplementary rf voltage applied to end caps), E is the scan of the rf voltage to eject sequentially and detect the daughter ions formed. (Adapted from reference 113.)

conventional rf amplitude ramp of the ring electrode. A schematic of the sequence of events necessary to perform an MS/MS experiment with an ITMS is shown in Figure 2-18. Multiple stages of MS/MS (MS^n) can be carried out by insertion of additional ejection and excitation steps into the pulse sequence.

Like FT–ICR, ITMS has only been used to perform daughter ion scans. Parent ion and neutral loss data can be extracted from the complete MS/MS data matrix that results when the daughter ion MS/MS spectra of all ions in the mass spectrum are obtained. Also, like the FT–ICR, since the entire experiment takes place in the same region in space, there is the possibility of unwanted side reactions occurring, such as ion/molecule association reactions, charge exchange, or other reactions between parent and/or daughter ions with neutral analyte or background contaminants. Again, injection of externally formed ions or time-delay experiments would help to diminish or identify such processes.

2.7. Reaction Regions

2.7.1. Kiloelectron-Volt Ion Kinetic Energy Reaction Regions

Perhaps the most important part of the MS/MS experiment (once the sample is ionized) is the change in mass and/or charge that occurs in the reaction region. In early studies of metastable ions using sector instruments (and for most metastable studies today), the reaction region is the region between analyzers, or after the acceleration region and prior to the first analyzer. These regions have no magnetic or electric fields (other than fringing fields) and were thus called field-free regions. The dimensions of these regions can range from a few centimeters to more than a meter in length.

The use of a collision gas to induce dissociations[9–11] led to the advent of collision cells[114] to increase the gas pressure in a defined, localized region within the mass spectrometer. A typical collision cell is about 1 cm long, with its entrance slit normally located at an ion optical focal point. Ideally, the collision cell is differentially pumped to minimize the gas load in the rest of the mass spectrometer. A disadvantage of a conventional collision cell is that the only exit for the gas is through the ion entrance and exit slits. While the cell itself is only 1 cm long, the target gas effusing from the cell makes the actual high-pressure region longer and less well defined.

As an alternative to this conventional arrangement, McLafferty et al. designed a molecular beam that was perpendicular to the ion beam.[52] A much simpler version of this idea was demonstrated by Glish and Todd; a hypodermic needle was inserted at the focal point normal to the ion axis and the gas flow from this needle directed into a diffusion pump.[115] Both of these methods gave improved performance over conventional collision cells and have the added advantage of not requiring the insertion of additional slits into the ion beam path. This latter fact is important in cases in which a collision region is added to an existing instrument. In such modifications, the mechanical alignment of a conventional collision cell is critical.

One other aspect of high ion energy collision regions is of notable importance. It is desirable to be able to float electrically the collision cell in the reaction region at some potential different from that of the field-free region in general. This allows the daughter ions that result from CID to be separated from daughter ions formed in metastable processes. The latter ions are formed throughout the length of the field-free region and thus have a different energy from the CID ions formed within a collision cell reaction region held at a different potential.

To illustrate this point, assume that a parent ion of m/z 100 dissociates to a daughter ion of m/z 60 in the second reaction region of a BE geometry instrument. If the ion source is held at 8 kV and the analyzers are at ground potential, the daughter ions in a conventional MS/MS experiment would have an ion energy of 4.8 keV, whether formed by CID in the collision cell or by metastable dissociation. However, if the collision cell is floated at 6 kV, parent ions undergoing CID would have a collision energy of 2 keV. The m/z 60 daughter ion would then have an ion energy of 1.2 keV from the conservation of momentum of the fragmentation, plus an additional 6 keV of acceleration upon exiting the collision cell. This gives a total daughter ion energy of 7.2 keV. The difference in kinetic energy is used to differentiate between ions formed in the two processes.

An alternative to CID is photodissociation. There are two possible geometries for photodissociation studies of ions with high kinetic energies: crossed photon and ion beams or coaxial photon and ion beams. Since the number density of ions in a beam is low, the latter method is advantageous in that it has a substantially longer effective path length, although some work has been done with crossed-beam configurations in which the reaction region was part of the laser cavity itself.[116,117]

Photodissociation work in the field of MS/MS has been pioneered by Beynon and co-workers using coaxial photon and ion beams.[118-120] As more powerful lasers have become available, crossed-ion and photon beam experiments in which the reaction region is not in the laser cavity have been performed.[121] For photodissociation experiments in general, even with the longer path lengths of the coaxial geometries, the cross section for photodissociation is such that the abundances of the photodissociated daughter ions are quite low. The photon beam is generally chopped so that a phase-sensitive detection method can be used to differentiate the products of photodissociation from those of metastable ion fragmentations, which are often more abundant.

Another means by which ions can be excited is by interaction with electrons. Recently, two different methods for this experiment have been proposed: electron-induced dissociation[122,123] and electron-capture-induced dissociation.[124] The latter experiment can only be applied to multiply charged positive ions, and there are some problems in the deconvolution of the signal originating in this process. Additional study of both of these methods is needed to assess their utility and practicality.

2.7.2. Electron-Volt Ion Energy Reaction Regions

For low-collision-energy CID work, conventional collision cells have often been used. However, since ions with a few electron volts of kinetic energy have relatively low ion velocities, it is possible to focus ions scattered by collision from the central axis of the ion path back toward this axis by the use of a strong focusing device. A quadrupole operated in the rf-only mode is such a strong focusing device and can be used as a reaction region for either CID or photodissociation. As noted in Section 2.2.3, quadrupoles operated in the rf-only mode pass all ions above a certain cutoff mass with equal efficiency, but this is not strictly true.[125] When an rf-only quadrupole is used as a reaction region, mass discrimination in ion transmission is a problem that must be considered. In general, there are two possible modes of operation in the measurement of MS/MS spectra using an rf-only quadrupole as a reaction region: constant parent ion transmission or constant daughter ion transmission.[17]

Constant parent ion transmission involves setting the rf amplitude of the rf-only quadrupole at a fixed fraction relative to the rf level that would be necessary to pass the parent ion if the quadrupole was operated in the mass filter mode. As the mass of the parent ion increases, the rf amplitude on the rf-only quadrupole enclosing the reaction region is also increased. In this mode of operation, lower rf amplitudes give a more uniform transmission of ions over the entire mass range of daughter ions, but at a lower overall efficiency of transmission. Conversely, if the rf amplitude is set at too high a level, daughter ions of lower mass will not be efficiently transmitted through the quadrupole, and in extreme cases, may not be observed in the daughter ion MS/MS spectrum.

The constant daughter ion transmission mode of operation involves scanning the rf amplitude of the reaction region quadrupole. The rf level is varied with the setting of the final mass analysis quadrupole so that each daughter ion is passed through the rf-only quadrupole with an equal efficiency. The problem with this mode of operation is that the parent ion transmission through the rf-only quadrupole will vary. This will lead to increased parent ion losses as the daughter ion mass is decreased and to discrimination against low-mass daughter ions in the MS/MS spectrum. Means of correcting for this discrimination effect have been proposed.[17]

Recently, a new type of reaction region has been demonstrated for MS/MS that uses collisions with a surface to cause surface-induced dissociations (SID).[126] Parent ions impinge upon the surface at low collision energies, and the reflected parent ions and the daughter ions formed as the result of the collision are then extracted and analyzed. Typically, the angle between the incident beam and the emerging beam is 90°. This type of reaction region has several advantages. It minimizes the pumping needed in the reaction region since no extra gas is introduced into this region as is the case with collision with a neutral target gas. The SID experiment is simple, with no mechanical, electrical, or optical components such as those involved with photodissociation and with electron-induced dissociation processes. Finally, a solid surface placed in the path of the incident ion beam provides a 100% interaction efficiency, although that does not necessarily mean that SID will have a higher overall cross section than does CID. The chemistry and physics of SID and remaining activation techniques are discussed in Chapter 3.

Reactions in MS/MS

3.1. Introduction

Chapter 2 described in detail instrumentation for MS/MS and discussed the relative merits of the different types of mass spectrometers used for MS/MS experiments. This chapter emphasizes the variety of reactions that can occur and be detected in an MS/MS instrument after the ions leave the ion source (or after the first stage of mass analysis in an ion-trapping instrument) and the fundamental information that these reactions can provide.

The most commonly observed reaction type is that of unimolecular dissociation, which can occur either spontaneously from metastable ions or from initially stable ions rendered unstable following an activation reaction. The pathways through which an ion dissociates are dependent on the ion structure and therefore can be useful in ion identification. The daughter ion experiment is especially useful in that it directly provides parent ion/daughter ion relationships that may not be obvious from the conventional mass spectrum. Of all reactions possible for polyatomic molecules, next to ionization itself, unimolecular dissociation has and will continue to be most useful.

A second general type of reaction in MS/MS is that of activation of ions in a reaction region. This is pursued to induce unimolecular dissociation and can be accomplished by several means. The most widely used method is collisional activation, either at ion translational energies of a few tens of electron volts or a few thousand electron volts. Fragmentation resulting from collision is referred to as collision-induced dissociation (CID). Excitation of ions by photon absorption, for instance, via laser irradiation, has also been widely used. The resulting fragmentation is commonly referred to as photodissociation. Recently, experiments have been described in which ions are excited by collision with a solid surface, giving rise to surface-induced dissociation. Finally, excitation of polyatomic ions with an electron beam has been used. Details of the methods for the activation of ions are particularly important in MS/MS because, in most cases, the ionization method is chosen to impart as little energy to the ion as possible and to minimize fragmentation in the ion source. Once the energetically "cold" ions are mass-selected in the first stage of MS/MS, it is necessary to induce them to fragment in the reaction region so that structural information can be obtained by a second stage of mass analysis. Important characteristics of an activation method are the amount of energy that can be imparted to an ion, the narrowness with which the range of this energy can be defined, the ease with which the amount of energy added can be varied, and the cross section of the activation reaction (i.e., the probability that it will occur).

If conditions of collision energy and frequency are so selected, associative as well as dissociative collisions can occur in a reaction region. In an ion/molecule collision, one or more atoms can be exchanged between the collision partners (the mass-selected ion and the neutral target gas molecule) in a short-lived collision complex. This type of reaction typically occurs at collision energies of less than about 20 eV and is most commonly observed in high-pressure ion sources and in ICR mass spectrometers. In these cases, the reactions occur at, or near, thermal energies. Ion/molecule reactions in MS/MS, at collision energies of several electron volts, have been observed in the reaction regions of triple-quadrupole and hybrid mass spectrometers, as well as in ICR instruments. The reactions that occur are, of course, characteristic of the chemical reactivity of the ion and therefore the functional groups within the ion. As with dissociation reactions, association reactions can be used to provide additional information about an ion in an MS/MS analysis.

A final class of ion/molecule reactions described in this chapter are charge permutation reactions, defined as reactions in which the ion and/or neutral collision partner undergo a change in charge. An exothermic charge transfer reaction can occur at any collision energy. However, some charge permutation reactions are highly endothermic and require keV collision energies for their initiation and observation. Certain charge permutation reactions are analytically useful and, in some cases, such reactions have been used to obtain fundamental information about ion structures

In this chapter, each of the reaction types is discussed in greater detail, with particular emphasis on how each reaction is related to an MS/MS experiment. Unimolecular dissociation, for example, is described only insofar as necessary to make clear those factors that apply to the MS/MS experiment. The activation reaction central to most MS/MS experiments is discussed in greater detail, particularly the reaction of collisional activation. Associative ion/molecule reactions and charge permutation reactions are then described. Illustrative examples are included throughout the chapter. Further examples can be found in Chapter 4, which is devoted to the application of MS/MS to fundamental studies such as ion structure differentiation, reaction mechanism elucidation, and thermochemical determinations.

3.2. Unimolecular Dissociation

Unimolecular reactions are widely studied in chemistry since the dissociations of small, isolated ions provide fundamental information about the rates of dissociation reactions, the partitioning of energy, and the validation of models developed to describe such reactive systems. The theory of unimolecular reactions, referred to both as quasiequilibrium theory and RRKM theory (after Rice, Ramsberger, Kassel, and Marcus), has been described in detail elsewhere.[127-130] Mass spectrometry has been a useful tool in testing the assumptions of the theory because reactions can take place in collision-free conditions (isolated reactants and products), and the ionic species can be selected and identified with high specificity. In general, the theory has been successful in the rationalization of conventional electron impact ionization mass spectra of simple polyatomic molecules and has therefore been adopted as a useful statistical theory of mass spectra. Several versions of the theory have been described with various degrees of stringency in the basic assumptions.[127-130] The extent to which the basic assumptions are obeyed by real systems is still an active area of research,[131] particularly for those seeking the elusive goal of state-selective chemistry. For the purposes of this text, the four basic assumptions of the theory are as follows:

1. The time required for dissociation of a polyatomic ion is long compared to the time required for its formation and excitation.
2. The rate of dissociation of an ion is much slower than the rate of redistribution of excitation energy over all internal modes of the ion.
3. The observed dissociation products result from a series of competing and consecutive reactions.
4. Ions achieve a condition of internal energy equilibrium in which energy is distributed over all internal states with equal probability.

The assumption that energy is randomized throughout the ion prior to reaction has important implications for mass spectrometry. Under conditions normally attained in a mass spectrometer, weak bonds are preferentially broken, providing fragment ions that reveal the structure of the parent ion. This behavior is consistent with the assumptions outlined previously, which implies that fragmentation depends upon the total energy in the ion system, but not in how or where the energy was deposited. In single-stage mass spectrometry, the energy for fragmentation is deposited into the system during ionization, and the statistical theory assumes that the ion has no memory of how it was formed. In MS/MS, energy can be deposited into the ion both during ionization *and* in a subsequent activation reaction. The assumptions therefore require that the ion has no memory of how it was formed, nor of how or where it was subsequently excited. This point will be further addressed in the section on activation methods. Provided the basic assumptions of the statistical theory are obeyed, what follows is a qualitative description of the factors that determine the daughter ion MS/MS spectrum.

The daughter ion MS/MS spectrum of a mass-selected polyatomic parent ion, like the mass spectrum of a compound, is determined by the internal energy of the parent ion(s), the microscopic rate constants for each possible dissociation, and the time during which the network of reactions is allowed to occur. For a population of ions with a distribution of internal energies, observed over a range of ion lifetimes, the resultant daughter ion MS/MS spectrum provides a fragmentation pattern integrated over time and internal energy.

Central to the theory is the internal-energy-dependent rate constant $k(\epsilon)$ that characterizes each possible dissociation. For illustrative purposes, the simplified form of the rate constant equation is given by:

$$k(\epsilon) = \nu\left[(\epsilon - \epsilon_0)^{(s-1)}\right]/\epsilon \qquad (3\text{-}1)$$

where ϵ is the internal energy of the ion, ϵ_0 is the critical energy for a reaction, s is an effective number of oscillators, and ν is the ratio of the product of the vibrational frequencies of the activated complex (i.e., the transition state ion structure) to that of the parent ion. The term ν is often referred to as the frequency factor and is a measure

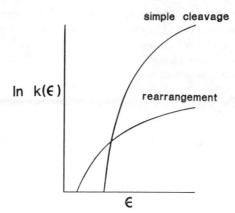

Figure 3-1 ■ Hypothetical $\ln k(\epsilon)$ vs. ϵ curves for a simple cleavage fragmentation and a dissociation reaction involving rearrangement.

of the entropic constraints on the reaction. When $\epsilon \gg \epsilon_0$, the value of $(\epsilon - \epsilon_0/\epsilon)$ approaches unity, and the rate constant approaches the frequency factor. When ϵ is only slightly greater than ϵ_0, the rate constant is substantially less than the frequency factor. The detailed rate constant expression includes the ratio of the sum of vibrational and rotational states in the activated complex with an energy less than or equal to $(\epsilon - \epsilon_0)$ to the density of states in the parent ion in the interval $(\epsilon + d\epsilon)$. The rate constant is also determined by the number of equivalent ways through which the activated complex can be reached from the parent ion and is constrained by conservation of angular momentum.[131-133]

A plot showing behavior typical of $\ln k(\epsilon)$ versus ϵ for a dissociation involving a rearrangement, and a simple cleavage dissociation, is given in Figure 3-1. Rearrangement reactions generally have low critical energies and low frequency factors due to the unfavorable entropy change necessary to reach the transition state. Simple cleavages often have higher critical energies and higher frequency factors than do rearrangement reactions. Ions that can dissociate via competitive rearrangement and simple cleavage reactions generally dissociate via the rearrangement process at low internal energies (due to a lower critical energy), whereas at higher internal energies, the simple cleavage reaction tends to dominate due to a higher frequency factor.

Several aspects of the unimolecular dissociations of polyatomic ions are exemplified with the dissociation reactions of the tetramethylammonium cation. The cation dissociates to two predominant daughter ions over a relatively wide range of internal energies. These daughter ions correspond to the loss of a methyl group, and loss of methane, viz:

$$(CH_3)_4N^+ \rightarrow (CH_3)_3N^{+\cdot} + CH_3^{\cdot} \tag{3-2}$$

$$(CH_3)_4N^+ \rightarrow (CH_3)_2NCH_2^+ + CH_4 \tag{3-3}$$

Experimental results indicate that the latter product ion arises from two mechanisms, a stepwise loss of a methyl group and hydrogen radical and a concerted loss of methane.[134] The kinetic scheme for the dissociation of this ion is relatively simple,

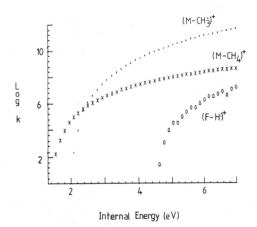

Internal Energy (eV)

Figure 3-2 ■ The $\log k(\epsilon)$ vs. ϵ curves calculated for the dissociation reactions of $(CH_3)_4N^+$ indicated in equations (3-4) and (3-5).

containing two competitive reaction channels and one consecutive reaction, viz:

$$(CH_3)_4N^+ \xrightarrow{k_1(\epsilon)} (CH_3)_3N^{+\cdot} + CH_3^\cdot \xrightarrow{k_2(\epsilon)} (CH_3)_2NCH_2^+ + H^\cdot \qquad (3\text{-}4)$$

$$(CH_3)_4N^+ \xrightarrow{k_3(\epsilon)} (CH_3)_2NCH_2^+ + CH_4 \qquad (3\text{-}5)$$

A set of calculated $\log k(\epsilon)$ versus curves for these three reactions is shown in Figure 3-2.[134] The concerted loss of methane was modeled as a rearrangement reaction while the other losses were assumed to be simple bond cleavages. As the figure indicates, methane loss is the only process that occurs at energies near the threshold for dissociation. Soon after the critical energy for methyl loss is reached, the rate for this reaction surpasses that of methane loss. At still higher internal energies, the further loss of a hydrogen atom proceeds rapidly. With the appropriate kinetic equations, the respective $k(\epsilon)$ values, and a time over which the reactions can occur, the breakdown curve can be calculated. The breakdown curve shows the relative abundances of the fragment ions as a function of internal energy in the dissociating ion. The breakdown curve for the tetramethylammonium cation is shown in Figure 3-3, which indicates the two processes that give $(CH_3)_2NCH_2^+$ separately. The breakdown curve shows that ions with internal energies in the range of (0.5–2.5 eV) fragment exclusively by loss of methane via a concerted mechanism. At intermediate energies, only methyl loss occurs. At the highest energies used in the calculation, loss of methyl and hydrogen dominates. In this particular example, only over very narrow internal energy ranges (2.5–3.0 eV and 4.8–5.2 eV) do products from two processes appear simultaneously.

Rarely are ions with a narrowly defined distribution of internal energies formed in mass spectrometry.[135] It is rarer still for ions sampled in MS/MS to be characterized by a narrow internal energy distribution, particularly after an activation reaction. The mass spectrum or MS/MS spectrum, therefore, is a product ion distribution resulting from the dissociations of parent ions with a relatively wide range of internal energies. To reproduce the final spectrum accurately, the breakdown curve must be multiplied by the parent ion internal energy distribution. The daughter ion MS/MS spectra of the

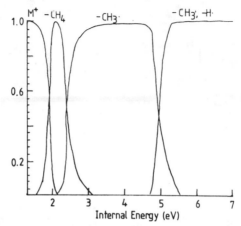

Figure 3-3 ■ Calculated breakdown curve for $(CH_3)_4N^+$ based on the reactions of equations (3-4) and (3-5) and the $\ln k(\epsilon)$ vs. ϵ curves of Figure 3-2.

tetramethylammonium cation predicted for Boltzmann parent ion internal energy distributions characterized by temperatures of 1500 and 2400 K are shown in Figure 3-4. Contributions from both mechanisms to methane loss are indicated. As is the case in these examples, when the internal energy distribution is broad, simultaneous contributions to the final spectrum can be observed from many or all of the ions that appear in the breakdown curve up to the highest internal energies involved. Rigorously, the Boltzmann distribution is not usually a realistic internal energy distribution for ions in mass spectrometry, since the ions that fragment are rarely in thermal equilibrium. The actual distribution is usually unknown, but is typically quite broad, as is the Boltzmann distribution.

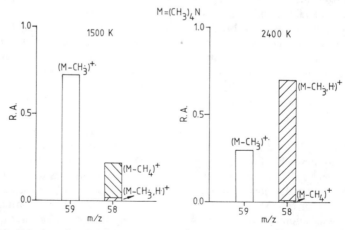

Figure 3-4 ■ Daughter ion spectra of $(CH_3)_4N^+$ predicted for Boltzmann internal energy distributions characterized by temperatures of 1500 and 2400 K.

The background discussion given here for unimolecular dissociation is meant to illustrate that daughter ion spectra in MS/MS, like mass spectra, are determined by the energy-dependent rate constants of the possible reaction channels, the time over which the reactions can take place, and the internal energy distribution of the sampled ions. Particular emphasis is given to the effect of internal energy by describing the breakdown curve and pointing out the importance of the internal energy distribution. It is important to be able to vary the internal energy of the ions to maximize the amount of information obtainable from the mass spectrum. Methods for energy resolution are described in detail in Section 3.3.

Time resolution is another useful technique to obtain more detailed information on an ion. Since ion lifetimes are inversely related to the rates of dissociation, acquisition of mass spectra as a function of ion lifetime provides information very similar to that contained in the breakdown curve. A number of methods have been used to obtain time-resolved mass spectra in various ion lifetime windows. For example, field ionization kinetics can provide mass spectra as a continuous function of ion lifetime from 10^{-11} to 10^{-9} s.[136,137] Metastable ions represent dissociations of ions with lifetimes on the order of microseconds. Lifshitz and co-workers have obtained electron ionization[138,139] and photoionization[140] mass spectra as a function of time up to a few milliseconds. Ion cyclotron resonance spectrometers and ion traps are capable of storing ions for many seconds and are therefore capable of studying ions with essentially infinite lifetimes.[141] Lifshitz et al. have also used ion trapping in the space charge of an electron beam in conjunction with the time-resolved magnetic dispersion MS/MS method of Enke[44] to make time-resolved kinetic energy release measurements of metastable ions.[142-144] Very little effort has been made to develop methods to obtain daughter ion spectra even crudely as a function of ion lifetime following an activation step. One paper by Morgan et al.[145] reports ion lifetimes following collisional activation. By electrically floating the collision cell in the second reaction region of a BE spectrometer, dissociations within and outside of the collision cell could be distinguished. For daughter ions to be formed within the collision cell, rate constants of 2×10^7 s^{-1} or greater were required. For the ions studied, it was noted that more than 99% of the parent ions fragmented within the collision cell following keV collision energy collisional activation. Clearly, it is desirable to obtain detailed ion lifetime data at the much shorter times typical for highly excited parent ions. In most MS/MS instruments, ions typically have at least a microsecond after the activation reaction to dissociate before they enter the next stage of mass analysis. It has proved to be experimentally easier to vary the amount of internal energy deposited into the ion in the activation reaction than to vary the time window of study. A number of experiments that provide this former capability are described in Section 3.3

Metastable Ions. Ions in a mass spectrometer can be classified rather arbitrarily as stable, unstable, or metastable, depending upon when, or if, they fragment during their passage through the instrument. Stable ions are those that have lifetimes longer than the time of passage through the instrument. Unstable ions are those that fragment within the ion source, and metastable ions dissociate outside of the ion source but before detection, without the benefit of an activation reaction separate from ionization. Figure 3-5 shows a hypothetical internal energy distribution for ions formed in the source of a mass spectrometer. Ions with high internal energies can fragment with rate constants greater than about 10^6 s^{-1} and therefore do so before leaving the ion source. Ion source residence times can vary markedly with the ionization method and instru-

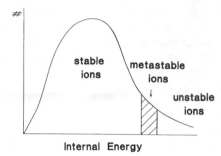

Internal Energy

Figure 3-5 ■ Hypothetical internal energy distribution $[P(\epsilon)$ vs. $\epsilon]$ indicating regions that lead to stable, metastable, and unstable ions on the time scale of a typical beam-type mass spectrometer.

mental parameters. For example, source residence times in electron ionization sources are on the order of microseconds,[135] whereas in chemical ionization, residence times can be up to 0.1 ms[146] due to the higher pressures of reagent gases used in these sources. In most MS/MS instruments, the ion transit time is from tens of microseconds to a millisecond, depending upon the mass of the ion, its kinetic energy, and the distance it must travel. Metastable ions have intermediate internal energies and therefore dissociate with intermediate rate constants. Although metastable ions can be the result of electronic and rotational predissociation and tunneling through a potential barrier,[147] most metastable polyatomic ions are vibrationally excited ions with internal energies slightly above the threshold for dissociation. This mechanism of metastable ion formation has been referred to as vibrational predissociation.[147] Much can be learned from the study of metastable ions, as discussed in a book and several recent reviews.[12, 148, 149] The discussion here is limited to those aspects of metastable ions that are useful in demonstrating key points regarding MS/MS.

The first modern MS/MS instruments, BE instruments, are frequently used to study the dissociations of metastable ions. Only dissociations in two well-defined regions of these instruments can be studied, viz the first and second reaction regions. Figure 2-8 shows a schematic diagram of such an instrument, and equation 2-2 gives the velocity of an ion in meters/second for a given accelerating voltage and mass and charge of the ion. For ions with 5000 eV of kinetic energy and masses less than about 2500 daltons, the velocities result in flight times within the instrument of a few tens of microseconds. Only ionic products of dissociations that take place within reaction regions 1 and 2 (see Figure 2-8) can be mass-analyzed. Ions spend only a few microseconds in each of these regions, and this short residence time narrowly defines the range of dissociation rate constants that can be sampled. This range, therefore, also narrowly defines the range of ion internal energies that are sampled. For most metastable ions in the BE type of spectrometer, the range of internal energies falls within 0.1 to 1 eV above the threshold for decomposition.

In the dissociation of a singly charged polyatomic ion, represented by:

$$m_p^+ \rightarrow m_d^+ + m_n \tag{3-6}$$

some fraction of the internal energy in the reaction coordinate in excess of that of the ground-state products is partitioned into the kinetic energy of separation of the fragments. This so-called *kinetic energy release*, symbolized by T, leads to a spread in

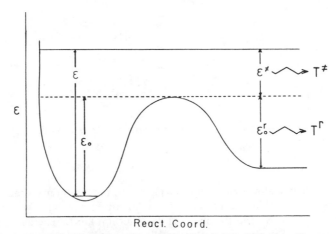

Figure 3-6 ■ Origin of kinetic energy release, T, for the case in which a significant reverse critical energy for the reaction exists.

the kinetic energy of the daughter ion, since the daughter ion is ejected isotropically in the center-of-mass frame of reference. The excess internal energy can be viewed as consisting of two components, the reverse critical energy and the nonfixed energy. The reverse critical energy is the energy difference between the ground state of the products and the critical energy for the fragmentation. The nonfixed energy is the internal energy in excess of the critical energy. Figure 3-6 shows schematically the origin of T for a fragmentation with a large reverse critical energy, i.e., a reaction in which the ground state of the products is significantly lower than that of the activated complex. The observed kinetic energy release for this reaction includes a fraction of the reverse critical energy ϵ^r and a fraction of the nonfixed energy $\epsilon^{\neq}$ such that:

$$T = T^{\neq} + T^r \tag{3-7}$$

where $T^{\neq}$ is the contribution from $\epsilon^{\neq}$ and T^r is that from ϵ^r. When r is negligible, the value of T is largely due to $T^{\neq}$. In reactions for this case, a linear relationship has been observed between $\epsilon^{\neq}$ and $T^{\neq}$ for a number of fragmentations,[150] viz:

$$T^{\neq}/\epsilon^{\neq} = 1/(\alpha\Phi) \tag{3-8}$$

where Φ is the number of degrees of freedom and α is an empirical parameter with an average value of 0.44. [A more accurate, though less simple, relationship has been given.[151] Equation (3-8) has been reassessed recently and is found to be valid within certain limits.[152]] No simple relationship between T^r and ϵ^r has been observed, but the fraction T^r/ϵ^r has been observed to vary from 0.1 to 1.0.[153] The factors that determine how energy is partitioned between T and internal modes upon fragmentation, particularly in cases where ϵ^r is large, are not well understood. Qualitatively, large T^r/ϵ^r values are observed when the energy in the reaction coordinate is not efficiently coupled to the other degrees of freedom in the activated complex.[154,155]

When dissociation occurs in the second reaction region of a BE instrument, the ionic products can be analyzed for their kinetic energy by a scan of the electric sector plate voltages. In the absence of a kinetic energy release, the observed width of the

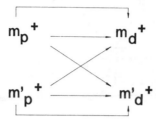

Figure 3-7 ■ Schematic indication of possible origins of non-Gaussian peak shapes for dissociations of metastable ions.

daughter ion peak would be equal to the daughter ion/parent ion mass ratio multiplied by the energy spread of the parent ion. For keV kinetic energy ions, the kinetic energy release T leads to a spread in the observed daughter ion kinetic energies that is much larger than T itself. The kinetic energy release is amplified in the direction of ion flight, an effect associated with the conversion from the center-of-mass reference frame to the laboratory frame of reference.[156] In practice, the observed energy spread arises from a distribution of T values. This distribution of released kinetic energies follows from the statistical nature of unimolecular dissociation. Peak shapes in the kinetic energy spectrum of daughter ions from metastable ions are heavily influenced by the magnitude and distribution of released kinetic energies. In general, the peak shape for a dissociation proceeding via a single mechanism with an average kinetic energy release of up to few hundred millielectron volts is Gaussian. In practice, peaks are often observed not to be purely Gaussian. This usually indicates mixtures of parent ions, mixtures of product ions, or that several mechanisms are involved.[12,148,149] These possibilities are depicted schematically in Figure 3-7. Large kinetic energy releases can result in the so-called flat or dish-topped peaks. Dishing is a discrimination effect due to finite instrumental slit height in the z direction.[12,148,149]

The BE geometry mass spectrometer is a powerful tool for kinetic energy spectroscopy due to its use of kinetic energy analysis with the electric sector and the amplification of kinetic energy release in the laboratory frame of reference. However, this amplification and the use of an electric sector as the final analyzer degrade the attainable mass resolution in daughter ion mass analysis. Alternative geometries of MS/MS instruments have been constructed to avoid this degradation (see Chapter 2). These spectrometers are much less sensitive to kinetic energy release and therefore have better mass-resolving power. They do not provide the peak shape and position information available from kinetic energy analysis.

Metastable ions typically have lifetimes of several tens of microseconds and internal energies in excess of the critical energy for the lowest energy dissociation of several tenths of an electron volt. Metastable ions therefore typically produce few product ions in their dissociative reactions, and those product ions that do form result from the lowest energy dissociation mechanisms. These dissociations are characteristic of the so-called reacting configuration of the ion. This configuration may or may not also characterize the structure of stable ions in the mass-selected beam used in MS/MS experiments. Figure 3-8 shows a two-dimensional energy diagram to illustrate this point. If a rearrangement reaction has a critical energy less than that of the lowest energy dissociation reaction, the rearrangement can occur *before* the ion fragments.

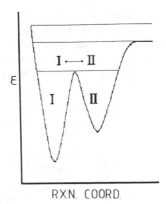

RXN. COORD.

Figure 3-8 ■ Energy diagram indicating a situation in which a metastable ion can rearrange prior to fragmentation.

The possibility of rearrangement and the fact that there are usually few peaks in the metastable ion spectrum, together restrict the general use of metastable ions in the differentiation of stable ion structures. Furthermore, the mechanisms for the dissociation of many metastable ions often involve rearrangements in conjunction with fragmentation, since rearrangement reactions generally have lower critical energies than simple cleavages. Rearrangement reactions are structurally less informative than simple bond cleavages, and, since they are low critical energy reactions, they are more sensitive to differences in ion internal energy.[157-159] Fragmentations of metastable ions are therefore strictly only characteristic of the reacting configuration.[148] The role of metastable ions in ion structure studies is discussed further in Section 4.2.1.1. Means of exciting the stable ions in a mass-selected beam have been developed to differentiate between stable ion structures and are discussed in the following section.

3.3. Activation Reactions

In most MS/MS applications, it is desirable to minimize fragmentation in the ion source to concentrate charge in one ion representative of each neutral component introduced into the source. Soft ionization methods (see Chapter 5) are therefore preferred. To obtain additional information from the ion with a second stage of mass analysis, some change in mass and/or charge of that ion must occur. Relatively few metastable ions can be observed in the mass-selected population of ions observed in a BE instrument. Usually only the lowest energy dissociation reactions are observed for metastable ions, and these reactions may not characterize the stable ions that make up the majority of the mass-selected ion population, as discussed in the preceding section. Methods of excitation for ions have been developed to increase the amount of their internal energies. Such excitation increases the absolute number of ions that dissociate and increases the likelihood for structurally diagnostic simple cleavage reactions.

Several methods for the excitation of ions in a reaction region have been developed. The most widely used method is the collision of a high-velocity ion with a target gas [referred to as collisional activation (CA)]. As discussed in Chapter 2, different MS/MS instruments allow for CA at ion kinetic energies in excess of a few kiloelectron volts, at kinetic energies less than 100 eV, or both. Other methods of ion activation include interaction of the ion with an intense beam of photons, a beam of electrons, or via collision of the ion with a surface. Each of these methods will be discussed, with particular emphasis given to collisional activation. Important characteristics used to evaluate the various techniques include:

1. the amount of energy that can be deposited into the ion;
2. the distribution of energies deposited (i.e., how closely the amount of energy can be defined);
3. how readily the deposited energy can be varied; and
4. the cross section for the activation reaction (i.e., the probability that it will occur).

3.3.1. Collisional Activation

The collisional activation of polyatomic ions has been observed in mass spectrometers since the earliest days of the development of the technique. Indeed, due to the poor vacuum in the earliest mass spectrometers, collisions between ions and residual neutral atoms and molecules in the ion flight path were unavoidable. Anomalous signals due to dissociations resulting from these collisions (and concurrently from metastable ion dissociations) were correctly identified by Aston as arising from reactions in a field-free region.[160] Until the explicit experiments of MS/MS, however, collisions in field-free regions (reaction regions) were artifacts to be reduced with better instrumentation, and the detailed study of the ionic products of dissociations in these regions was largely ignored. Until the early 1970s, only a few studies regarding the collisional activation of organic ions appeared.[9-11,161] McLafferty and co-workers described early research that recognized collisional activation as a tool in organic mass spectrometry for the differentiation of ion structures.[162,163] Initial studies involved collisions of keV energy projectile ions with neutral collision gas targets.[22,23] Multiquadrupole instruments were developed for MS/MS at a later time, and collisional activation of lower kinetic energy ions (eV range) resulting in fragmentation was then also pursued.[63,64,164] In preservation of the original distinctions, most MS/MS instruments currently in use are configured to study either keV kinetic energy projectile ions for CA or ions with kinetic energies less than 100 eV. A few papers have described studies that use the intermediate collision energy region of 100 to 1500 eV; such studies are described later in this section.

There are two primary reasons why CA was the first activation method to be used for analytical MS/MS and why it is currently the most widely used method. First, the cross sections for collision-induced dissociations of polyatomic ions are generally 10 to 200 Å^2. These cross sections, which indicate the probability for the reaction, are several orders of magnitude larger than those for other activation reactions such as photodissociation. Second, the number density of the target gas is easily variable over several orders of magnitude. In an instrument with a differentially pumped reaction region, a target gas pressure sufficient to attenuate completely the main beam can be maintained without significant compromise in the subsequent mass analysis of daughter ions. As in any type of beam experiment, a Beer's law relationship applies to the typical MS/MS experiment, i.e.:

$$I = I_0 e^{-n\sigma l} \tag{3-9}$$

where I_0 is the main beam current in the absence of the target gas, I is the attenuated main beam current, n is the target gas number density, l is the path length, and σ is the cross section for all processes that contribute to the loss of main beam current, of which CID is one. The other major competing loss processes are charge transfer to the target gas and scattering of the ions out of the collection region of the second mass analyzer.[165,166] The importance of these loss processes can be minimized with the proper choice of the target gas and its number density. With a large cross section for CID (σ_{CID}), and a readily variable target gas density (n), parent ions can be induced to fragment in a relatively simple collisional activation experiment. It is also likely that when the target gas density increases to the point at which multiple ion/target collisions are probable, the cross section for CID of the ion that has already undergone one collision is significantly larger than for an ion that has not. Experiments that exploit this phenomenon are only now being completed. In beam-type MS/MS instruments, the path length of the reaction region is not generally variable. In Fourier transform–ion cyclotron resonance mass spectrometers (FT–ICR, Section 2.6), however, the total path length of the ion can be on the order of meters and can be readily varied by controlling the time between initial mass selection and the second mass analysis. This makes FT–ICR particularly useful in experiments based on activation methods with low cross sections.

In most applications of collisional activation, the masses and relative abundances of the daughter ions are used to identify the structure of an ion, or, in analytical applications, to differentiate one isomeric ion structure (or mixture of structures) from another. In this latter application, very careful comparison of CID spectra is required (see also Section 4.2.1.2). It has been demonstrated by van Tilborg and van Thuijl that the relative CID cross sections can provide an additional test in differentiating ion structures.[167–169] Although this approach has been rarely used for CID, measurements of cross sections for photodissociation are commonplace (see Section 3.3.2). Relative cross sections for scattering and for charge exchange are also expected to be ion-structure dependent.[162]

The overall mechanism for collision-induced dissociation (CID) in both low- and high-collision-energy regions is generally accepted as proceeding in two steps.[170–172] The sequence involves collisional activation of the parent ion in the first step and unimolecular dissociation in the second step, i.e.:

$$m_p^+ \xrightarrow{N} m_p^{+}* \longrightarrow m_d^+ + m_n \tag{3-10}$$

where $m_p^{+}*$ is the activated parent ion. The overall net equation of the CID process includes mass and energy balance:

$$q + m_p^+ + N = m_d^+ + m_n + N' + T \tag{3-11}$$

where q is the endothermicity of the collision (i.e., the amount of energy converted from the translational energy of the collision partners into internal energy), N' is the target molecule in its postcollision state, and T is the kinetic energy liberated in the unimolecular dissociation. Provided that no photoemission occurs, the internal energy of $m_p^{+}*$ appears as T and the internal energies of m_d^+ and m_n. Figure 3-9 shows a two-dimensional energy diagram of $m_p^{+}*$ indicating the fraction of q deposited into the ion and the fraction released as kinetic energy (compare with Figures 3-6 and 3-8).

The second step of CID, namely unimolecular dissociation, has previously been discussed in Section 3.2. What follows is a discussion of the kinematics and dynamics

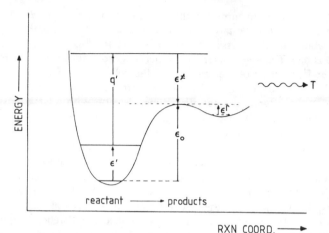

Figure 3-9 ■ Two-dimensional energy diagram indicating an arbitrary fraction of q that appears in the reaction coordinate, symbolized here as q', and the origin of T for the resulting fragmentation. ϵ' is the level of energy present in the reaction coordinate from the initial internal energy of the ion. (Adapted from reference 171.)

of the CA reaction. The mechanisms of excitation are discussed for polyatomic ions in the collision energy range of 1 to 10,000 eV. The discussion emphasizes collision energy effects, but also considers scattering-angle selection and the effects of target gas and target gas pressure.

3.3.1.1. Kinematics. The possible pre- and postcollision vector velocities of the collision partners in collisional activation are constrained by the conservation of energy and momentum. A kinematic description of the collision gives the possible postcollision vector velocities following from the specified initial velocities and the conservation laws. The collision endothermicity, q, is determined from collision dynamics (see the next section), as are the most probable scattering angles of the collision partners. Several important considerations in the use of collisional activation in MS/MS, however, follow directly from the conservation laws and, hence, the kinematics of collision. This section provides a description of the collision through use of a Newton diagram,[170] which is applicable over the entire collision energy range.

In the laboratory frame of reference, the complete description of a binary collision requires that six coordinates be specified (two particles, each with three position coordinates). To simplify matters, it is useful to convert from the laboratory frame to the center of mass (CM) coordinate system. For any binary system with no external forces acting upon it, the kinetic energy of the center of mass of the collision pair is conserved and, therefore, can be factored out of the descriptive equations. Reducing the problem to one in which the CM coordinate system is at rest allows for a complete kinematic description of the collision. This description specifies the relative velocities of the particles before and after collision and the angle between the velocity vectors. Prior to collision, the ion of mass m_p and target of mass N move with a velocity v_i relative to one another and with velocities relative to the CM of U_{mp} and U_N, respectively, so that:

$$v_i = U_{mp} - U_N \qquad (3\text{-}12)$$

where

$$U_{mp} = Nv_i/(m_p + N)$$ (3-13)

and

$$U_N = m_p v_i/(m_p + N)$$ (3-14)

The relative velocity after collision, v_f, at an angle θ_{CM} with respect to v_i, and the velocities U_{mp}' and U_N' have the same relationships as above, i.e.:

$$v_f = U_{mp}' - U_N'$$ (3-15)

where

$$U_{mp}' = Nv_f/(m_p + N)$$ (3-16)

and

$$U_N' = m_p v_f/(m_p + N)$$ (3-17)

These quantities are shown schematically in the Newton diagram given in Figure 3-10. In an elastic collision, $v_i = v_f$, by definition, with the net effect of the collision being restricted to a rotation of the line connecting the ion and target in the Newton diagram. In an endothermic inelastic collision, in which translational energy is converted into internal energy, $v_i > v_f$, and the relative velocity vector may also be rotated. The elastic and inelastic circles for both the ion and target, as shown in Figure 3-10, indicate the magnitudes and possible directions of v_f for an elastic collision and for an endothermic inelastic collision.

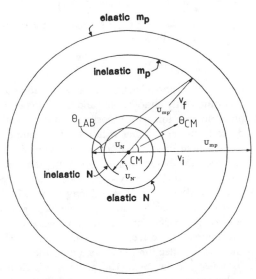

Figure 3-10 ■ Newton diagram of an ion of mass m_p colliding with a target gas N. The diagram indicates all of the possible final velocities of the ion and target in an elastic collision and in an inelastic collision. The laboratory and center-of-mass scattering angles for an arbitrary final velocity following an inelastic collision are indicated.

For some practical situations in MS/MS (e.g., when low collision energies are used), it is important to recall that, since the kinetic energy of the CM is conserved, not all of the laboratory kinetic energy is available for conversion to internal energy of the collision partners. Only the *relative* kinetic energy of the collision partners is available for conversion to internal modes. Conservation of energy requires that the difference between the magnitude of the initial relative kinetic energy of the collision partners $[m_p Nv_i^2/2(m_p + N)]$ and the final relative kinetic energy $[m_p Nv_f^2/2(m_p + N)]$ equals q[173]:

$$m_p Nv_i^2/2(m_p + N) - m_p Nv_f^2/2(m_p + N) = q \qquad (3\text{-}18)$$

The value of q reaches maximum value when the final relative kinetic energy of the system is 0, i.e., when the collision partners stick together. The initial relative kinetic energy E_{REL} therefore represents an upper limit on the value of q. In terms of the laboratory kinetic energy E_{LAB} of a high-velocity ion impinging on a target assumed to be at rest:

$$E_{REL} = (N/N + m_p) E_{LAB} \qquad (3\text{-}19)$$

E_{REL} increases with target mass, making more of the kinetic energy of the projectile available for conversion into internal energy. This is particularly important in low-energy collisions in which a relatively large fraction of E_{REL} is converted to internal energy (see Section 3.3.1.2). In practice, there is a spread in E_{REL} determined by the spread in E_{LAB}, designated as W_1, and by the thermal motion of the target, W_2. The latter spread (expressed as the full-width at half-maximum value) for a monoenergetic particle beam impinging on a thermal target is[174]:

$$W_2 = \left[11.1m_p kT^0 E_{REL}/(m_p + N)\right]^{1/2} \qquad (3\text{-}20)$$

where T^0 is the temperature of the target gas and k is the Boltzmann constant. The total FWHM spread in E_{REL}, W_0, is given by:

$$W_0 = \left[(W_1)^2 + (W_2)^2\right]^{1/2} \qquad (3\text{-}21)$$

Figure 3-10 shows only the final possible relative velocities for the parent ion and the target. If the collision is sufficiently endothermic and the ion is excited to an energy above its dissociation threshold, the ion can dissociate. On dissociation, the charged and neutral fragments recede from one another with a relative kinetic energy of T, the kinetic energy release. The effect of dissociation on the Newton diagram is shown in Figure 3-11 for the case in which the dissociation occurs in the same plane as the initial deflection. Superimposed upon the final velocity vector of the parent ion (the center of mass of the dissociated ion) are possible relative velocity vectors of the charged and neutral fragments, shown here for a single-valued T. Since, for polyatomic ions, the orientation for fragmentation is generally isotropic, the final velocities of the fragments relative to one another are opposite in direction, and the vectors fall on spheres (or circles if coplanar with the initial deflection). (For legibility, the elastic circles for m_p and N are omitted, as is the circle for m_n.) The angle at which a daughter ion is experimentally observed, therefore, may not accurately reflect the angle at which the parent ion was scattered, particularly when T is large and when the charged fragment is low in mass. Furthermore, a distribution of T values exists when the ion population fragments statistically[175,176] (see also Section 3.2). The final angular distribution of

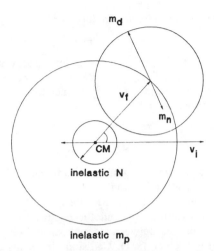

Figure 3-11 ■ Newton diagram showing an inelastic collision of an ion of mass m_p with a target N that results in the fragmentation of the ion into a daughter ion and neutral fragment. This diagram shows the special case when the fragmentation occurs coplanar with the collision plane. The kinetic energy released in the fragmentation gives an additional velocity vector, all possible orientations of which are indicated by the m_d circle, that adds to the final velocity of the daughter ion. The angle at which a daughter ion is observed can be largely determined by collisional scattering or kinetic energy release, or a combination of both.

daughter ion vectors following CID is therefore determined by a complicated function of the collision dynamics, kinematics, and mechanisms of dissociation.[177,178] The angular distribution of daughter ions following CID of polyatomic ions is discussed further in Section 3.3.1.2.

The magnitude of q is not directly measurable except in coincidence experiments that measure the postcollision scattering angles and kinetic energies of the projectile and target, respectively.[179-181] The amount of internal energy deposited in the ion is, however, reflected in its CID MS/MS spectrum, and under some circumstances, q is closely approximated by the translational energy change of the projectile, ΔE_{LAB}. In the laboratory frame of reference, q is equal to the difference in the kinetic energy changes of the ion and the target, i.e.:

$$q = \Delta E_{LAB} - \Delta E_N \tag{3-22}$$

where ΔE_N is the kinetic energy change in the target molecule, assumed initially to be at rest. With this assumption, a kinematic analysis of the energy loss of the projectile ion yields[182]:

$$\Delta E_{LAB} = q + \left[\left(m_p/(m_p + N)\right)\right](1 - \cos\theta_{CM})\left[2NE_{LAB}/(m_p + N) - q\right]$$
$$+ 1/4(m_p/N)(q^2/E_{LAB})\cos\theta_{CM} \tag{3-23}$$

The last two terms of equation (3-23) represent the kinetic energy imparted to the target on collision. Under conditions in which negligible scattering occurs, and when

$E_{LAB} \gg q$ and $N \gg m_p$, this relationship reduces to[183]:

$$\Delta E_{LAB} \approx q \qquad (3\text{-}24)$$

so that the measured kinetic energy loss of the ion approximates the value of q. A useful measure of how closely ΔE_{LAB} approximates q at negligible scattering angles is given by the disproportionation factor D[170,184] defined as the ratio of the changes in the kinetic energies of the projectile and the target, i.e.:

$$D = \Delta E_{LAB}/\Delta E_N$$

$$\approx 4NE_{LAB}/m_p q \qquad (3\text{-}25)$$

In BE geometry mass spectrometers with good angular collimation, set to accept a narrow range of laboratory scattering angles ($\pm 0.1°$), the energy loss of the projectile, when $N \gg m_p$, can be used as a direct measure of q.[171,183] Beynon and co-workers have used this approach to measure the endothermicities for various types of reactions including, for example, collisions between atomic ions and targets and charge-changing reactions[12,156,185–188] (see Section 4.4.1). Collisions involving polyatomic ions, which have high densities of states, usually involve a range of q values centered on a value of a few electron volts. Energy loss measurements for polyatomic ions are therefore more difficult. Energy losses associated with parent ions that dissociate can be evaluated by kinetic energy analyses of the daughter ions.[170] For some organic ions, the energy loss evaluated by this method correlates with the energy deposited in the ion as judged by the fragmentation pattern.[189] Conversely, Dawson et al. have used the measurement of daughter ion energy distributions with assumed q values to infer the scattering-angle distributions in a triple quadrupole instrument in which ions from a wide range of scattering angles are transmitted.[172,190–195]

3.3.1.2. Dynamics. Collision dynamics involves changes on the molecular level that take place as the collision partners approach and then recede from one another. There are three classes of collision: elastic, when there is no net change in the kinetic energy of the system (and hence $q = 0$); inelastic, when $q > 0$, and there is a net decrease in the kinetic energy of the system with a concomitant increase in internal energy of the collision partners; and superelastic, when $q < 0$, and the collision partners experience a net increase in kinetic energy and decrease in internal energy. Collision-induced dissociations of polyatomic ions involve inelastic collisions.

Figure 3-12 is a generalized depiction of the collision event for a fixed target and a projectile ion approaching with a kinetic energy E_{LAB}. For a given E_{LAB}, the outcome of the collision is determined by the closeness of the collision, the intermolecular interaction potential $V(r)$, and the probabilities under these conditions of transitions between initial and final states of the collision pair.[196] A useful measure of the closeness of collision is the impact parameter b, defined as the distance of closest approach between the collision partners in the absence of an interparticle force. Figure 3-13 gives a two-dimensional depiction of the potential energy between a fixed target and a projectile ion as a function of the distance of separation between the two (assuming that this potential is independent of the orientation of the ion and target as they approach each other). Low-impact parameter collisions refer to short-range interactions that tend to probe the repulsive part of the internuclear potential. Large-impact parameter collisions refer to long-range interactions that tend to probe the attractive part of the potential. As a general rule, low-impact parameter collisions result in

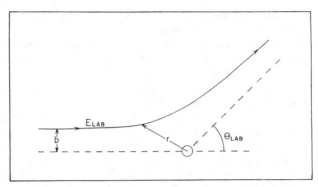

Figure 3-12 ■ Generalized depiction of a collision between a high-velocity projectile and a stationary target in which the interaction potential, $V(r)$, is repulsive over the entire range of r. A $V(r)$ that is strongly attractive at some values of r could result in a more complicated trajectory depending upon the impact parameter and the collision energy.

significant particle scattering due to nuclear repulsion, whereas little or no scattering occurs for large-impact parameter collisions, which sample the relatively weak attractive part of the potential.[197,198]

A number of possible mechanisms for collisional activation have been identified. Many of these mechanisms have evolved from studies involving collisions of diatomics with atoms.[199] These studies are useful as a guide to the mechanisms likely to be predominant for the collisional activation of polyatomics at both eV and keV collision energies. Nevertheless, CA mechanisms of polyatomic ions are not yet well understood. A clear understanding of the mechanisms has been elusive due to the overall two-step mechanism of CID. As a rule, polyatomic ions subjected to collisional activation seem to fragment according to the tenets of quasiequilibrium theory[166,170,200,201] (Section 3.2). The excited ions have no memory of how or where energy was originally deposited. Ions dissociate from excited vibrational states of the ground electronic state.

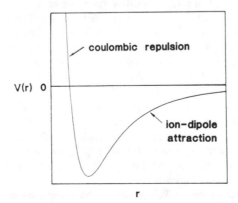

Figure 3-13 ■ Schematic of the interaction potential for an ion/target system. (Adapted from reference 196.)

MS/MS, however, has provided impetus for new detailed studies that promise to shed further light on the underlying mechanisms.[202]

An important factor for all types of excitations in inelastic collisions is the Massey parameter,[203] which is the ratio of the time of collision, t_c, to the characteristic period of the motion being excited, τ. The value of t_c is determined by an interaction distance (usually taken as 4–7 Å[204]) divided by the relative velocity of the collision partners. The value of τ is equal to Planck's constant divided by the difference in energy levels involved in the transition leading to the excitation. When t_c/τ is much less than 1, the collision is fast and transitions are likely. The cross section for the transition reaches a maximum when t_c/τ is about equal to 1 and falls off slowly as the ratio decreases.[205] When t_c/τ is much greater than 1, the system can adjust to the perturbation of the collision while it is occurring so that transitions are unlikely. The collision time, therefore, plays an important role in determining the nature of excitation. For small values of τ, which correspond to large energy differences, the excitation tends to be electronic in nature since electronic energy levels are relatively widely spaced, and electron motion is fast. Larger values of τ correspond to vibrational and possibly rotational spacings. The following sections describe the most important possible mechanisms for collisional activation of polyatomics and contain a discussion of the experimental evidence for each mechanism in eV and keV collision energy experiments.

I. *Direct vertical electronic excitation.* If the interaction time between the target and projectile is short, relative to the nuclear motion within either particle, an electron can be promoted to a higher energy state via a curve-crossing mechanism[171,206] (Figure 3-14). If the collision occurs slowly (i.e., $t_c/\tau \gg 1$), the electrons can constantly adjust to maintain the condition of lowest overall system energy, resulting in no net excitation. If the collision is sufficiently fast (i.e., $t_c/\tau \ll 1$), a net crossing can occur, resulting in internal excitation. Once the excitation has occurred, the energy can then rapidly redistribute within the ion.

In analogy with electron excitation (and ionization) and photoexcitation (and photoionization), a Franck–Condon-type excitation can be induced by a rapidly changing electric field experienced by an electron, in this case from screened nuclei of the collision partners.[207,208] Again, for polyatomic ions, rapid radiationless transitions prior to fragmentation can take place. For some diatomic ions, collisional activation results in excitation of the ion to a repulsive state and resultant dissociation within the time of a vibrational period.[147] In these cases, large values of T are observed, since few other energy repositories are available. With the exception of the collisional activation of doubly charged ions, a situation in which the Coulombic repulsion of two charged fragments can result in T values of several electron volts,[12] polyatomic ions show relatively small values of T, the typical range being 10 meV to 1 eV. This is consistent with the statistical theory of mass spectra.

With these mechanisms, close approach of the nuclei of the ion and the target is not required for excitation, i.e., the nuclear repulsive part of the interaction potential need not be sampled. Large impact parameter collisions therefore result in negligible scattering. On the Newton diagram of Figure 3-11, m_p would appear after collision very near to $\theta_{CM} = 0$, and the angular distribution of the daughter ions would be determined primarily by T for each reaction.

The collisional activation of keV energy polyatomic ions is often attributed to a direct vertical electronic excitation mechanism. For 5-keV $H_2^{+\cdot}$ ions, assuming an interaction distance of 5 Å, t_c is about 7×10^{-16} s, which is much shorter than the

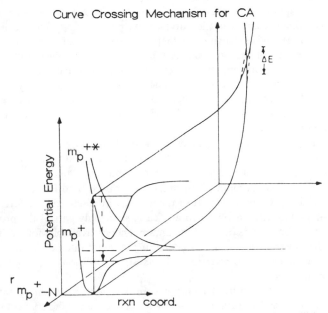

Curve Crossing Mechanism for CA

Figure 3-14 ■ Schematic depiction of the curve-crossing mechanism for collisional activation. (Adapted from reference 206.)

fastest vibrations in the ion (about 10^{-14} s), and on the order of electron motion (10^{-15} s). Direct vertical electronic excitation is therefore a likely mechanism for collisional activation. For a 5-keV ion of m/z 200, assuming the same interaction distance, t_c is about 7×10^{-15} s. As the mass of the projectile ion increases at a fixed keV energy, the likelihood for vertical electronic excitation decreases and that for vibrational excitation, as described in the next section, increases.

II. *Direct vibrational excitation.* Several mechanisms have been identified that lead to direct excitation of vibrations of the electronic ground state of an ion. A mechanism that acts at relatively long range is based on the induction of vibration and/or rotation by the rapidly changing polarization forces during the collision.[209] This mechanism, often referred to as the Russek mechanism, has been used successfully in rationalizing experimental results for the CID of HeH$^+$.[147,209] This mechanism depends on the long-range part of the interaction potential and, hence, does not necessarily result in scattering of the collision partners. This type of mechanism is therefore usually associated with relatively small projectile energy losses at scattering angles near 0° that correspond with the transfer of one or a few vibrational quanta. The importance of the Russek mechanism with respect to CID of polyatomics is therefore probably limited to situations that involve the dissociation of ions already close in energy to the dissociation limit.

An important mechanism for direct vibrational excitation is the "impulse" model[177,210] in which an atom or group of atoms of the target undergo an elastic collision with part of the ion, while the remainder of the ion acts only as a spectator. If the collision is rapid enough (i.e., faster than the vibrational frequency), the nuclei cannot adjust adiabatically (vibrationally) to the perturbation of the collision during the interaction time, and a transition is therefore likely. In the impulsive collision, the nuclear-repulsive part of the potential is sampled, and the collision can result in an appreciable transfer of momentum between particles. Impulsive collisions lead to large angle scattering in the center-of-mass frame. In general, smaller impact parameters (closer approach of the collision partners) result in larger angle scattering.[211] Closer approach of the collision partners results in greater repulsion and, therefore, greater momentum transfer.

Since the impulsive collision involves an elastic collision with only part of the ion, the maximum energy transferable from translation to internal modes and the relationship between the laboratory energy loss and q must be modified in the kinematic analysis to take this into account.[210] In practice, the mass used for the ion is not easily chosen since it depends upon the part of the ion that interacts and the force constant(s) of the vibration(s) directly involved. In any case, since only part of the projectile takes part in the impulsive collision, the maximum energy available for conversion into internal modes is less than would be the case if the target interacts with the entire ion. Nevertheless, large energy transfers are observed for collisions reasoned to be of the impulsive type.

The formation of a long-lived collision complex (with a lifetime greater than the time required for several vibrations) is another mechanism for the efficient excitation of an ion. If there is a deep potential well in the energy surface of the ion/target combination, a "sticky" collision can take place in which a relatively long-lived collision complex is formed. If the complex is sufficiently long-lived, E_{REL} tends to be shared statistically among all the degrees of freedom of the collision complex.[212–214] With a polyatomic ion, the percentage of E_{REL} transferred to the internal modes of the ion when the complex breaks up can be very large. When the ion and target react chemically within the complex, other decomposition channels from the complex become competitive with the reversion of the complex to the ion and target. This situation can occur in ICR spectrometers and in high-pressure ion sources in which ion/molecule reactions are commonly observed. In these cases, the collision partners generally have thermal energies. Reactive collisions also occur at superthermal energies in MS/MS and are discussed at more length in Section 3.4 If the lifetime of the complex is longer than several rotational periods, the complex will break up randomly with respect to the initial beam direction, and the angular distribution of the products will be uncorrelated with the initial direction of the beam. An example of an ion/molecule reaction in which this is apparently the case is the basis of Figure 3-15, which shows an intensity contour map for O_2D^+ resulting from the collision of O_2^+ onto a D_2 target at $E_{LAB} = 24.8$ eV ($E_{REL} = 2.76$ eV). At a higher collision energy of $E_{LAB} = 49.2$ eV, the O_2D^+ angular distribution becomes weighted more heavily toward the initial direction of the beam, indicating that the complex has a shorter lifetime due to the increased internal energy of the complex.[176] In general, the lifetime of any collision complex decreases with increasing collision energy, just as ion lifetimes are inversely correlated with internal energy. Excitation mechanisms that involve formation of a complex are likely to be important only in low-collision-energy MS/MS, and their importance is likely to decline as the collision energy increases. The scant experimental evidence that exists for the dynamics of low-collision-energy CID of polyatomic ions tends to support this generalization (see Section 3.3.1.4).

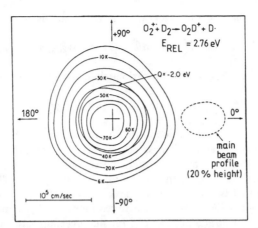

Figure 3-15 ■ Intensity contour map, plotted in the center-of-mass frame of reference, for O_2D^+ resulting from the collisions between $O_2^{+\cdot}$ and D_2 at $E_{LAB} = 2.76$ eV. (Adapted from reference 176.)

III. *Electronic excitation in conjunction with vibrational excitation.* Vibrational excitation can occur, of course, with vertical electronic excitation if the Franck–Condon factors are appropriate. Vertical electronic transitions often occur to vibrationally excited states of the excited electronic state, since the bond length of the excited electronic state of the ion is often longer than in the ground state of the ion. The discussion here refers to excitation of vibration that is not the direct result of a vertical electronic transition. If the impact parameter is sufficiently small, an impulsive collision can occur in conjunction with electronic excitation.[199,215] This can be viewed qualitatively as a combination of a direct electronic excitation mechanism (I) with the impulsive scattering mechanism described previously (II). Evidence for such "oblique" transitions (as opposed to vertical transitions) has been presented.[216,217] An important distinction between this mechanism and that of vertical electronic excitation is that the former is expected to result in scattering of the ion, whereas the latter results in negligible scattering. Ions that have undergone an impulsive mechanism with or without electronic excitation, therefore, can be preferentially sampled by selecting ions at off-zero scattering angles. The Russek mechanism can also occur in conjunction with electronic excitation.[215] This mechanism is not expected to result in significant scattering since both interactions of the mechanism occur at large impact parameters.

3.3.1.3. Kiloelectron-Volt Collision Energy Collisional Activation. The first MS/MS instruments of the modern era (BE instruments) excite and analyze ions with keV kinetic energies. The initial CID experiments, therefore, employed keV collision energies. It is widely regarded that the primary excitation mechanism in this collision energy region is electronic (process I mechanisms).[170,206] Most high-collision-energy instruments accept ions only within a very small range of laboratory scattering angles (a few tenths of a degree, typically), and therefore discriminate against those mechanisms that can involve scattering at larger angles. Nevertheless, few ions at keV energies are scattered beyond 1° in the laboratory frame of reference. The relationship between the laboratory scattering angle, θ_{LAB}, and the center-of-mass scattering angle,

θ_{CM}, assuming that the target interacts with all of the ion, is:

$$\theta_{LAB} = \tan^{-1}\left\{ v_f N \sin \theta_{CM} / \left(v_f N \cos \theta_{CM} + m_p v_i \right) \right\} \tag{3-26}$$

If q is much less than E_{REL}, v_f is approximately equal to v_i so that:

$$\theta_{LAB} = \tan^{-1}\left\{ \sin \theta_{CM} / \left(\cos \theta_{CM} + m_p / N \right) \right\} \tag{3-27}$$

The latter is exact for an elastic collision in which $q = 0$. For an ion of mass 100 daltons impinging on a target of mass 40 daltons at a collision energy of 8000 eV, $E_{REL} = 2286$ eV. Values of q are very rarely in excess of 20 eV for polyatomic ions and are usually less than 5 eV. Using the preceding relation, a θ_{CM} of 4° results in θ_{LAB} of about 1.1°. (Note that two values of θ_{CM} give the same θ_{LAB}. Forward-scattered ions that have small values of θ_{CM} and backscattered ions with θ_{CM} approaching 180° both provide similar θ_{LAB} values. For example, for the preceding case, a θ_{CM} of 178° leads to a θ_{LAB} of about 1.3°. However, since little scattering past a few degrees is observed in the laboratory frame, i.e., intermediate values of θ_{CM} are not indicated. It seems unlikely that highly backscattered ions contribute to the low angle scattering for keV energy polyatomics.) As the mass ratio m_p / N increases, larger θ_{CM} are required to produce the same value of θ_{LAB}. For the illustration given previously with $N = 4$ instead of 40, $E_{REL} = 308$ eV, and a θ_{CM} of 4° leads to a θ_{LAB} of about 0.15°. For most of the polyatomic ions studied using keV collision energy CID (where v_i is on the order of 10^5 m/s), the parent ions are forward-scattered, indicating that most of the collisions are glancing. At lower collision energies and, possibly, for high mass ions at keV collision energies (i.e., for which v_i is on the order of 10^4 m/s), the scattering distributions are much broader (see the following discussion). There is other convincing evidence that direct electronic excitation takes place for many polyatomic ions at keV collision energies. For example, the highly endothermic charge permutation reactions that are commonly observed at keV collision energies (but not at eV collision energies) *require* electronic excitation (see Section 3.5). There are no quantitative measures, however, of the extent to which each of the possible mechanisms contributes to CID. The technique of angle-resolved mass spectrometry (ARMS, see the following discussion) is based on the presumption that at least one of the mechanisms that results in inelastic scattering is operative. The fact that the kinetic energies released in dissociations lead to different angular distributions for the daughter ions has led some to question the importance of scattering in ARMS data. Recent results have shown, however, that CID does occur for 5-keV acetone ions scattered out to laboratory angles of 0.8°.[218] Moreover, as the mass of the ion increases, its velocity decreases (and, for a fixed target mass, E_{REL} decreases), increasing the interaction time and thereby increasing the likelihood for vibrational excitation. Evidence that this is so for relatively high mass ions has been reported.[219-221]

Kim et al. have used the Massey criterion in conjunction with cross section data for charge transfer reactions to calculate roughly the distribution of energies deposited into an ion upon collision at keV collision energies.[166] Figure 3-16 shows typical results. Qualitatively, the distributions show a most probable energy transfer of a few electron volts, with a long tail extending out to very large energy transfers. Ions formed in highly endothermic reactions (> 5 eV) in CID MS/MS spectra generally originate from the decreasing part of the distribution. When the energy transfer distribution is

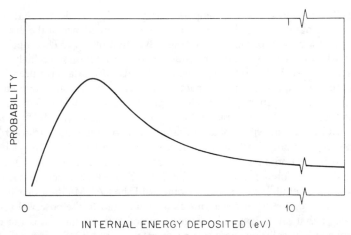

Figure 3-16 ■ Qualitative shape of the distribution of energies transferred in keV collision energy CA.

convoluted with the distribution of internal energies prior to collision, the final distribution of internal energies, it is argued, should have the same shape at high energies, i.e., a slowly decreasing function. This supports the contention that relative abundances of ions that result from processes that have high critical energies should be independent of parent ion internal energy following collisional activation (see Section 4.2.1.2). The calculated energy transfer distribution should be regarded only as a rough estimate of the true distribution, since the calculation assumes only a mechanism of direct electronic excitation and does not take into account Franck–Condon factors. Nevertheless, the shape of the distribution seems reasonable based on reported CID data.[222] The calculated curves have been used to conclude that energy transfer to high mass ions is small, due to the lower velocity of the higher mass ion. The CID of relatively high mass ions has been reported, however, and energy losses of up to 20 eV for such ions have been noted.[220] These highly endothermic collisions have been attributed to highly impulsive collisions (vibrational excitation) that are assumed not to occur in the calculation of Kim et al. Derrick et al. have reported trajectory calculations that model this vibrational energy transfer.[221] The fact that high mass ions do undergo collisional activation and dissociate is evidence for direct vibrational excitation of polyatomic ions at keV energy collision energies, as is expected when viewed on the basis of relative velocity. Bricker and Russell, however, have also observed large energy losses in keV energy collisions of high mass ions with several targets.[223] These data suggest that much of the energy loss can be attributed to ionization of the target gas.

In an attempt to obtain more detailed information from keV collision energy CID, Cooks and co-workers developed the technique of angle-resolved mass spectrometry (ARMS).[224-229] In this method, CID spectra are recorded as a function of θ_{LAB}. Many examples have been given in which the relative abundances of daughter ions obtained as a function of scattering angle have correlated with results obtained from independent techniques, such as charge exchange mass spectrometry[230] and photoion–photoelectron coincidence,[231,232] which give fragment ion abundances as a function of parent-ion internal energy. These comparisons[224–226,228,233] indicated that it might be possible to obtain, rather crudely, the internal energy dependence of the dissociation of a mass-selected ion by recording CID data as a function of θ_{LAB}. This capability might be particularly useful for the differentiation of ion structures, since higher scattering angles correlate with higher internal energies. The ARMS data have been interpreted based on the reasoning that small impact parameter collisions are more endothermic and result in greater scattering.[233] Such an interpretation implies that a momentum transfer excitation mechanism applies to ions observed at $\theta_{LAB} > 0.1°$. Evidence for electronic excitation and vibrational excitation due to momentum transfer collisions taking place concurrently for diatomic ions has been reported.[216,217] This is mechanism III discussed in Section 3.3.1.2. A major complication to ARMS data is the contribution to the daughter ion angular distribution from T, the kinetic energy released in the dissociation. Todd et al. have shown for some published ARMS data that the different T values for the competing dissociations alone could lead to the observed results.[234] It was concluded that an endothermic momentum transfer mechanism is not important for these ions at keV collision energies, and therefore that ARMS is not a useful means for the selection of ions of different internal energies. van der Zande et al.[235] have published a coincidence experiment in which both the charged and neutral fragments from the CID of ionized acetone were detected; in these experiments, the center of mass of the ion could be specified without any complication from T. These results showed that CID was associated with ion scattering out to $\theta_{LAB} = 0.8°$. A relationship between θ_{LAB} and the transferred energy could not be derived from these experiments. More recently, Boyd et al. have published several papers focusing on the theoretical[177,178] and experimental[236] foundations of ARMS. For some ions, but not all, they have observed a correlation between θ_{LAB} and T. The kinetic energy release T is expected to be sensitive to excess internal energy for dissociations without a large reverse critical energy (see Section 3.2). From the data acquired to date, it seems unlikely that an a priori relationship between θ_{LAB} and the transferred internal energy can be established. More detailed studies, such as some of those alluded to previously, are needed to establish those circumstances in which scattering is likely to be important for high-velocity polyatomic ions.

3.3.1.4. Electron-Volt Collision Energy Collisional Activation.

The collisional activation of polyatomic ions in the collision energy region of 1 to 100 eV became important to the mass spectrometry community with the introduction of multiquadrupole mass spectrometers.[237] This collision energy region is also accessible in hybrid MS/MS instruments, Fourier transform–ion cyclotron resonance mass spectrometers, ion traps and in some multiple-sector instruments (see Chapter 2). In this collision energy region, CID spectra can be highly sensitive to small absolute changes in the collision energy. A correlation between collision energy and internal-energy deposition has been observed repeatedly[79,172,190–195,238–243] and is interpreted as due to the relatively large changes in E_{REL} that can occur in this collision energy region and the relatively large fraction of E_{REL} transferred into internal energy. In general, this fraction decreases as E_{REL} increases.

Figure 3-17 illustrates several of these points. This figure compares a plot of the ratio of the abundances of the daughter ions at m/z 91 and m/z 92 from the molecular ion of n-butylbenzene obtained from low-energy CID as a function of E_{REL} (Figure 3-17a),[241] and from charge exchange experiments as a function of internal

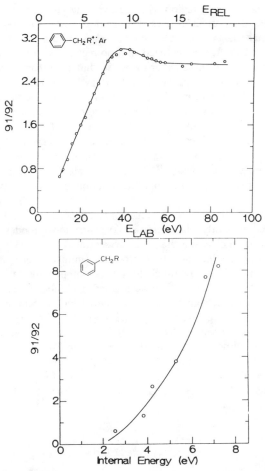

Figure 3-17 ▪ (a) Plot of the ratio of the abundances of the daughter ions at m/z 91 and m/z 92 from the molecular ion of n-butylbenzene obtained from low-collision-energy CAD as a function of E_{REL}. (Adapted from reference 241.) (b) Plot of the same ratio shown in (a) obtained as a function of internal energy via charge exchange mass spectrometry. (Adapted from reference 244.)

energy (Figure 3-17b).[244] This ratio has been used by a number of workers to study internal energy deposition in activation reactions in MS/MS (see Section 4.4.3.2). Several important characteristics of low-collision-energy CA are evident in these data. First, it is clear from the comparison that the internal energy of the parent ions after collision increases with collision energy up to E_{REL} of about 9 eV (E_{LAB} of about 40 eV). Second, relatively large fractions of E_{REL} are converted to internal energy of the ion. The maximum of the ratio reaches about 2.7, which corresponds to an internal energy (from Figure 3-17b) of about 4.6 eV. This indicates a maximum transfer of slightly over 50%. Third, the fraction of E_{REL} converted to internal energy *decreases* as the collision energy increases over the entire collision energy range tested. This is obvious in the collision energy region where the ratio levels off, but is also true over the range in which the ratio quickly rises. For example, at $E_{REL} = 9.2$ eV, the ratio indicates an internal energy of 4.6 eV, while at $E_{REL} = 3.7$ eV, the ratio indicates an internal energy of 3.2 eV. It was reported that no metastable ions were observed in the absence of the target gas in these experiments so that the internal energies of the ions prior to collision must be less than the critical energy for the lowest energy dissociation. The critical energy for formation of $C_7H_8^{+\cdot}$ from ionized n-butylbenzene, the lowest energy fragmentation for this ion, is 1.4 eV. Therefore, the ranges of possible energy transfers at $E_{REL} = 3.7$ eV and $E_{REL} = 9.2$ eV are 1.8 to 3.2 eV and 3.2 to 4.6 eV, respectively. These values correspond to percentage transfers of 48% to 86% at the lower collision energy, and 35% to 50% at the collision energy for maximum absolute energy transfer. Similar comparisons[238-243] also indicate that internal energy deposition increases with collision energy but that the fraction of E_{REL} converted to internal energy decreases.

It is widely accepted that eV collision energy collisional activation of polyatomic ions proceeds predominantly via excitation of vibrational modes of the electronic ground state. It has been observed that ions are scattered over a wide range of angles at eV collision energies. This indicates that either the impulsive mechanism of excitation is operative, energy transfer via relatively long-lived complex formation occurs, or both. Douglas and Dawson et al.[172,191-195] have studied a number of polyatomic ions in a triple-quadrupole mass spectrometer and determined the dissociation cross sections as a function of collision energy. The daughter ion yield is observed to increase rapidly once E_{REL} reaches the threshold for dissociation and tends to level off at an E_{REL} value of two to three times the threshold value. Figure 3-18 gives an example of this behavior, showing the yields, expressed as the dissociation cross section relative to the maximum observed dissociation cross section σ/σ_{MAX}, for the formation of $C_6H_5^+$ from $C_6H_5Cl^{+\cdot}$ using Ar and N_2 as target gases.[172] When the data are plotted as a

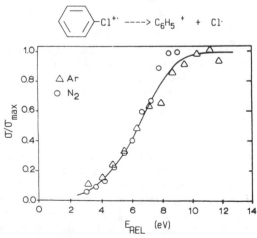

Figure 3-18 ■ Plot of the relative yields of $C_6H_5^+$ from $C_6H_5Cl^{+\cdot}$, expressed as the dissociation cross section divided by the maximum observed dissociation cross section σ/σ_{MAX}, as a function of E_{REL} using Ar and N_2 as targets. (Adapted from reference 172.)

function of E_{REL}, both target gases are seen to produce similar behavior. Douglas compared these data with behavior predicted by a model of complex formation with equilibration of the energy of the complex and with an alternative impulsive collision model.[172] The experimental data showed behavior intermediate between the predictions of the two models. The results were interpreted as evidence for complex formation with incomplete equilibration of energy. Toennies et al. have studied the energy loss of atomic projectiles resulting from collisions with polyatomic molecules (e.g., Li^+ incident on CH_4) at low eV collision energies and have developed a statistical model that accounts for the results.[212] Futrell et al. have reported Newton diagrams for $CH_4^{+ \cdot}/Ar$ and $C_3H_8^{+ \cdot}/Ar$ and report that the data can be rationalized on the basis of a statistical model.[202] The frequently made observation that energy transfer is highly efficient in eV energy collisions, particularly near the first threshold for dissociation, is good evidence that complex formation is fairly general for polyatomic ion/neutral collisions under these conditions, even with inert target gases such as argon. Under conditions in which large scattering angles are observed and the collision energy is greater than about 20 eV, Toennies has observed rather large energy losses that are attributed to impulsive collisions.[245] It seems reasonable that the lifetime of a collision complex should decrease with collision energy and that an impulsive collision mechanism should become more important as the collision energy increases.

The distribution of energies transferred to the ion in eV collision energy collisional activation is not well defined. At low collision energies, however, E_{REL} (in addition to the original internal energy of the target, which is generally small) can be a practical upper limit to the energy transfer. Kenttämaa et al. have used an empirical thermochemical approach to ascertain the shape of the energy transfer distribution for several types of reactions, including low-collision-energy CID (see Section 4.4.3.2).[246-248] Figure 3-19 shows the energy distributions calculated using this approach and the appropriate spectra for the triethyl phosphate molecular ion for 70-eV electron ionization, 7-keV collisional activation using air as the target gas, and low-collision-energy collisional activation using argon as the target gas at various collision energies and target gas pressures. The distribution obtained for 7-keV CID qualitatively reproduces the shape of the distribution expected from the work of Kim and McLafferty (Section 3.3.1.3). As expected, the low-collision-energy results indicate that the distribution is more heavily weighted to high energies as the collision energy is increased, and as the number of collisions is increased (via an increase in the target gas pressure). (The minimum observed between 3 and 4 eV for all of the activation methods is not observed for other ions studied by these workers.)

A few studies describe CID results for collision energies of a few tens of electron volts continuously up to several kiloelectron volts.[43, 95, 249, 250] Despite the difficulties in obtaining data relatively free of instrumental effects over such a wide range of collision energies, these studies illustrate several important points regarding the CID of polyatomic ions. A BB sector instrument with an electrically floatable collision cell and postacceleration of the daughter ions was used to study the CID of a number of parent ions over a range of about 10 to 6000 eV.[43, 249] For the molecular ion of methane, formed by 70-eV electron ionization, it was shown that the cross section for CID using helium as the target did not vary more than an order of magnitude (within experimental error) over the entire collision energy range investigated.[43] These results emphasize that the inherent efficiency for CID is *not* higher at eV collision energies than at keV collision energies. This observation has also been made for the excitation and dissocia-

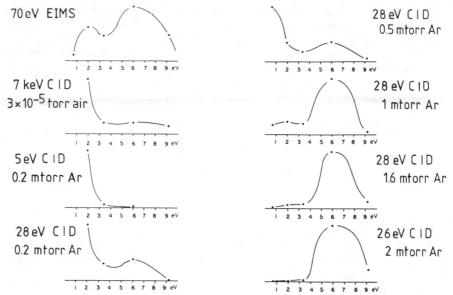

Figure 3-19 ■ Calculated internal energy distributions of triethyl phosphate molecular ions following 70-eV electron ionization, 7-keV collision energy CID, and low-collision-energy CID at various combinations of collision energy and target gas pressure. The high-collision-energy results and the low-collision-energy data obtained at a target pressure of 0.2 mTorr reflect predominantly single collisions. (Adapted from reference 246.)

tion of the molecular ion of bromobenzene and $C_8H_{17}^+$ formed from *n*-hexadecane in collisions with helium using an EBEB geometry instrument.[250] The fact that less detector amplification is required to observe daughter ions on the same scale as the parent ion in CID spectra acquired using triple-quadrupole instruments and eV collision energies, as compared with CID spectra obtained using BE instruments and keV collision energies, is due primarily to instrumental factors rather than to a higher cross section for reaction. Among these are the longer path length and lower ion velocity in a quadrupole collision cell, both of which increase the time window over which fragmentation products can be collected.[250] Furthermore, the relatively large angular acceptance of the quadrupole collision cell/mass filter combination, possible detector discrimination against daughter ions with lower kinetic energies than the parent ion in some BE instruments, and the fact that peaks are broadened in BE instruments by kinetic energy release (which reduces peak heights) also contribute to differences in daughter ion signals. It is also noteworthy that CID yields do not show a dramatic decrease in the intermediate collision energy region of 100 to 2000 eV, as might be expected if a sharp change in mechanism from direct vibrational excitation to electronic excitation occurs.

Another point illustrated in the experiments alluded to previously is that the *average* energy deposited in the parent ion can reach a maximum at a relatively low-collision-energy, whereas the distribution of energies transferred at keV collision energy is broader and includes much larger energy transfers of low probability (i.e., the high-energy tail discussed in Section 3.3.1.3). Evidence for the high-energy tail is the appearance, with relatively low abundances, of ions that arise from high-energy processes at keV collision energies.[249] These ions are virtually absent in CID spectra acquired at low collision energies. Similar observations have been made in a comparison of high- and low-collision-energy data for aza- and amino-polynuclear aromatic hydrocarbons.[251] Calculations of the shape of the energy transfer distributions of eV and keV CID data also show this behavior.[246-248]

3.3.1.5. Target Gas Effects. At low collision energies, the nature of the target gas can have a much greater effect on the appearance of the MS/MS spectrum than at high collision energies. This is due largely to the relatively long-lived collision complex mechanism for energy transfer. Since, at low collision energies, a large fraction of E_{REL} can be converted to internal energy (Section 3.3.1.4), target mass has a significant effect on the amount of energy transferred into the ion. Since E_{REL} increases with target mass, heavier mass targets lead to more extensive fragmentation of the parent ion at a fixed E_{LAB}. For example, Douglas has measured the relative dissociation cross section for $C_6H_5Cl^{+\cdot}$ incident on Ar and N_2 up to a collision energy of about 50 eV.[172] The dissociation cross section was observed to be significantly larger at all but the highest collision energies for Ar compared to that for N_2. When the data were plotted as a function of E_{REL}, however, the cross-section behavior was indistinguishable for the two targets. These data clearly demonstrate the significant kinematic effect of target mass in low-collision-energy CID.

Chemical effects of the target are also extremely important in determining the appearance of the low-collision-energy MS/MS spectrum. Such an effect is most dramatically demonstrated when products of a reactive collision, i.e., a reaction in which one or more atoms are exchanged between the ion and the target, appear in the spectrum. An example is given in Figure 3-20a, which shows a spectrum resulting from the molecular ion of propane at a laboratory collision energy of 6.1 eV colliding with an ethylene target. Major peaks above m/z 44 (the mass of the molecular ion) appear at m/z 69, 67, and 55 corresponding to $C_5H_9^+$, $C_5H_7^+$, and $C_4H_7^+$, respectively. Daughter ions from the molecular ion of propane are observed in the spectrum along with ions due to dissociations of adduct ions, as seen by comparison of this spectrum with that of ionized propane using argon as the target gas at the same E_{REL} and target pressure (Figure 3-20b). For example, ions at m/z 41 and 28 are enhanced in abundance with ethylene as the target gas. When a spectrum containing only dissociations of the parent ion is desired, such ion/molecule reactions are undesirable, and in these experiments, an inert target gas is the best choice. However, in some cases, bimolecular chemistry may be analytically useful, and this topic is taken up in more detail in Section 3.4.

Reactions such as proton or electron transfer from the ion to the target can also occur and also affect the efficiency of CID. Since these reactions are competitive with CID, it is usually desirable to minimize their contributions by the use of target gases with relatively high ionization potentials or low proton affinities. Insofar as ionization potentials are roughly inversely related to target mass, a compromise is often made between target mass and ionization potential to obtain sufficient energy transfer without a significant decrease in the total dissociation. For this reason, along with the desire to avoid ion/molecule association reactions, target gases such as argon and nitrogen are usually used for low-energy CID.

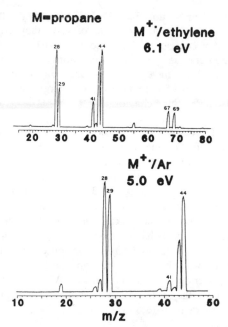

Figure 3-20 ■ Low-collision-energy spectra of the molecular ion of propane in collisions with (a) ethylene and (b) argon.

Due to the much shorter interaction times usually encountered at high collision energies and the much smaller fraction of a rather large E_{REL} transferred into internal energy, a change in the nature of the target has different effects on high-collision-energy MS/MS spectra than on the corresponding low-collision-energy MS/MS spectra. Kim and McLafferty[166] have shown that the relative magnitude of the CID reaction $CH_4^{+\cdot} \to C^{+\cdot}$ increases with the ionization potential of the target gas. This reaction, which requires an energy transfer of about 15 eV,[184] was argued to result primarily from direct electronic excitation. Higher ionization potential targets apparently tend to enhance CID in keV energy collisions via the direct electronic excitation mechanism. Helium, in this regard, is the target of choice. Other studies have confirmed this conclusion,[165,252] also showing that processes that compete with CID, such as scattering and charge transfer to the target, are minimized with selection of helium as the target gas. Most high-collision-energy CID studies, for these various reasons, use helium as the target gas.

At all collision energies, the effect of increasing the target gas pressure is to increase, at least initially, the total number of daughter ions and the relative abundances of product ions that arise from more highly endothermic reactions. These changes are a result of increasing the number of ions that undergo collision, as well as increasing the likelihood for each individual ion to undergo multiple collisions. A parent ion can undergo several collisions prior to dissociation and, at sufficiently high pressures, daughter ions formed after one or more collisions can themselves undergo CID. The two limiting cases that combine to contribute to the CID spectrum are represented below:

$$m_p^+ \xrightarrow{N} m_{d_1}^+ \xrightarrow{N} m_{d_3}^+ \xrightarrow{N} \text{etc.} \tag{3-28}$$

$$m_p^+ \xrightarrow{N} m_p^{+}* \xrightarrow{N} m_p^{+}** \xrightarrow{N} \text{products} \tag{3-29}$$

Reaction (3-28) represents the CID of the parent ion in a first collision, followed by the CID of a daughter ion in a second collision, and so on. Reaction (3-29) represents the stepwise excitation of the parent ion prior to dissociation. The overall effect in either case is to increase the total internal energy of the ions that are sampled when the target gas pressure is increased. Increasing the target gas pressure is, therefore, a convenient and frequently used means of increasing the degree of dissociation for higher mass ions (for which E_{REL} and the ion velocity are relatively low) and for ions that are particularly stable. For example, positive molecular ions of polynuclear aromatic hydrocarbons are particularly stable and produce very few ions in the CID spectra obtained using low collision energies (30-eV laboratory collision energies or less) under single collision conditions. It has been shown that more useful structural information is obtained with an increase in the collision gas pressure.[251]

Several studies have addressed target gas pressure effects on CID spectra both at high[252-255] and low[256] collision energies. Kim has applied probability theory to the sequence of multiple collisions.[254] With some simplifying assumptions, such as equal ion loss cross sections for parent and daughter ions, the probability for the detection of an ion P_{ion} (defined as the ratio of ions detected and ions that enter the collision cell) is given by the Beer's law relationship:

$$P_{ion} = e^{(-n_D \sigma_L l)} \tag{3-30}$$

where n_D is the target gas number density, σ_L is the total cross section for all processes that lead to ion loss, and l is the collision cell path length. Assuming that the parent ion internal energy prior to the first collision is unimportant, the probability that the detected ion is the parent ion can be determined, provided that the energy transfer distribution is known. Kim derives the expressions and uses several simple distributions to compare the predictions of the theory with experimental data. Despite the rather gross simplifications, reasonable agreement is obtained. Kim has since refined the model by incorporation of different ion loss cross sections for the daughter ions into the model.[255]

Another approach has been used by Dawson and Douglas to describe triple-quadrupole results,[256] and by Ouwerkerk et al. to describe keV collision energy data[252] acquired as a function of collision gas pressure. This approach uses the appropriate kinetic expressions for the important competitive and consecutive dissociations of the parent ion, incorporating a cross section for each. For example, Ouwerkerk et al. studied the CID spectra of the methane molecular ion and its carbon-containing daughter ions as a function of target gas pressure. Assuming that the cross sections for the dissociation steps indicated in the scheme of Figure 3-21 are the same regardless of whether the ion is formed in the collision cell or in the ion source, each indicated cross section was measured experimentally. Assuming further that the loss cross sections for each of the ions were equal, the relative abundances of the ions in the CID spectrum and the contributions of each dissociation route can be calculated from the measured cross sections. As an example, the authors calculated the relative contributions to the $CH_2^{+\bullet}$ ion current from $CH_4^{+\bullet}$ directly and from CID of the intermediate daughter

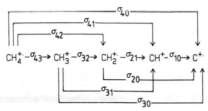

Figure 3-21 ■ Kinetic scheme for the dissociation of the molecular ion of methane.

CH_3^+, as a function of pressure for Xe, Ar, and He as collision targets.[249] An alternative application of the model is to measure the cross sections under single-collision conditions (as opposed to their determination by a best fit of the data acquired over a large pressure range) and to predict crudely the CID spectrum under multiple-collision conditions.

3.3.2. Photodissociation

Next to collisional activation, excitation by photon absorption is the most common means for activation of polyatomic ions. Photoexcitation holds several advantages over collisional excitation. In terms of the four criteria for an activation method, photoexcitation can, in principle, impart a wide range of energies, depending upon whether or not the ion can absorb a photon of the irradiating wavelength. The energy deposited into the ion is very well defined, and the energy can be varied insofar as the photon source wavelength is variable. The efficiency of photoexcitation depends upon the absorption spectrum of the ion; photodissociation cross sections are on the order of 10^{-2} Å^2. The very narrow and well-defined energy transfer distribution is a great advantage over the poorly defined collisional energy transfer characteristic of collisional activation experiments. The usefulness in the study of ion structures is greatly increased. An example of potential analytical advantage is the addition of a highly selective activation step in the neutral loss and/or parent ion MS/MS scan. The possibility for variation of the internal energy of the ion in a well-defined manner is also highly useful. The photodissociation spectrum of a parent ion, i.e., the extent of parent ion depletion via dissociation, acquired as a function of the irradiation wavelength, provide very useful information regarding ion structure. The relative abundances of the photofragments (analogous to the CID spectrum) provide further information. Note that the photodissociation spectrum refers to a wavelength spectrum, whereas the photofragment spectrum refers to an MS/MS spectrum acquired at a fixed wavelength. Additional details are available if the daughter ion translational energies can be measured. The major impediment to more widespread use of photodissociation for analytical purposes is much lower cross section for photodissociation relative to that for CID. This has required more intense light sources that have not been available until recently. This requirement is particularly limiting for most analytical mass spectrometers since these are beam instruments. For these cases, lasers are usually required to provide a sufficiently intense light source. For Fourier transform–ion cyclotron resonance mass spectrometers, less intense light sources can be used since the ion/photon interaction time can be lengthened appreciably.

Much of the present understanding of photodissociation of polyatomic organic ions has been derived from the pioneering work of Dunbar and co-workers.[106,257,258] These studies were carried out in an ion cyclotron resonance mass spectrometer in which ions (usually formed by electron ionization) were irradiated by light over a relatively long period of time. The experiment can be described in terms of Beer's law: the flux of photons absorbed, I_{ab}, which result in dissociation of the parent ion, is:

$$I_{ab} = I_0 n \sigma_{PD} l \qquad (3\text{-}31)$$

where I_0 is the initial photon flux, n is the number density of ions in the region irradiated by the photon beam, and l is the path length, representing the intersection of the photon beam with the orbits of the ions. In a typical ICR experiment, about 10^5 ions are created by the ionization pulse. This value gives an ion number density of roughly 10^6 ions/cm^3 in the volume intersected by the laser beam (the ions are concentrated along the electron beam path prior to rf excitation). Assuming a value of σ_{PD} equal to 5×10^{-2} Å^2, and a path length l of 0.2 cm (a longer path length can be obtained with an irradiation geometry coaxial with the electron beam), the product $(n_D l)$ is equal to 10^{-12}. Assuming that the population of ions is homogeneous and that these ions dissociate without complications (e.g., a reaction with background gas in the ICR cell), 10^4 ions can be dissociated in 1 s of irradiation by an I_0 of 10^{16} photons/s. This photon flux corresponds to a rather modest laser power of 5-mW radiation of wavelength 450 nm. The ability to trap ions for extended periods of time permits a group of ions to be exposed to large numbers of photons. This makes the ICR mass spectrometer particularly well suited for a wide variety of experiments using light. For example, the ICR has been used to study slow multiphoton dissociation of polyatomic ions using low-intensity continuous-wave infrared radiation,[259] as well as dissociation induced by high-power pulsed infrared radiation.[260] Two-photon experiments using either photons of the same wavelength or a combination of an infrared photon with a visible photon are also performed with ICR spectrometers.[257] To date, these studies have virtually all been aimed at obtaining fundamental information from gas-phase ions, such as the photodissociation spectrum, and for the differentiation of isomeric ion structures. Analytical applications, such as to complex mixture analysis, have lagged, but this is due in part to the relatively few ICR spectrometers in use as general analytical instruments. As the photodissociation processes of polyatomic gas-phase ions become better understood, as the use of lasers in chemistry and in mass spectrometry expands, and as the use of Fourier transform–ion cyclotron resonance mass spectrometers in analytical laboratories becomes more widespread, this situation will surely change.

The ion trap mass spectrometer is also capable of storing ions. This device has only recently been used for MS/MS but promises to become more widely used in this capacity. The capability for long storage times makes this device, like the ICR mass spectrometer, suited to photodissociation studies. Louris et al. have demonstrated photodissociation in an ion trap using a fiberoptic waveguide.[261]

Laser light sources make photodissociation of ions in beam-type spectrometers practical. Indeed, the first triple-quadrupole mass spectrometer was constructed for the study of photodissociation of organic ions in the rf-only quadrupole.[62] In recent years, several commercial and other sector mass spectrometers have been modified to allow for the laser irradiation of keV translational energy organic ions.[118–121,262–271] Beynon et al. have published a number of papers that describe the use of photodissociation in the second field-free region of a BE spectrometer[120,156,258,265-269] and in the ion source of a sector mass spectrometer[272-275] (the latter is not an MS/MS experiment). As reported by Harris and Beynon for a BE mass spectrometer modified for photodissociation studies,[276] an ion current of 5×10^{-10} A of a 6-keV ion of m/z 100 yields an ion number density of roughly 2×10^3 ions/cm^3 over a path length of 30 cm in the second field-free region of the instrument. Assuming that σ_{PD} is 5×10^{-2} Å^2, the product $(n\sigma_{PD}l)$ is about 3×10^{-13}. To obtain a photofragment ion yield of 6.25×10^4 ions/s $(10^{-14}$ Å), a photon flux of about 2×10^{17} photons/s is required, which corresponds to a laser power of 88 mW at a wavelength of 450 nm. Harris and Beynon point out that the ion and laser beams may not be optimally aligned, and that the laser beam must be chopped to correct for the daughter ion contribution from metastable ions. The practical requirement for photon flux is therefore higher. Also, practical values of σ_{PD} are often much lower than those assumed here; therefore, Beynon uses a minimum laser power of 750 mW. Under these conditions (viz, a 6-keV ion of m/z 100 with a path length of 30 cm irradiated by 750-mW continuous wave, of 450-nm photons), an individual ion spends roughly 3 μs in the field-free region; the ion is exposed in that time to roughly 5×10^{12} photons. Furthermore, ion velocities in a triple-quadrupole instrument are roughly one order of magnitude lower than in sector instruments, but the path length is slightly shorter than 30 cm, so that the number of photons available for absorption by any particular ion is not generally more than a factor of 10 greater than for a similar experiment performed using a sector instrument.

The types of information most frequently obtained from photodissociation experiments in an ICR spectrometer are the optical photodissociation spectrum and the relative abundances of charged photofragments. The photodissociation spectrum is obtained simply by irradiation of the ions for a fixed time at a series of wavelengths and comparison of the number of ions measured with the light on and with the light off. Under fixed experimental conditions, the relative cross section is $-\ln(\text{signal light-on})/(\text{signal light-off})$. An example of a photodissociation spectrum is given in Figure 3-22, which shows the spectrum for the molecular ion of phenetole.[277] This type of spectrum is not the true optical absorption spectrum of the ion, since absorption

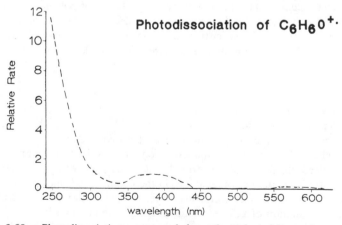

Figure 3-22 ■ Photodissociation spectrum of phenetole. (Adapted from reference 277.)

that does not result in fragmentation is not detected. Relaxation of the ion via fluorescence may also compete with fragmentation. This phenomenon has been noted in several cases[257] but, in general, dissociation rates are higher. The photodissociation spectrum is useful as an ion fingerprint and, therefore, as a basis upon which to distinguish isomeric ion structures. A comparison of relative photofragment abundances has been used less frequently to distinguish ion structures in ICR photodissociation studies; however, these experiments should prove to be very useful for the same purpose, particularly when spectra are obtained at several wavelengths. Most photodissociation experiments to date, however, use photon energies of 1 to 2 eV; such energy is generally sufficient to induce only the lowest energy dissociations and the least structurally diagnostic reactions (see Sections 3.2 and 3.3). Furthermore, the ions sampled in the ICR photodissociation experiments are generally very long-lived, with lifetimes on the order of milliseconds, implying that they are of low internal energy. Higher-energy ions would be more likely to dissociate, to fluoresce, or to undergo stabilizing collisions if the pressure is sufficiently high. The total postirradiation ion internal energies tend to be low in photodissociation experiments. Possible methods to increase the internal energies of the ions are to use shorter wavelength photons or to use multiphoton excitation.[257]

The triple-quadrupole instrument can also be used to obtain the relative abundances of photofragments created from a mass-selected ion. However, the photodissociation spectra obtained in a triple-quadrupole mass spectrometer are not generally obtained in the same experiment as performed in an ICR spectrometer. Only a small fraction of the parent ions are typically photodissociated in the triple-quadrupole mass spectrometer, so that the ratio (signal light-on)/(signal light-off) is always nearly unity, and a small difference between two large signals must be measured. With beam instruments (sectors and quadrupoles), the contribution to the daughter ion signal from the dissociation of metastable ions and from CID resulting from collisions with residual gas in the instrument can, together, be of a magnitude comparable to that from photodissociation. A method of correction for these contributions must therefore be designed into the experiment. McGilvery and Morrison, with a triple-quadrupole instrument, used an up–down counting system in which ions were counted and added both during and then shortly after a laser pulse. After a delay of 1 ms, background dissociations were counted for an equal period of time and subtracted from the previous total to give the corrected number of counts due to photodissociation.[62] Harris et al. and other groups using sector instruments have used mechanical chopping of the laser beam and phase-sensitive detection to differentiate between the continuous background signal and the periodic photodissociation signal.[120] If only one photofragment is formed over the range of wavelengths investigated, a simple experiment that measures the relative photodissociation cross section on a triple-quadrupole instrument sets the first quadrupole to pass the parent ion, sets the final quadrupole to pass the photofragment, and scans the wavelength of light that is passed through the rf-only quadrupole.

Despite the low signal levels and added experimental complexity, much can be learned about the unimolecular chemistry of gas-phase ions using a beam instrument, particularly in experiments in which daughter ion kinetic energies can be measured. Harris et al. and others have completed a number of elegant studies of photodissociated ions in sector instruments. The studies of Harris have primarily involved the photodissociation of ions in the second field-free region of a BE mass spectrometer and translational energy analysis of the daughter ions. In this instrument, the laser beam is colinear with the ion beam in the field-free region. A very similar approach has been used by Wagner-Redeker and Levsen[262] with a BE spectrometer. Bowers et al. have used a crossed-beam arrangement in the second field-free region of a BE instrument. The laser beam is directed along the z direction,[264] i.e., in the nonfocusing direction; maximum overlap between ion and photon beams is therefore obtained for this crossed-beam arrangement. Harris has also modified an EB mass spectrometer to allow a laser beam to pass through the ion source and into the first field-free region.[118,265-269] Russell et al. have also reported such a geometry, as well as a crossed-beam arrangement with the laser beam focused into the first field-free region.[121] In colinear ion/photon beam experiments, photodissociation can be studied either in the field-free region or in the ion source itself. The latter experiments are not defined as MS/MS since the absorbing species is not mass-selected. Nonetheless, there are interesting applications for this type of experiment.

Most photodissociation MS/MS studies have been devoted to the differentiation of isomeric ion structures. One common approach has been to measure relative photodissociation cross sections as a function of wavelength for each ion structure (see Chapter 4). Complementary information is obtained by comparison of the relative abundances of the photofragments. An additional experiment involves photodissociation of ions in the second field-free region of a BE spectrometer and a comparison of T values. Such information is unavailable from experiments with ICR and triple-quadrupole mass spectrometers. Measurement of T values has been used in the differentiation of isomeric xylene ions.[118] The ability to deposit a known, narrowly defined energy in the activation step of MS/MS allows for more detailed studies of the unimolecular dissociation of polyatomic ions than is possible with collisional activation. For example, T values have been measured as a function of photon energy for alkylbenzene cations.[268] A similar experiment studied the effect of ion source temperature on T at a fixed photon wavelength for ionized n-alkylbenzenes.

3.3.3. Electron Excitation

Electron/ion collisions can be used to excite ions, although this process has only rarely been used to induce the dissociation of polyatomic ions. Relatively little is known of the capabilities or limitations of electron excitation as an activation method for MS/MS, but some preliminary conclusions can be drawn from the studies reported to date. With regard to the four criteria for an activation method (see the introduction to Section 3.3): (i) in analogy to electron ionization, the amount of energy that can be imparted to an ion via bombardment by electrons with tens of electron volts of energy can be quite large (over 10 eV); (ii) the distribution of transferred energies is expected to be broad and poorly defined, with the upper limit being the energy of the bombarding electron itself; (iii) the *average* transferred energy can be conveniently varied (though not precisely controlled) by variation of the energy of the bombarding electrons, and (iv) although too few cross sections have been measured to give a typical value, an estimate of 6 $Å^2$ has been made for ionized cyanobenzene,[110] and values on the order of a few tenths of an angstrom squared have been measured for the electron-impact-induced dissociation of $D_2O^{+\cdot}$, H_3O^+, and D_3O^+.[120]

An early application of electron bombardment as an excitation process for polyatomic ions was reported by Cody and Freiser; the experiments were performed in a trapped-cell ICR instrument.[110,278,279] Just as with photodissociation, an ICR mass spectrometer is well suited for this experiment due to its ability to trap ions for a time sufficient for reaction to occur. In the initial experiments of Freiser, ions were formed by a pulse of electrons with an energy greater than the ionization potential of the sample and were subsequently excited by lower energy electrons. The major variables in these experiments were the trapping time and the electron energy. The daughter ion abundances, both absolute and relative, were shown to be very sensitive to the electron energy. Variations in the absolute daughter ion abundances reflect changes in the cross section for electron-induced dissociation. The maximum observed cross section for the cyanobenzene molecular ion was estimated to be 6 $\mathring{A}^2$ at an electron energy of about 7.5 eV. (The energy of the electrons used for exciting the ions was held below the ionization potential of the sample, since the sample was present throughout the experiment.) The variation of daughter ion relative abundances with increasing electron energy was consistent with increasing energy deposition into the ion. It was also noted that increased trapping time increased the probability that the ion would undergo electron-induced excitation and that a daughter ion could also undergo excitation. Due to the high background pressure, however, contributions from ion/molecule reactions were seen in the spectrum. Many of the complications encountered in the early studies can be avoided with newer methods. The newer FT–ICR mass spectrometers, for example, are equipped with pulsed valves that allow the sample pressure to be increased during the ionization pulse and quickly reduced prior to electron excitation. The proper timing of the experiment should therefore reduce contributions from ion/molecule reactions and allow for higher-energy electrons to be used for excitation, i.e., values greater than the ionization potential of the sample. Electron excitation of polyatomic ions in a dual-cell FT–ICR mass spectrometer has recently been demonstrated.[111] Ions were formed in the source region during a fixed period. Parent ions were then selected by sweeping out higher and lower mass ions and were allowed to pass into the analyzer cell. Electron excitation of the ions was performed in the analyzer cell. The authors point out that several advantages are gained by use of electron excitation rather than CID in the dual-cell FT–ICR mass spectrometer. For example, parent ions are not excited to larger orbits as with the CID experiment, making daughter ions formed in the source region more readily transferable through the conductance aperture between the cells. Elimination of the need for admission of a collision gas via a pulsed valve is another advantage. The authors also point out, however, that electron excitation is not selective. Collisional activation in the FT–ICR experiment requires that only a particular m/z be translationally excited, whereas the electrons can interact with any ions in their path. The mass selection of the parent ion must therefore be "cleaner" with the use of electron excitation in the FT–ICR mass spectrometer. Another important point is that since the ions must be trapped in the electron beam to be excited, the method is restricted to positive ions.

Electron-induced dissociation in the second reaction region of a BE spectrometer has been reported for the molecular cations of nitrobenzene and dimethyl malonate.[122] An electron energy of 40 eV was used with an electron current on the order of milliamperes. Under these conditions and with a background pressure of about 5×10^{-6} Torr, the signals obtained with the electron current on were comparable in intensity to those from the dissociations of metastable ions and from collisionally activated ions (from the background pressure) observed with the electron current off. As expected based on the increase in the internal energy of the ions, the relative abundances of ions resulting from simple cleavages increased relative to the abundances of those resulting from rearrangement and the kinetic energy release for NO˙ loss was seen to increase. Care must be taken in these experiments, since an increase in the background pressure in the second reaction region (and hence, an increase in the number of ions excited by collisional activation) may arise from heating of the filament. Modulation of the energy of the electron beam and use of a lock-in amplifier for detection might be used to discount the effect of filament heating.

Electron-induced dissociation cross sections for some polyatomic ions (admittedly small) have been measured with a specialized beam instrument.[123] In this instrument, ions formed in an electron cyclotron resonance ion source are mass- and charge-selected and their paths are then crossed with an electron beam. The charged products are momentum- and charge-analyzed with a magnetic sector. The electron beam is chopped to allow for the correction of dissociation events from background reactions (pressure in the instrument is about 5×10^{-9} Torr). Cross sections have been measured for the reactions $H_3O^+ + e^- \rightarrow O^+ \cdot + \text{neutrals}$, $H_3O^+ + e^- \rightarrow OH^+ + H_2$, and $D_3O^+ + e^- \rightarrow D_2O^+ \cdot + D \cdot$. The range of electron energies investigated was ten to several hundred electron volts. Maximum cross sections, which tend to occur at 100 eV or less, are measured to be 0.1 to 0.7 $Å^2$.

The excitation mechanism for electron bombardment at these electron energies is expected to be of the Franck–Condon type, i.e., a vertical electronic transition that occurs without a change in the position of momenta of the nuclei. This type of excitation, of course, also takes place with UV photoionization, a process in which a photon traverses 1 nm (molecular dimensions) in about 3×10^{-17} s; the process also occurs in 70-eV electron ionization in which the electron traverses 1 nm in about 2×10^{-16} s. In Franck–Condon excitation, electronic transitions are induced in the target (in this case an ion) by the rapidly changing electric field experienced by the ion due to the fast-moving electron. It might be expected that the effects of irradiation of ions with fast electrons will be similar to irradiation by photons.[207] The frequency spectrum of the electric field experienced by an ion when bombarded with a fast electron beam is relatively constant below the frequency corresponding to $1/t_c$, where t_c is the time of the electron/ion collision. The passage of fast electrons is similar in effect to irradiation by white light with the exception that some optically forbidden transitions may be more likely with electron bombardment. The wide range of frequencies has important implications with regard to the distribution of energies transferred to the ion as the electron energy is varied. It seems likely that the distribution of energies transferred to the ion will be increasingly broad as the electron energy is raised, just as it is with electron ionization. More studies are needed on the electron excitation of polyatomic ions to see how far the analogy with electron ionization can be carried.

It seems unlikely that electron excitation of polyatomic ions will supplant collisional activation as a general analytical tool, or that electron excitation will replace photodissociation in more detailed studies. However, it offers some advantages over each and will probably be useful in some specialized applications. As with photodissociation, vacuum-pumping requirements are reduced relative to those for collisional activation, scattering of the ion can be neglected since the "light" collision partner has a negligible effect on the center of mass of the ion, and charge transfer to a target gas is not a concern. The cross sections appear to be larger than those for photodissociation, especially when the ion lacks a good chromophore. Greater energies are accessible with electron excitation than with photons from current lasers (at the expense of a well-defined energy transfer distribution), and the electron energy is easily varied. Since the excitation mechanism is of the Franck–Condon type, electron excitation may be particularly well suited to the formation of doubly charged ions from mass-selected singly charged ions.

3.3.4. Surface-Induced Dissociation

A new and very promising means for excitation of polyatomic ions is based on collision of the ion with a surface. A few early reports in the mid-1970s indicated that a beam of polyatomic ions bombarding a solid surface could be dissociated. Cooks et al. reported the phenomenon in a mass spectrometer of EB geometry in which keV energy ions impinged upon a surface at glancing angles.[280,281] However, McLafferty et al., in a similar type of experiment with a number of different surfaces, did not observe dissociations that could be attributed to collisions with the surface.[166] The low ionization potentials of the metal surfaces employed apparently resulted in predominant neutralization of the ion. Gandy et al. reported surface-induced dissociation (SID) of mass-selected polyatomic ions in which the primary beam with kinetic energies of several tens of electron volts impinged directly onto the surface, and daughter ions produced were analyzed by time-of-flight mass spectrometry at right angles to the primary beam.[282] No other studies on the SID of polyatomic ions were reported until 1985, when Cooks and co-workers reported the SID of a number of polyatomic ions in a modified hybrid geometry mass spectrometer.[126,283] Briefly, the Cooks experiment involves bombardment of a rotatable surface by mass-selected ions with translational energies in the range of 20 to 150 eV. The angle of the incident ion beam with respect to the surface normal is usually about 25°, and daughter and reflected ions are extracted from the surface into a broad band-pass 45° energy filter and subsequently into a quadrupole mass filter. Since the first report, Cooks has reported a number of applications of the technique and several studies of the fundamental characteristics of SID. A stainless steel surface has been used in the studies reported to date but it may be significant that at the pressures employed in the experiments the surface is covered with impurities.

Most of the general characteristics of SID as an activation method for MS/MS are already apparent. The Cooks group has demonstrated that: (i) large energy transfers are possible (average transfers of almost 8-eV have been observed), (ii) the distribution of energy transferred is relatively narrow compared with collisional activation and electron excitation, and (iii) the energy transferred is easily controlled by variation of the translational energy of the incident ions. Cross sections (or other measures of the probability of the reaction) have not been reported, but a "typical" value of the fraction of ion current hitting the surface that reaches the final detector in Cooks' instrument is about 0.5%. Given that charge transfer to the surface (thereby neutralizing the ion) competes with SID and that only a fraction of the ions are likely to be extracted, transmitted, and detected, this conversion ratio of incident-to-detected ions indicates that SID is an important reaction channel, at least for the ions investigated. The generality of this conclusion remains to be established.

That the energy transferred in SID is readily varied and can be quite large was apparent in the first publication describing the technique.[126] For example, Figure 3-23 shows SID mass spectra of ionized d_3-acetophenone (m/z 123) acquired at parent ion energies of 22, 30, 40, and 60 eV. At 22 eV, the molecular ion reflected intact from the surface is the base peak in the spectrum, and only one daughter ion of significant

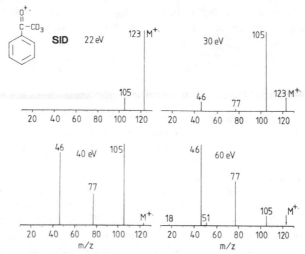

Figure 3-23 ■ Surface-induced dissociation daughter ion MS/MS spectra of the molecular ion of d_3-acetophenone. (Adapted from reference 126.)

abundance is observed. This daughter results from loss of CD_3·, probably the lowest energy fragmentation. As the impact energy increases, the daughter ion relative abundances clearly show an increasing energy transfer, as reflected by an enhancement of the low mass daughters at higher impact energies. The rapid decrease in the intensity of the parent ion also probably reflects increasing energy transfer, although a dependence of the cross section for charge transfer (neutralization) on impact energy may also influence the parent ion intensity. The SID spectrum of ionized phenetole at 60-eV indicates a much larger energy transfer than evident for the 70-eV electron ionization mass spectrum, the 7000-eV collision energy CID spectrum, or the 22-eV collision energy CID spectrum using argon as the target gas under multiple-collision conditions. Other similar examples have also been described. The approach discussed in Section 3.3.1.4 for the determination of the energy transfer distribution was used for the SID of $Fe(CO)_5^+$:[283] These results were compared with those obtained using keV and-eV collision energy collisional activation. The energy transfer distribution for SID was relatively narrow compared with the ion/gaseous target experiments (see Figure 3-19), and the average energy transfer was observed to increase from slightly less than 2-eV at an impact energy of about 20-eV to almost 8-eV at an impact energy of about 130 eV. In retrospect, it is not surprising that collisions with a surface give a narrower energy transfer distribution than collisions with a gas, since it seems likely that a much narrower range of impact parameters would be involved in SID. All collisions with the surfaces tend to be direct, whereas glancing collisions are likely with a gaseous target.

Unlike the collision between a fast projectile and an essentially stationary target gas atom or molecule, the center of mass collision energy in a collision with a surface is not easily determined. Two limiting cases are apparent: (i) the mass of the target is that of a single atom on the surface, be it a metal atom or a surface impurity, and (ii) the mass of the target, N, is that of a collection of atoms large enough so that the fraction $N/(N + m_p)$, is nearly unity, and the center of mass collision energy is essentially the laboratory collision energy. The former case assumes that the projectile strikes one surface atom that can recoil freely, i.e., the forces holding the atom in place at the surface are extremely weak. In the second case, the bonds at the surface are extremely rigid.

It is significant that SID spectra show the same daughter ions observed in MS/MS spectra generated with other activation methods. This similarity suggests that the overall mechanism for SID is a two-step mechanism like that for CID, viz, the first step involves an inelastic collision and the second step is unimolecular dissociation. However, the mechanism for energy transfer in the collision between a polyatomic ion and a surface is not yet clear. The possibilities that exist for collision with a target gas also exist here, e.g., impulsive collisions and complex formation. Some evidence has been given for chemically reactive collisions on the surface, which indicates that in some cases a complex may form with a lifetime sufficient for atom transfers to occur.[284] It seems unlikely, however, that complexes on the surfaces allow an approach to statistical equilibration of energy, considering the large internal energies deposited into the ions.

Research into the use of SID is still preliminary, and its ultimate role in MS/MS cannot be predicted. Certainly the high energy transfers observed, the ready variability of energy transfer, and the reduction of gas load are very attractive features. Provided cross sections are not too low, SID could supplant CID in many MS/MS applications. The degree to which cross sections for SID can vary relative to other mechanisms for different types of ions and surfaces is as yet unknown. Applications of SID to negative ions are only now being reported. New methods of SID that involve transmission of parent ions through an array of microchannels appear promising, and investigations of SID will be a large area of MS/MS activity in the next few years.

3.4. Reactive Collisions

The preceding sections of this chapter are devoted either directly to unimolecular dissociation or to techniques that ultimately lead to the unimolecular dissociation of a mass-selected ion. In most applications of MS/MS, it is the unimolecular chemistry of the ion that is used for its identification. A less well-exploited approach is to use the bimolecular chemistry of mass-selected ions to identify and to distinguish them. The approach in this case is to use a target gas that is likely to undergo a reactive collision with the mass selected ion, i.e., a collision in which atoms are transferred between the ion and the target. (Another technically reactive collision is one that results in electron transfer, i.e., charge exchange. A distinction is made here between these types of reactions, and charge exchange is included as a member of the family of charge permutation reactions discussed in Section 3.5. It should be recognized, however, that charge exchange, usually considered to be a competitive reaction channel, may accompany an ion/molecule reaction.) In order to be analytically useful, the reaction must result in a change in mass and should be specific for some chemical functionality of the ion. The relative kinetic energies of the collision partners must be low so that a long-lived complex can be formed that allows atom transfers to occur. This condition generally does not hold in high-collision-energy reaction regions. Atom transfers via a direct mechanism can occur at keV collision energies, but these reactions are characterized by much lower cross sections than are low-energy ion/molecule reactions.[285]

Reactive collision experiments are therefore generally restricted to multiquadrupole, hybrid, ion cyclotron resonance, and ion trap mass spectrometers. Such reactions can also occur in multisector instruments with additional deceleration into and acceleration out of the reaction region.

The study of ion/molecule reactions is hardly new. In 1916, Dempster observed, and correctly identified, H_3^+ produced by the now familiar reaction of $H_2^{+\bullet}$ with H_2.[286] Nevertheless, comparatively little attention was given to ion/molecule reactions until the 1950s. During that decade, most reactions studied were those induced in the ion source of a mass spectrometer. Subsequently, to define more accurately reaction conditions and to measure with higher accuracy the characteristics of a reaction such as its rate and its cross section, more sophisticated experiments were devised. These included drift-tube studies,[287,288] beam experiments,[180] and ion cyclotron resonance.[141] Most of the crossed- and merged-beam experiments and the beam/stationary target gas studies were performed using a variety of tandem mass spectrometers constructed especially for these purposes. From these studies, more detailed information has been obtained for ion/molecule reactions than for any other type of bimolecular reaction in chemistry. A number of books and reviews have been published in the field.[285,289-295]

Some 50 years after the first observation of an ion/molecule reaction product, the development of chemical ionization[296] began the use of the ion/molecule reaction in analytical mass spectrometry. Furthermore, the introduction of the triple-quadrupole instrument in MS/MS made possible the use of ion/molecule reactions in a reaction region of a tandem mass spectrometer for analytical purposes. Further impetus for the investigation of the analytical utility of ion/molecule reactions has come with the development of the FT–ICR and its MS/MS capabilities. The use of ion/molecule association reactions in MS/MS has been concentrated in the last few years.

For a detailed discussion of the present understanding of ion/molecule reactions in the range of collision energies of 0 to 50 eV, the reader is referred to the many excellent treatments of this topic.[285,289-295] Only a brief discussion of the subject is given here, with particular emphasis on aspects relevant to MS/MS. Most ion/molecule reactions are rationalized as proceeding either via a direct mechanism or via a collision complex in which the entire ion/target combination plays a role in the reaction. In a direct mechanism, pairwise interactions occur only between *portions* of the ion and/or the target, with the remainders of the molecules as spectators. Reactions deemed to proceed, at least in part, via a direct mechanism tend to involve simple atom transfers, as in the reaction $Ar^{+\bullet} + D_2 \rightarrow ArD^+ + D^{\bullet}$.[297] However, most reactions of polyatomic ions with polyatomic targets are thought to proceed via complex formation.

An example of an ion/molecule reaction deemed to proceed via a relatively long-lived complex is the reaction of ionized ethylene with ethylene.[131,298-301] This reaction, represented as:

$$C_2H_4^{+\bullet} + C_2H_4 \rightarrow C_4H_8^{+\bullet*} \rightarrow \text{products} \qquad (3\text{-}32)$$

occurs over a relatively wide range of collision energies. The intermediate complex is generally not observed unless it is relaxed by collision with a third body, as occurs in a high-pressure ion source. Figure 3-24 shows the partial MS/MS spectrum of products from the reaction of mass-selected $C_2H_4^{+\bullet}$ with ethylene in the first reaction region of a QEB hybrid instrument at a collision energy of 9.0 eV. This spectrum shows products

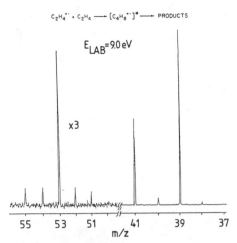

$C_2H_4^{+\cdot} + C_2H_4 \longrightarrow [C_4H_8^{+\cdot}]^* \longrightarrow$ PRODUCTS

$E_{LAB} = 9.0\,eV$

x3

55   53   51   41   39   37

m/z

Figure 3-24 ■ Partial MS/MS spectrum obtained from collisions of ionized ethylene with neutral ethylene at a collision energy of 9.0 eV ■ (Adapted from reference 94.)

that arise from two major reaction channels:

$$C_4H_8^{+\cdot}{}^* \rightarrow C_4H_7^+ + H^{\cdot} \rightarrow \text{products} \qquad (3\text{-}33)$$

$$C_4H_8^{+\cdot}{}^* \rightarrow C_3H_5^+ + CH_3^{\cdot} \rightarrow \text{products} \qquad (3\text{-}34)$$

The transient intermediate dissipates some of its internal energy through dissociation via one of these pathways. The primary products, $C_4H_7^+$ and $C_3H_5^+$, dissipate energy further by loss of hydrogen(s). The relative abundances of the various ionic products in each reaction path are very sensitive to collision energy. Plots of these abundance variations for each reaction path are shown in Figure 3-25. In each case, the trend is toward greater dissociation (the lower mass ions increase in relative abundance) as the collision energy is increased. As the collision energy increases, more internal energy is present in the collision complex. This result might be expected considering that E_{REL} increases with collision energy. Indeed, endothermic reactions that proceed through a long-lived collision complex can be driven by the collision energy. However, as the lifetime of the complex decreases, as it should with increasing internal energy, the likelihood that all of the energy of the collision would be statistically partitioned between internal and translational modes decreases. Statistical theories have been applied to ion/molecule complexes with some success.[302] Comparisons of the fragmentation behavior of the $C_4H_8^{+\cdot}{}^*$ complex with that observed in electron ionization and photoionization mass spectra of butene isomers showed very similar behaviors.[298-301] Comparisons of the behavior of an ion/molecule complex thought to have the structure of an ion formed by electron ionization are complicated, however, by differences in the distributions of internal energy and angular momentum.[132] Another important point (not evident in Figure 3-25) is that the absolute intensity of ion/molecule reaction products decreases with the collision energy. This is a common observation and is consistent with ion/molecule interaction potential models (ion-induced dipole attraction and variations thereof) that predict decreasing reaction cross section with increasing collision energy.[302] The $C_2H_4^{+\cdot}/C_2H_4$ example illustrates an important

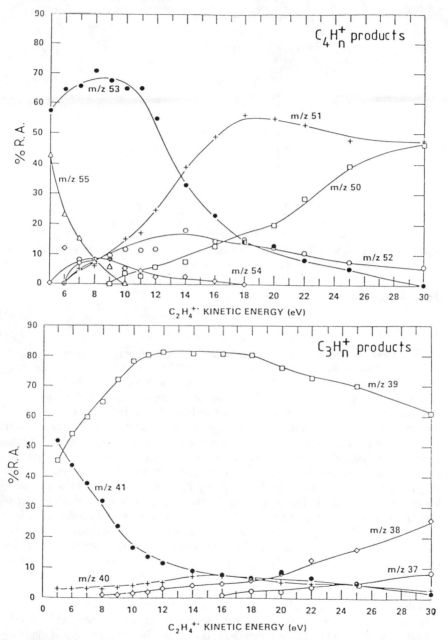

Figure 3-25 ■ Plots of the relative product ion abundances from the reaction of ionized ethylene with neutral ethylene obtained as a function of collision energy. (Adapted from reference 94.)

practical point in the use of a chemically selective target gas in that the cross section for the reaction is highest at low collision energies under which condition dissociation is minimized.

In many ways, the most powerful tool for studying ion/molecule reactions over the last two decades has been the ICR spectrometer in which collisions occur at near thermal energies. The MS/MS and high mass resolution capabilities provided by the application of Fourier transform methods to the ICR technique has made it even more powerful for this type of study.

3.5. Charge Permutation Reactions

A major class of reactions that commonly occur in MS/MS instruments involves changes in charge and are referred to collectively as charge permutation reactions. When such a reaction occurs with polyatomic molecules, dissociation of the projectile or target, or both, can follow. The most commonly encountered charge permutation reactions are listed below:

$$m_p^{+\cdot} + N \rightarrow m_p + N^{+\cdot} \tag{3-35}$$

$$m_p^{-\cdot} + N \rightarrow m_p + N^{-\cdot} \tag{3-36}$$

$$m_p^{2+} + N \rightarrow m_p^{+\cdot} + N^{+\cdot} \tag{3-37}$$

$$m_p^{-\cdot} + N \rightarrow m_p^{+\cdot} + N + 2e^- \tag{3-38}$$

$$m_p^{+\cdot} + N \rightarrow m_p^{2+} + N + e^- \tag{3-39}$$

$$m_p + N \rightarrow m_p^{+\cdot} + N + e^- \tag{3-40}$$

These reactions are written with odd-electron ions and even-electron targets, but they are not necessarily restricted to this combination. In the reactions discussed here, all reactants are assumed to be in their respective ground electronic states. Reactions (3-35) through (3-37) are examples of a charge exchange process between the ion and the target and can be observed at all collision energies used in MS/MS. Reactions (3-38) through (3-40) are classified here as examples of collisional ionization reactions. These reactions are typically only observed at keV collision energies. Each class of reactions, as it relates to MS/MS, is discussed in turn.

3.5.1. Charge Exchange Reactions

3.5.1.1. $m_p^{+\cdot} + N \rightarrow m_p + N^{+\cdot}$. Charge exchange is one of the most commonly observed and extensively studied types of ion/molecule reaction. The most common charge exchange reaction involves the transfer of an electron from a neutral species to a positive ion [reaction (3-35)]. Lindholm and co-workers pioneered the practical use of this reaction to define ion internal energies in mass spectrometry and to obtain breakdown curves.[230] The internal energy with which the ion is formed in a charge exchange reaction is given by:

$$\epsilon = RE(m_p^{+\cdot}) - IP(N) \tag{3-41}$$

where $RE(m_p^{+\cdot})$ is the recombination energy of $m_p^{+\cdot}$, and $IP(N)$ is the vertical ionization potential of N. (In this experiment, N is the species of interest.) If m_p is formed in the ground electronic state, the recombination energy is the negative value of its vertical ionization potential.[230] A breakdown curve can be generated for an ion by use of several reagent ions of different recombination energies. These types of studies have been performed in an ion source[303] and in mass spectrometers designed for this purpose. Two geometries of tandem instruments have been used: longitudinal[304] and perpendicular.[230] In longitudinal tandem mass spectrometers, the products of charge exchange are extracted in the same direction as the primary beam, whereas in the perpendicular spectrometers, the charged reaction products are extracted at 90°. In longitudinal mass spectrometers, among which most MS/MS instruments would be classified, the charge exchange spectra can be sensitive to collision energy, although much less so than are low-collision-energy CID mass spectra. The lesser collision energy dependence for charge exchange mass spectra measured with a longitudinal mass spectrometer may be due to the nature of the charge exchange reaction. Electron transfer does not require the close approach of nuclei or the formation of a long-lived collision complex. Momentum transfer can occur in conjunction with charge exchange, although it is not a prerequisite. To minimize the contribution to the mass spectrum of collisions that involve momentum transfer, perpendicular mass spectrometers have been used.

In MS/MS, it is usually desirable to minimize charge exchange since it competes with CID. Target gases with relatively high ionization potentials (e.g., He, Ar, N_2) are therefore generally used. Endothermic charge exchange reactions can still occur, however, with the additional energy supplied by the translational energy of the ion. The extent to which charge exchange occurs cannot generally be directly measured in keV collision energy MS/MS spectrometers, since the ion(s) from the target gas are not formed with high translational energies and are not therefore detected. Charged products from charge exchange can, however, be detected in many low-collision-energy MS/MS instruments (since they are very similar to longitudinal tandem instruments constructed for this purpose),[305] although not necessarily with the same efficiency as CID products.

3.5.1.2. $m_p^{-\cdot} + N \rightarrow m_p + N^{-\cdot}$. Reaction (3-36) represents a charge exchange reaction between an anion and a target; in this case, the electron is donated by the projectile to the target. The relative electron affinities of m_p and N determine whether or not the reaction is exothermic. Many molecules will not take part in this reaction since they have negative electron affinities and do not form a stable negative ion. As with reaction (3-35), if this reaction occurs in the reaction region of a keV collision energy beam instrument, the charged products are not detected because of their low kinetic energy, but they can be detected in a low-collision-energy instrument.

A major difference between reactions (3-35) and (3-36) is that electron detachment from $m_p^{-\cdot}$ competes with electron transfer.[203] For stable anions, electron detachment requires an external perturbation such as a collision. The neutralization reaction:

$$m_p^{-\cdot} + N \rightarrow m_p + N + e^- \qquad (3\text{-}42)$$

is termed electron detachment, and along with autodetachment, is a charge permutation reaction that is not directly observable with most MS/MS instruments [unlike electron detachment from a positive ion (reaction (3-39))]. Electron detachment can also occur at low collision energies (i.e., at E_{REL} less than the electron affinity of m_p) via the formation of a short-lived $m_p N^-\cdot$ complex that ejects the electron.[208] Electron detachment from $m_p^-\cdot$ competes with CID of negative ions, along with reactions (3-36) and (3-38) and can be the most important reaction channel for some anions. Electron affinities of most molecules that form stable anions are typically less than a few electron volts and are often less than or comparable to the lowest energy dissociation reactions, so that electron detachment may be faster than fragmentation. In many MS/MS daughter ion spectra of molecular anions, only a few dissociation reactions are observed, and these are generally reactions with low critical energies. It is possible, although there is no direct evidence, that the anions that undergo more highly inelastic collisions eject the extra electron rather than dissociate with retention of the charge. In cases in which structurally diagnostic daughter ions are not observed in the daughter ion MS/MS spectrum, reaction (3-38) has been studied to enhance the specificity of the spectrum.

3.5.1.3. $m_p^{2+} + N \rightarrow m_p^+\cdot + N^+\cdot$. A common reaction for doubly charged positive ions is the capture of one electron from the target molecule to form two positive ions. The reaction is exothermic if the ionization potential of N is less than the recombination energy of m_p^{2+} to form m_p^+, which is often the case. Either product can be internally excited after collision and may dissociate. This reaction has been studied with BE spectrometers, in which the $m_p^{1+}\cdot$ ions formed in reaction region 2 have mass/charge ratios twice as large as singly charged ions formed in the ion source. The $m_p^+\cdot$ ions formed by reduction can therefore be analyzed free of interference from source ions by setting the electric sector plate voltages at twice their normal values ($2E$). The resulting spectrum is referred to as a $2E$ mass spectrum and was first described by Beynon et al.[306] The intensities of the ions in the $2E$ mass spectrum are determined by a number of factors, among which are the intensities of m_p^{2+} ions that reach the collision cell, the relative cross sections for reaction (3-37), and competitive processes such as double electron transfer, fragmentation, and scattering. Most recently, the $2E$ mass spectrum method has been used by Moran and co-workers using an EB geometry instrument and reaction region 1 to obtain fundamental information regarding polyatomic doubly charged ions.[307-312]

If the singly charged ion formed by electron capture is sufficiently excited, it can dissociate. This process has recently been referred to as electron-capture-induced decomposition (ECID)[124] in analogy to collision-induced dissociation. In a BE spectrometer, the products from this reaction in reaction region 2 can be analyzed by a scan of the electric sector voltage. Most daughter ions are of a kinetic energy within the $2E$ to E voltage window. Spectra measured in this kinetic energy window do not contain contributions from unimolecular dissociations or from CID of singly charged ions formed in the ion source (these ions are transmitted in the 0 to E voltage region). Interfering ions formed in other processes can still be passed to the detector in the $2E$ to E voltage region, however. Principal among these is the charge separation reaction:[313]

$$m_p^{2+} \rightarrow m_{d1}^+ + m_{d2}^+ \tag{3-43}$$

this may be either unimolecular or induced by a collision.[12,314] Charge separation is a commonly observed reaction for doubly charged ions, but has not been explicitly classified here as a charge permutation reaction, since neither the mass-selected ion nor the target undergoes a change in charge before dissociation. In a BE mass spectrometer, the charge separation reaction is usually easily identified by a very large value of T, which predominantly arises from the Coulombic repulsion of the fragments.[12] The ECID reaction should prove useful in the study of doubly charged ions including, for example, the differentiation of isomeric ion structures and in the determination of double ionization energies.[124] The fate of the newly formed ion from the target molecule, $N^{+\cdot}$, has not been studied using MS/MS instruments. With a BE spectrometer, some information regarding the state of the target may be reflected by the translational energy change of the projectile, but the newly formed ion $N^{+\cdot}$ itself is not detected. These ions can, however, be detected in low-collision-energy instruments.[315]

3.5.2. Collisional Ionization Reactions

3.5.2.1. $m_p^{-\cdot} + N \rightarrow m_p^{+\cdot} + N + 2e^-$. Reaction 3-38, often referred to as charge inversion, involves the removal of two electrons from an anion to form a singly charged positive ion. The other products are the target molecule and two free electrons. This situation is probably the usual case, although electron capture by the target to form an anion and one free electron also occurs. The relative contributions of the two processes depends on the collision conditions. The reaction is highly endothermic, requiring a minimum energy (assuming that the reactants are in their respective ground states) equivalent to the sum of the electron affinity and ionization potential of the molecule m_p.[316,317] For most polyatomic molecules that form stable anions, this sum is on the order of 10 eV. Assuming the ionization potential is known, the electron affinity can be obtained using peak position measurements to determine the energy loss associated with the inversion reaction (see Sections 3.3.1.1 and 4.4.1). Energy loss measurements associated with charge permutation reactions are discussed further in Section 3.5.2.2 and 4.4.1.

Reaction (3-38) is analytically useful because the daughter ion MS/MS spectra of $m_p^{-\cdot}$ ions often show little, if any, fragmentation, and the daughter ions that are observed generally arise from low-critical-energy reactions. The charge inversion spectra, i.e., the daughter ion MS/MS spectra of the positive analog of the parent ion, generally contain a variety of fragmentations, including products from highly endothermic reactions.[318,319] For example, Zakett et al. observed virtually no negatively charged daughter ions in the daughter ion MS/MS spectra of the anions of polynuclear aromatic hydrocarbons. However, the charge inversion spectra contained sufficient fragmentation information to allow the isomers of these compounds to be distinguished.[320]

The charge inversion spectrum of $m_p^{-\cdot}$ usually resembles the daughter ion spectrum of $m_p^{+\cdot}$ formed in the ion source, but often reflects a greater energy deposition in $m_p^{+\cdot}$ by exhibiting more extensive fragmentation and greater relative abundances of daughter ions formed from highly endothermic reactions. This may seem surprising in light of the fact that about 10-eV are required to form initially $m_p^{+\cdot}$ from $m_p^{-\cdot}$.[316,317] This effect may be attributable to major differences between the Franck–Condon factors in the formation of $m_p^{+\cdot}$ from $m_p^{-\cdot}$ and in the direct excitation of $m_p^{+\cdot}$. It should also be recognized that the structure of $m_p^{-\cdot}$ may well be quite different than that of ground state $m_p^{+\cdot}$.

As with all of the collisional ionization reactions discussed, the cross section for charge inversion is generally very low at low collision energies. As expected based on the Massey parameter (Section 3.3.1.2), charge inversion is enhanced at the shorter collision times encountered in high-collision-energy instruments. Only one report of charge inversion occurring in a triple-quadrupole instrument has appeared.[321] A collision energy of 180 eV (E_{REL} 70 eV) was used with SF_6 as the target gas under single-collision conditions. A relatively low reaction cross section of about 1 $\mathring{A}^2$ was observed. In general, charge inversion is not observed using triple-quadrupole or other low-collision-energy instruments.

The other possible charge inversion reaction has also been observed [322-324], i.e.

$$m_p^+\cdot + target(s) \rightarrow m_{(pordj)}^-\cdot + products \qquad (3\text{-}44)$$

where the target species and products other than the anion are unspecified. This reaction can be classified among the charge transfer reactions since it involves the transfer of two electrons to the parent ion. This reaction has been little studied due to its low cross-section. At thermal collision energies, the reaction is extremely unlikely in that the transfer of the second electron requires the removal of an electron from the target when the electron affinity of m_p is likely to be several electronvolts less than the ionization potential of N. At keV collision energies, however, the cross-section for the reaction is enhanced.

3.5.2.2. $m_p^+\cdot + N \rightarrow m_p^{2+} + N + e^-$. Reaction (3-39) is commonly known as charge stripping and is most often recognized when it occurs following a collision in the second reaction region of a BE instrument, since the doubly charged ion that is formed falls at one-half the kinetic energy/charge ratio of the singly charged parent ion. Indeed, it was the development of the BE instrument that spurred interest in charge stripping as a means for studying dications. Since the first reports of charge stripping via ion kinetic energy spectrometry,[325-328] applications of the reaction have been described in areas such as ion thermochemistry, ion structure differentiation, and mixture analysis. A number of factors determine if charge stripping is observed. These include: (1) the stability of m_p^{2+}, (2) the experimental conditions such as collision energy and the nature of the target, and (3) the cross sections of competing processes such as CID, scattering, and charge transfer relative to that for charge stripping.[329]

The main experimental factors that determine the extent to which charge stripping occurs are the nature of the target and the collision energy. Since an electronic transition usually must occur, the collision energy must be as high as most MS/MS instruments can access, in order to have a more favorable Massey parameter for the reaction (see Section 3.3.1.2). In contrast to high-collision-energy collisional activation, helium is not the most efficient target gas for the promotion of the charge-stripping reaction. A number of studies have indicated that oxygen is much more efficient than helium.[185,330-334] This effect is attributed to the possibility for resonant electron capture by O_2[335] (electron capture by O_2 has been shown to occur in another type of charge-stripping reaction described in Section 3.4.2.3). Both NO_2[336] and NO[333] have also been observed to be unusually efficient targets in the promotion of charge-stripping reactions.

Assuming ground-state reactants, the minimum energy required for reaction (3-39) is the energy required to remove an electron from $m_p^+\cdot$, i.e., its second ionization potential. The reaction is usually considered to proceed via a vertical transition so that the vertical second ionization potential applies. Under certain conditions, the second ionization potential can be obtained by measurement of the value of the minimum energy loss in the electric sector scan. With no conversion of translational to internal energy, the doubly charged cation formed in reaction (3-39) would appear at exactly one-half the kinetic energy/charge ratio of the main beam. Under the proper conditions, which are discussed in Section 3.3.1.1, the difference between this and the observed peak position can be used to determine the endothermicity of the reaction. The leading edge of the peak gives q_{min}, the minimum endothermicity of the reaction, which is taken to be the second ionization potential. This technique has proven to be superior to the electron ionization technique in measurement of second ionization potentials. The two methods do not strictly measure the same value. The latter method gives the onset for the reaction of m_p forming m_p^{2+} and two electrons, while charge stripping involves the reaction of $m_p^{+\bullet}$ to form m_p^{2+} and one electron. The sum of the first ionization potential and q_{min} is not necessarily the energy required to form m_p^{2+} from m_p directly.[329]

An important application of the charge-stripping reaction to gas-phase polyatomic ion chemistry has been in the differentiation of isomeric ion structures. When the data are acquired with a BE mass spectrometer, charge-stripping products can usually be distinguished from CID products (if they are not buried by signals that result from the dissociation of m_p^+ into daughter ions with one-half the mass) by their relatively narrow peak shapes. The doubly charged parent ion peak is not broadened by kinetic energy release as are the CID products of $m_p^{+\bullet}$, and the peaks due to losses of hydrogen that often accompany the charge-stripping process are only slightly broadened by kinetic energy release, since the detected ion product is usually far more massive. Dissociation reactions of the doubly charged ion into two singly charged products also occurs, and the products of such reactions can be distinguished by their relatively large T values; these ions are not usually used in the differentiation of isomeric ion structures. Despite the fact that many fewer peaks are observed in charge-stripping spectra than in most CID MS/MS spectra, the former often differ significantly for isomeric ions when the CID MS/MS spectra do not.[329] It has therefore been concluded that the relative abundances of the products of charge-stripping reactions are more sensitive to structural differences than those resulting from CID. This may be due to the fact that charge stripping is such a highly endothermic process (typically requiring greater than 12-eV of energy) that a higher fraction of stable, nonisomerizing parent ions are sampled than is the case with collisional activation. With the smaller energy transfers encountered with collisional activation, ions near the threshold for dissociation (those that may have isomerized over a low barrier for rearrangement) are preferentially sampled.[151,337] This argument is similar to that used to justify the fact that collisional activation MS/MS spectra are superior to the spectra of metastable ions in the differentiation of ion structures (see Sections 3.3 and 4.2.1). It has also been argued that if the structures of $m_p^{+\bullet}$ and m_p^{2+} are sufficiently different, the Franck–Condon factors will also be different, and a highly excited m_p^{2+} will be formed. Structurally diagnostic dissociations from the highly excited ions are then likely to occur.[319]

As with all methods for the characterization of ion structures, the question of the effect of the parent ion internal energy on the spectrum has been raised. Several studies have reported that no significant effect of parent ion internal energy on the spectrum has been noted when the ions studied ($C_3H_6^{+\bullet}$ and $C_5H_8^{+\bullet}$ isomers) could not rearrange at energies less than the lowest energy fragmentation.[334,338,339] Jarrold et al. have measured a significant effect of the chemical ionization source pressure on the charge-stripping peaks arising from nonisomerizing $C_2H_8N^+$ and $C_2H_7O^+$ ions.[340] It has been argued that since the charge-stripping reaction is likely to be a vertical transition, the Franck–Condon factors can be highly sensitive to the vibrational energy of $m_p^{+\cdot}$, thereby resulting in the formation of m_p^{2+} ions of significantly different energy from precursors of different initial internal energy. Nevertheless, the number of cases in which the indications of the charge-stripping peaks are consistent with other methods for the characterization of ion structures seems to support the argument that charge stripping is sensitive to the ion structure and not overly sensitive to ion internal energy.[329]

A relatively new method utilizes reaction (3-37) to obtain spectra of charge-stripping products free of interferences from CID products. In the BE spectrometer, the CID of $m_p^{+\cdot}$ can produce broad peaks at, or near, one-half the kinetic energy of m_p^+ when the daughter ions are one-half the mass of the parent ion or nearly so. These ions can interfere with, or obscure, the peaks due to ions that result from charge-stripping processes. Kingston et al. have used a triple-sector mass spectrometer of BEE geometry and reaction (3-37) to eliminate the CID interferences.[341] A target gas is admitted into the third reaction region (i.e., between the electric sectors). The doubly charged ions formed in reaction region 2 can undergo charge transfer with the target gas in reaction region 3 to give a singly charged ion with a kinetic energy of approximately twice that of a product that results from a CID process. The lower kinetic energy ions can therefore be filtered out by the second electric sector. With the appropriate linked scanning of the two electric sectors, all doubly charged ions formed in reaction region 2 that subsequently undergo the charge transfer reaction in reaction region 3 can be detected without interference. The spectrum that results is not necessarily one that reflects the abundances of doubly charged ions formed directly from charge stripping, since the cross sections for reaction (3-37), as well as other reactions possible in reaction region 3, vary with the ion. Preliminary work indicates that isomeric structures can be distinguished with this experiment, and that the differences in ion structure persist through the reaction sequence.[342]

3.5.2.3. $m_p + N \rightarrow m_p^{+\cdot} + N + e^-$. Reaction (3-40) represents the ionization of a high-velocity neutral species m_p upon collision with a target N. The reaction is included in the discussion of commonly observed charge permutation reactions in MS/MS since it has been used in conjunction with reaction (3-35) in a new technique referred to as neutralization–reionization mass spectrometry (NRMS). This experiment, introduced by McLafferty and co-workers for the study of polyatomic ions and neutrals,[52,342] involves the formation of neutrals from a mass-selected ion beam via reaction (3-35) (and also via the decomposition of metastable ions, CID, or electron detachment from anions), the deflection of any residual charged particles from the beam, and the collision of the fast moving neutrals with a target gas. The neutral molecules are ionized by electron detachment, as indicated, or in some cases, by electron transfer to the target,[343] i.e., $m_p + N \rightarrow m_p^+ + N^{-\cdot}$ The McLafferty group[344] and Burgers and Holmes et al.[345] have shown this technique to give unique ion structural information about organic gas-phase ions and to be useful in the formation and study of highly unstable and reactive neutral species. For example, several isomeric ions could be readily distinguished via NRMS but could not be distinguished with

CID.[346] The NRMS method was successful because the neutral species are not likely to isomerize prior to dissociation, whereas such a process did occur after collisional activation but before dissociation when the ions were examined.

Danis et al. have studied the parameters that optimize each of the two charge permutation reactions of NRMS.[343,347] For the reionization reaction, oxygen was found to be the most efficient target in terms of the fraction of neutrals converted into ions [as has also been the case for reaction (3-39)]. By impinging fast neutral O_2 onto molecular targets (CH_3Cl and CH_3COCH_3) and observing O_2^- and O^- (in combined abundance greater than that of the corresponding cations), these workers demonstrated that electron capture by O_2 is an important reaction channel that may account for the enhanced performance of oxygen as a reionization agent.[343] In terms of the degree to which fragmentation occurs upon reionization, oxygen was found to be the "softest" target and helium the "hardest." To minimize fragmentation, oxygen is recommended for general use at pressures chosen to minimize multiple collisions. For greater fragmentation, higher target pressures are used to increase the number of collisions, or helium can be used instead of oxygen. The reionization efficiency can vary substantially with the nature of the species studied as well, but for any given sample projectile, the propensity for reionization did not vary by more than a factor of 3 over the collision energy range of 3 to 10 keV in collisions with oxygen. These data indicate that the cross section for reionization is fairly flat over this collision energy range. The cross section for collisional ionization of Rb, K, or Na by N_2, O_2, or Cl_2, on the other hand, has been observed to drop dramatically in the range of 1000 to 150-eV of collision energy.[343] High collision energies are therefore recommended to complete experiments in the flat portion of the cross section curve and to maximize transmission of ions and neutrals through the mass spectrometer.

In the neutralization step of the NRMS experiment, reaction (3-35) is used to produce fast neutral species. The neutral beam formed in this reaction may consist of the intact uncharged counterpart of the mass-selected ion and neutral daughter species derived from the dissociations of the ion. Such species exist in the beam along with the fast neutral products formed from the decomposition of metastable ions and from CID processes. Since the neutral products of reaction (3-35) are of interest in this experiment, the choice of conditions for the neutralization reaction may be very different from those used when charged products are analyzed, as in the experiment emphasized in Section 3.5.1.1. Danis et al. have shown, for example, that the efficiency for neutralization of high-energy ions is greater for metal vapor targets than for rare gas or molecular targets.[347] The highest efficiencies for all targets are observed at a pressure of target sufficient to attenuate the parent ion beam to about 30% of its original intensity. The proportion of charge transfer to competitive processes of CID is greater for the metal vapor targets. More of the neutral species present prior to reionization result from reaction (3-35) than is the case for helium as a target, for example. The energy imparted to the newly formed neutral species via reaction (3-35) can be varied by using targets of different ionization potentials. If a high-purity neutral beam is desired, a metal atom target with an ionization potential near that of the ion should be used under single-collision conditions. If greater structural information is desired, a low ionization potential target can be used under multiple-collision conditions to enhance the degree of fragmentation from the neutral species.

Application of MS/MS to Fundamental Studies

4.1. Introduction

As pointed out in Chapter 1, the initial studies performed using MS/MS instruments (specifically, BE spectrometers) in the early 1970s were directed toward obtaining fundamental chemical information from gas-phase ions. These studies focused primarily on metastable ions and keV collision energy collisional activation and collision-induced dissociation. Although the analytical utility of mixture analysis with MS/MS has provided the major impetus to the growth of the technique, the number and variety of studies aimed at obtaining fundamental chemical information using MS/MS have also grown rapidly. This chapter describes the major types of fundamental studies for which MS/MS has been used and presents examples of a general nature. Reference has already been made to many specific studies in Chapter 3 to illustrate the types of reactions that can occur in an MS/MS instrument.

MS/MS has played a particularly important role in assigning *structures* to ions in the gas phase. The majority of ion structural studies using MS/MS have utilized keV collision energy CID. This application is emphasized here along with MS/MS in conjunction with newer activation methods such as photoexcitation. Another important application of MS/MS has been in the elucidation of reaction mechanisms. The MS/MS method has been particularly useful in this regard both for unimolecular and bimolecular reactions, in many cases using comparisons of the dissociations of isotopically labeled ions. A final major area discussed here regards the use of MS/MS in obtaining thermochemical information from gas-phase species. These data have been acquired using several approaches, including measurements of energy loss, kinetic energy release, and relative fragment ion abundances.

Most chemical research experiments using MS/MS fall into one of these three classes, viz, ion structures, reaction mechanisms, and thermochemistry. This chapter is divided along these lines, giving examples of typical studies using the various reactions discussed in Chapter 3 as illustrations.

4.2. Ion Structures

The structures of the gas-phase ions formed in the mass spectrometer are of fundamental interest to the mass spectrometrist. A number of excellent reviews have been written recently on this topic.[135,171,348,349] Several approaches to obtaining ion structural information, some utilizing MS/MS methods, have been developed. These approaches fall into the following categories:

1. theoretical calculations,
2. ion thermochemistry,
3. unimolecular chemistry, and
4. bimolecular chemistry.

The ion structural information obtainable from the experimental methods is generally limited to constitution, i.e., how the atoms in the ion are bonded to one another and the locations of the likely charge site and radical site (if the ion has an odd number of electrons). With the exception of theoretical calculations, current ion structure characterization methods are insensitive to bond angles, bond lengths, and charge distribution, although some microwave spectroscopy data are available.

Details of bonding in ions has long been the province of theoretical calculations. Both molecular orbital and ab initio approaches have been applied to the study of gas-phase ions. A practical guide to these types of calculations along with numerous citations has recently been published.[350] Ion thermochemistry refers to the measurement of the heats of formation of ions via ionization energy and appearance energy measurements. The role of ion thermochemistry in ion structure determination has been reviewed by Holmes.[332] Accurate measurements (0.05 eV) are usually only obtainable with photoionization mass spectrometers or instruments with electron ionization sources with narrowly definable electron energy.[351,352] MS/MS instruments that are sensitive to kinetic energy release, however, are useful in evaluation of the accuracy of an appearance energy measurement and can be used to measure the appearance energy of a metastable ion. These points are discussed in Section 4.4.2, which describes kinetic energy release measurements in thermochemical determinations. Heats of formation are almost always used in conjunction with information regarding the unimolecular decomposition chemistry of the ions. The latter information is most conveniently obtained using MS/MS with an activation method or from a study of metastable ions. The role of metastable ions in ion structure determination is discussed in Section 4.2.1.1, and the role of collisional activation and other activation methods is described in Section 4.2.1.2. The bimolecular or ion/molecule reaction chemistry of isomeric ions has also been used to differentiate ion structures, primarily using ion cyclotron resonance mass spectrometers. The role of MS/MS in the study of ion/molecule reaction chemistry for the purpose of differentiating ion structures is discussed briefly in Section 4.2.2. A more complete discussion of bimolecular chemistry and MS/MS is given in Section 4.3. Finally, a brief discussion of neutralization–reionization mass spectrometry and radical trapping is given in Section 4.2.3. These techniques provide structural information on neutral species that are otherwise difficult to study.

4.2.1. Unimolecular Chemistry

For many years the most common basis for distinguishing ion structures was the interpretation of the electron impact ionization mass spectra of isomeric neutral species. This approach has significant limitations, however, many of which are overcome by the use of MS/MS. For example, in electron ionization mass spectra, the parent ion/daughter ion relationship is unclear, the technique is limited to odd-electron parent ions, and the mass spectrum does not reflect T, the kinetic energy released in a dissociation. MS/MS allows the parent ion/daughter ion relationships to be established, it can be used to study the dissociation behavior of any ion that can be formed in the gas phase (provided it fragments or can be made to do so), and, finally, with the proper instrumentation, MS/MS can provide information on T.

Much of the background for the study of the unimolecular chemistry of gas-phase ions for ion structure characterization is given in Section 3.2, which is a brief discussion of unimolecular dissociation. Section 3.2.1 discusses metastable ions in particular. The following sections discuss the roles of metastable and activated ions in ion structural studies. As a rule, these studies should be used in conjunction with other information, e.g., heats of formation and theoretical calculations. The general procedure is to form parent ions using a variety of means that presumably give different structures or mixtures of structures and then to compare the respective metastable ion and CID daughter ion spectra. Relative fragment ion abundances as well as peak shapes (T values) and the nature of the products of charge permutation reactions are then compared to assess the differences between the ions.

Figure 4-1a shows a hypothetical two-dimensional energy diagram of an ion formed in the ion source of an MS/MS instrument. This particular example shows that a rearrangement of the ion can occur at internal energies less than the lowest critical energy for decomposition. The region directly above the critical energy for dissociation shows the range of internal energies that give rate constants leading to dissociation in an arbitrary reaction region of the mass spectrometer. Ions that are formed in the ion source within this internal energy region are metastable. In the internal energy region lying between the barrier for isomerization and the barrier for dissociation, the ions can freely interconvert between the two isomeric forms. An activation reaction that results in dissociation predominantly samples initially stable ions (see Section 3.3). If most of the initially formed ions have internal energies insufficient for isomerization the activation reaction will tend to sample the population of ion structures present at ionization. This is an important facet of the study of unimolecular chemistry for ion structure characterization. Metastable ion dissociations reflect the structure of parent

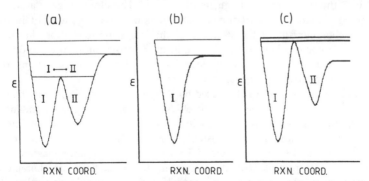

Figure 4-1 ■ Two-dimensional energy diagrams intended to depict commonly encountered situations in ion structural studies. (a) The case where the ion can rearrange at internal energies less than lowest critical energy for fragmentation. (b) The case where any barrier to rearrange is significantly higher than the lowest-energy fragmentation. (c) The case where a rate-determining isomerization can occur prior to rapid dissociation.

ions with internal energies near to the lowest energy dissociation and therefore not necessarily the stable ion structure. Nevertheless, the study of metastable ions can be very useful in an overall investigation of a set of isomeric ions.

Figures 4-1*b* and 4-1*c* are energy diagrams intended to describe two other commonly encountered situations. Figure 4-1*b* shows the situation for which any barrier for isomerization is much higher than the critical energy for the lowest energy decomposition. In this case, the fragmenting ion has the structure of the ion formed at the threshold for ionization. Even at internal energies sufficient for isomerization, the rates for dissociation are generally much faster. Figure 4-1*c* shows the situation where a rate-determining isomerization can occur prior to rapid fragmentation. Here, the dissociations may not be characteristic of the initial structure of the fragmenting ion, but isotopic labeling can be used to gauge the extent of the rearrangements that occur, as discussed later.

4.2.1.1. Metastable Ions.

Literature reviews of metastable ion studies give excellent discussions on the role of metastable ion dissociations in ion structure differentiation.[12,148,149] Structure differentiation can be based on the comparison of relative fragment ion abundances if two daughter ions of different mass are formed, and can also be based on comparison of peak shapes. The measurement of T from metastable ions is also useful in obtaining the heat of formation of an ion when its appearance energy has been measured. Each of these applications will be discussed in turn.

As discussed previously and in Section 3.2.1, the fragmentation of a metastable ion is not necessarily characteristic of the ion structure formed at the ionization threshold. The range of internal energies that give rise to metastable ions is narrow, but all values within the range are higher than the energy of the ion at ionization threshold. Therefore, if the ion can rearrange at energies less than, or near to, the lowest critical energy for dissociation, metastable ion spectra may not reflect the initial stable ion structure. A classic example where this is the case is the $C_4H_8^{+\cdot}$ isomers. Each of these isomeric ions has the structure of its neutral counterpart at ionization threshold, but isomerization can readily occur with each ion at energies below the critical energy for loss of $CH_3^{\cdot}$, the lowest energy fragmentation.[353,354] Therefore, in general, neither the abundances of fragment ions from metastable ions nor the T values obtained can be regarded as characteristic of the structure of the ion at its ionization threshold. The question then becomes whether or not these measurements can be used to differentiate between the structures that directly lead to the observed dissociations (i.e., the structures of the relatively energetic metastable ions).

The abundance ratio test was first used by Rosenstock et al.[355] and by Shannon and McLafferty.[356] It was assumed that closely similar daughter ion abundance ratios indicate that the metastable parent ions have the same structure. It was later shown that an abundance ratio from metastable ions can be very sensitive to internal energy,[157,158] varying by up to a factor of 5 for the same structure metastable ion. This situation is likely to prevail when the competing fragmentations are characterized by $k(\epsilon)$ versus ϵ curves of widely different slopes. If two ratios are closely similar, it can be concluded that the metastable parent ions very likely fragment from the same structure. Conversely, if the ratio is different, but the metastable parent ions show the same daughter ions, it *cannot* be concluded that the parent ion structures are different. In either case, further evidence from ion thermochemistry and CID is necessary to determine if the structure of the metastable parent ion is also the structure of the stable parent ion.

Isotopic labeling is also often very helpful in determining ion structures (see also Section 4.3). For example, Levsen and McLafferty have studied the metastable ion spectra of five $C_3H_8N^+$ isomers.[357] Four distinct spectra were observed. It was proposed that metastable ions initially formed as $CH_3CH=N^+HCH_3$ and $(CH_3)_2N^+=CH_2$ rearrange to a common structure or mixture of structures prior to fragmentation. The three other isomers were concluded to fragment from distinct structures. When the ions initially formed as $(CH_3)_2N^+=CH_2$ were specifically labeled with deuterium or ^{13}C, the metastable ion spectrum showed complete randomization of all deuterium and carbon atoms prior to loss of ethylene. Labeling of $CH_2=NHC_2H_5^+$, one of the isomers concluded to fragment from a distinct structure showed no atom randomization in the metastable ion spectrum.

Another test of ion structure is the comparison of kinetic energy released in the fragmentations in metastable ion spectra. If two possibly isomeric metastable ions show the same daughter ions and the peak shapes associated with the daughters are the same, these ions probably fragment from the same structure or mixture of structures. If there is a substantial difference in the measured T values (a factor of 2) it can be safely concluded that the metastable parent ions fragment via different structures. As discussed in Section 3.2.1, the kinetic energy released in a dissociation arises from the internal energy in excess of the critical energy for the fragmentation and from the so-called reverse critical energy. If two ions have the same structure, they have identical potential surfaces so that the reverse critical energies are the same, and the fractions of the reverse critical energy partitioned into translation are the same. Likewise, the kinetic shifts, i.e., the energy in excess of the critical energy required to drive the dissociation at a rate sufficient to be observed on the metastable time scale, are identical. The remainders of the respective excess internal energies are not necessarily equal since the ions are formed via different means. If parent ion internal energy has a major effect on the observed peak shape, it will therefore most likely be observed in cases where the reverse critical energy is small. Since the excess energy is generally small for metastable ions, the observed kinetic energy release is expected to be relatively insensitive to the means of forming the ion. The existing experimental evidence suggests that this is the case.[358-361] Metastable ion abundance ratios are more sensitive to internal energy because they are determined by the relative rates of competitive fragmentations that may be highly sensitive to internal energy.

If the isomeric ions do not interconvert at internal energies below the lowest critical energy for dissociation, the energy diagram of Figure 4-1b applies, and the peak shapes should be sensitive to ion structure. The metastable ions will also have the structures of the ion at the respective ionization thresholds. If the ions can freely interconvert at energies less than the critical energy for decomposition, an energy diagram similar to that of Figure 4-1a applies, and one or more structures may contribute to the observed peaks. Nevertheless, if the same structure or mixture of structures are formed, the shapes of the metastable peaks should be very similar. If, however, a rate-determining isomerization can occur prior to dissociation, as depicted in Figure 4-1c, further information will be required to distinguish this situation from two distinct ion structures. An ion initially formed as structure I, but with sufficient energy to rearrange to structure II, may have a large excess energy that may lead to a larger kinetic energy release when it rearranges and fragments than the kinetic energy release exhibited by a metastable ion initially formed as structure II. The abundance ratios may also be very different. This situation can be distinguished from two different ion structures by measurement of the appearance energies of the fragment ions. If the fragment ions all have the same appearance energy and that energy is higher than those calculated from established thermochemical values, a rate-determining isomerization is indicated.

4.2.1.2. Activated Ions. Two major drawbacks to the use of metastable ions in the characterization of ion structures are the sensitivity of metastable ion spectra to parent ion internal energy and the fact that metastable ions may not necessarily have the structure of the ion formed at ionization threshold. In order to sample stable ions, an activation method, usually CA, is used in conjunction with MS/MS. The idea is to excite the stable, non-isomerizing ions indicated in the energy diagrams of Figure 4-1. If they are sufficiently excited, they will tend to fragment quickly via structurally diagnostic cleavage reactions before extensive rearrangement occurs.

Collisional Activation. McLafferty and co-workers pioneered the use of high-collision-energy CID as a structural tool for polyatomic ions.[162,163,200] In general, these studies, like many studies of metastable ions, involve comparing the CID spectra of possibly isomeric ion structures. An important conclusion drawn from the early work is that CID spectra are relatively insensitive to the internal energy of the parent ion prior to collision,[162,206] and less sensitive than metastable ion abundance ratios. Differences observed in the CID daughter ion spectra of isomeric ions can be attributed to structural differences and not to different methods of ion preparation that may lead to different distributions of parent ion internal energy. This conclusion is valid when certain caveats, articulated by McLafferty et al., are heeded.[162] Principal among these is that peaks that can result from low-energy dissociation pathways must be excluded from the analysis in comparing CID spectra. When the peaks that also appear in the corresponding metastable ion spectra are excluded, parent ion internal energy changes show no appreciable effect on the relative intensities of the CID product ions. Ions that appear in the metastable ion spectra are excluded since the processes leading to these products are most sensitive to differences in internal energy.[162,359-361] Furthermore, parent ions that give daughter ions that also appear in the metastable ion spectrum are likely to have relatively low internal energies, so that the internal energy of the ion prior to collision can make up a large fraction of the postcollision internal energy. By focusing attention on daughter ions that arise from more highly endothermic processes, the possible effect of the initial internal energy is reduced.

CID has been used extensively in ion structural studies since the early 1970s.[135,171] Many cases have been noted in which the CID spectra of isomeric ions differed when the corresponding metastable ion spectra did not. The $C_3H_8N^+$ ions mentioned in the preceding section are an example. The metastable ion spectra indicated that three of the five possible isomers tested fragmented from distinct structures (or mixtures of structures). The ions presumed to be initially formed as $CH_3CH=NHCH_3^+$ and $(CH_3)_2N=CH_2^+$, on the other hand, were shown to interconvert when the parent ions were metastable. This was concluded both from isotopic labeling studies and metastable ion spectra. CID indicated, however, that the two structures were distinct stable structures.[357] CID spectra of these ions, omitting the daughter ions also present in the metastable ion spectra, generated from several different compounds are summarized in Table 4-1. This example demonstrates a case in which an energy diagram similar to that of Figure 4-1a applies. A barrier to isomerization exists below the lowest critical energy

Table 4-1 ■ Daughter Ion MS/MS Spectra of $C_3H_8N^+$ Isomers Formed
from Various Precursors

Compound	Structure	m/z	Relative abundances[a]								
			15	27	28	29	41	42	43	54	55
N-methylethyl amine	$CH_3CH{=}NHCH_3^+$		5	10	24	15	12	16	25	3	2
2,5-dimethyl piperazine	$CH_3CH{=}NHCH_3^+$		6	11	28	13	14	19	17	4	3
N-methyliso-propylamine	$CH_3CH{=}NHCH_3^+$		6	10	26	13	13	18	23	3	2
trimethylamine	$(CH_3)_2N{=}CH_2^+$		7	3	6	8	<6	52	21	2	1
tetramethyl-diaminomethane	$(CH_3)_2N{=}CH_2^+$		9	4	6	7	<6	62	19	2	1
dimethylamino-acetone	$(CH_3)_2N{=}CH_2^+$		8	4	6	7	<6	49	22	2	1

[a] Relative abundances calculated with respect to the summed abundances of all daughter ions except that at m/z 41.

for dissociation that metastable ions can surmount. The structures can therefore interconvert prior to dissociation, leading to dissociation from the same structure or mixture of structures. CID also samples ions formed with internal energies that may be insufficient for isomerization. If this is the case, the CID spectra can indicate that distinct stable structures exist at ionization threshold. The height of the isomerization barrier and the distribution of internal energies of the parent ions before CA, of course, have a major effect on the fractions of interconverting and noninterconverting ions in the sampled beam. Only minor differences in the CID spectra may be observed if the isomerization barrier is low or when the isomeric parent ions give the same daughter ions.

A special case of the energy diagram of Figure 4-1a is one in which ions can isomerize before dissociating, but do so only via one ion structure. This situation can exist when the lowest critical energy for dissociation for one of the isomers is much higher than that for the other. Metastable ions can interconvert many times but only dissociate from the ion structure that fragments most readily. Isotopic labeling can sometimes be used to differentiate this situation from the simpler case of Figure 4-1b. CID can also identify this situation if each of the isomeric forms of the ion can be formed with internal energies below the threshold for ionization.

When the isomerization barriers between isomeric ions are high, the energy diagram of Figure 4-1b applies, and the interpretation of both metastable ion and CID daughter ion spectra are generally straightforward. Ion thermochemistry and labeling experiments should give supporting evidence that isomerization prior to fragmentation does not occur. Metastable ion data alone, however, may be misleading if an energy diagram similar to that of Figure 4-1c should apply. This is the case such that a rate-determining isomerization of structure I precedes a fast dissociation, as mentioned in Section 4.2.1.1. The classic example of this situation was first described by Hvistendahl and Williams in a study of the isomeric oxonium ions $CH_3CH_2O{=}CH_2^+$ and $CH_3CH{=}OCH_3^+$.[362] The metastable ion spectrum of $CH_3CH_2O{=}CH_2^+$ shows the loss of H_2O (70% of daughter ion abundance) and loss of C_2H_4 (30%), whereas the metastable ion spectrum of $CH_3CH{=}OCH_3^+$ shows losses of H_2O (1.4%), C_2H_4 (77%), and CH_2O (22%). Furthermore, the average T values for loss of H_2O and C_2H_4 from metastable $CH_3CH{=}OCH_3^+$ are higher than for the corresponding losses from $CH_3CH_2O{=}CH_2^+$.[362] Both metastable ion tests would seem to indicate that the metastable ions fragment from distinct structures. It was proposed, however, that the ion $CH_3CH{=}OCH_3^+$ undergoes a slow isomerization to $CH_3CH_2O{=}CH_2^+$ followed by rapid dissociation. The $CH_3CH{=}OCH_3^+$ ions with sufficient energy to isomerize have energy well in excess of that required to dissociate from $CH_3CH_2O{=}CH_2^+$ (and well in excess of metastable $CH_3CH_2O{=}CH_2^+$), accounting for the differences in the daughter ion abundance ratios and average T values. Strong evidence for this interpretation came from the measurement of the appearance energy values for the losses of H_2O, C_2H_4, and CH_2O from $CH_3CH{=}OCH_3^+$. These values were all the same and were all higher than the corresponding values for $CH_3CH_2O{=}CH_2^+$. CID spectra indicate that the two isomeric forms are stable species.[135,171] The major differences observed in the metastable ion data can be attributed to large differences in the internal energies of the ions that fragment in the reaction region; initially formed metastable $CH_3CH_2O{=}CH_2^+$ having relatively low internal energy and $CH_3CH_2O{=}CH_2^+$ ions formed by the rearrangement of $CH_3CH{=}OCH_3^+$ having relatively high internal energy. CID predominantly samples stable ions so that differences in the spectra can be more confidently attributed to structural differences than to differences in internal energy.

The preceding discussion and the bulk of the CID ion structural studies have assumed that differences in the parent ion internal energy distributions have only a minor effect, if any, on CID spectra. This conclusion, however, has recently been called into question.[340,363,364] Jarrold et al. have reported CID data acquired from ions formed via chemical ionization as a function of ion source pressure.[340] Ion source pressure is expected to be inversely correlated with the internal energy of the parent ion and also, possibly, its angular momentum. (Theoretical and experimental evidence have been presented that show that the angular momentum of the parent ion can have a significant effect on its dissociation pattern.[131-133]) The higher collision frequency resulting from the pressure increase tends to relax the nascent ions formed from an exothermic proton transfer reaction. As expected, the data showed that daughter ions resulting from low-critical-energy fragmentations showed the greatest variation in relative abundance with ion source pressure. In several cases, however, some reactions that did not show products in the metastable ion spectra also showed variation of more than 10% in the CID spectra with ion source pressure changes. These variations could not be explained by rearrangement of the ions prior to collision since parent ions with high barriers to rearrangement were chosen (energy diagrams similar to that shown in Figure 4-1b). The results were attributed either to differences in parent ion internal energy or angular momentum. Parent ions in the early CA studies were generally formed by electron ionization and therefore the effect of parent ion internal energy was tested by varying the electron energy. Parent ions formed by electron impact ionization are not expected to differ widely in angular momentum. The authors suggest, therefore, that the observed sensitivity of the CID spectra to ion source pressure might be largely due to differences in angular momentum. Even more striking sensitivity to ion source pressure was observed for the relative abundances of the peaks due to charge stripping for some of these ions (see the following discussion). Porter et al. have also reported a significant internal energy effect on the CID spectrum of the benzoyl cation.[363] In a later study of the same ion, however, McLafferty et al.[360] observed no such effect.

The consensus on internal energy effects on CID spectra is that the relative abundances of daughter ions produced from high-critical-energy reactions are, as a rule, insensitive to parent ion internal energy. In any case, an observed variation in daughter ion abundances with source conditions should not be attributed to an internal energy effect until it can be confirmed that different mixtures of ion structures can result from changes in the ion internal distribution. Such a situation can arise when an energy diagram similar to that of Figure 4-1a applies. To sample stable, noninterconverting parent ions and to minimize the possibility of a parent ion internal energy effect, ions of low internal energy should be formed. This can most simply be done by using low ionizing electron energies in electron ionization. Another approach is to form ions at relatively high pressures in a chemical ionization source using either a proton transfer reaction or, in the case of odd-electron ions, a charge transfer reaction. At high pressures, a thermal distribution of internal energies and angular momenta is approached.[340]

Charge Stripping. Several of the charge permutation reactions described in Section 3.5 have been used to characterize ion structures. By far the most commonly studied reactions are the charge-stripping reaction for cations:

$$m_p^{+\cdot} + N \rightarrow m_p^{2+} + N + e^- \qquad (4\text{-}1)$$

and the charge inversion reaction for anions (see Section 3.5.2.1):

$$m_p^{-\cdot} + N \rightarrow m_p^{+\cdot} + N + 2e^- \qquad (4\text{-}2)$$

The first report in which the products of charge stripping were used in an ion structure characterization study appeared in 1975 in which $C_6H_6^{+\cdot}$. isomers were distinguished.[365] Since that time, numerous other examples have been published, many of which have shown cases where the peaks due to charge stripping could be used to distinguish between isomers when the CID spectra could not. This situation can arise with systems with energy diagrams similar to Figure 4-1a. If the barrier to isomerization is relatively low, many of the ions formed in the source can freely interconvert. These ions are nearer in energy to the critical energies for fragmentation than are the noninterconverting ions. Since the most probable energy transfer in keV collision energy collisional activation is a few electron volts (see Section 3.3.1.3) the interconverting ions are preferentially sampled. Charge-stripping products arise from much larger energy transfers (in excess of 10 eV, see Section 3.5.2.2) so that the stable ions are more uniformly sampled. The charge-stripping spectra, therefore, better reflect the structures of the noninterconverting ions than do the CID spectra. Another argument, independent of which ions are sampled, is that if the geometries of the singly and doubly charged ions are significantly different, the Franck–Condon factors for the transition from $m_p^{+\cdot}$ to m_p^{2+} will lead to highly excited doubly charged ions. The fragmentations from these ions are then reasoned to be very fast and therefore highly structural specific.[319] Charge-stripping products usually show the doubly charged ion and doubly and singly charged fragment ions. Most of the fragmentation products are often obscured by CID products, so that ordinarily only the relative abundances of the doubly charged parent ion and doubly charged daughter ions formed by losses of hydrogens are used in comparisons. For example, the $C_2H_5O^+$ isomers, CH_3CHOH^+, $CH_3OCH_2^+$,

Table 4-2 ■ Charge-Stripping MS/MS Spectra for $C_2H_5O^+$ Isomers.

Structure	m/z	Relative abundance				
		22.5	22	21.5	21	20.5
CH_3CHOH^+		0.5	4	83	11	1
$CH_3OCH_2^+$		—	98	—	2	—
$CH_2CH_2OH^+$		10	3	76	10	1
$CH_2=CHOH_2^+$		80	6	11	3	0.5

$CH_2CH_2OH^+$, and $CH_2=CHOH_2^+$ give the charge-stripping spectra summarized in Table 4-2.[366] The spectra are obviously different, but only two distinct CID spectra were obtained for these ions: one from the $CH_3OCH_2^+$ ion and the other from any of the other three isomers.

As with all methods for the characterization of ion structures, the question of the effect of parent ion internal energy on charge stripping has been raised. Several studies have been reported in which no significant parent ion internal energy effect on charge stripping was observed when the ions studied ($C_3H_6^{+\cdot}$ and $C_5H_8^{+\cdot}$) could not rearrange at energies less than the lowest energy fragmentation.[334,338,339] Another study has demonstrated a significant effect of the chemical ionization source pressure on the charge-stripping peaks arising from nonisomerizing $C_2H_8N^+$ and $C_2H_7O^+$ ions.[340] It has been argued that since the charge-stripping reaction occurs via a vertical transition, the Franck–Condon factors can be highly sensitive to the vibrational energy of $m_p^{+\cdot}$, thereby resulting in significantly different energy m_p^{2+} ions formed from parent ions of different energy. The latter study correctly points out that caution should be employed when using charge-stripping peaks alone to distinguish between isomeric ions. Nevertheless, the number of cases in which the indications of the charge-stripping peaks are consistent with those of other methods for characterizing ion structures supports the general conclusion that charge stripping is relatively insensitive to ion internal energy effects.[329]

Photodissociation. As pointed out in Section 3.3.2, photodissociation has advantages over collisional activation as an ion structural tool in those cases where the cross section for photodissociation is not too low. These advantages stem primarily from the very closely defined energies deposited into the ion and the ability to vary that energy. Several approaches to comparing photodissociation behavior can be taken. These include:

1. photodissociation spectra;
2. relative abundances of daughter ions;
3. kinetic energies released during dissociation; and
4. multiphoton dissociation.

The comparison of daughter ion relative abundances following photoexcitation is similar to a comparison of CID spectra. The other approaches also have collisional activation analogs. The photodissociation spectrum gives a measure of the relative cross section for photodissociation as a function of wavelength. van Tilborg and van Thuijl have shown that relative CID cross sections can provide additional information in comparing ion structures.[167–169] However, although CID cross sections as a function of energy deposition into the parent ion may be obtainable over a limited range at low collision energies, the values obtained are coarse. The kinetic energies released in the CID process can, in principle, be used as an ion structural tool just as they are with metastable ions. However, the experiment is better done with photoexcitation since the broad range of internal energies deposited during collisional activation, coupled with a range in translational energy losses experienced upon collision, compromise the measurement of T. The collisional activation analog to multiphoton dissociation is CID under multiple-collision conditions. Again, however, the CID experiment allows rather poor control over the number of collisions and the energy transferred in each collision.

The photodissociation spectrum is the most commonly obtained data in ion structure studies using lasers as the source of excitation energy. The general procedure for obtaining these spectra is described in Section 3.3.2. A good example of this approach is an ion cyclotron resonance study of $C_6H_6O^+ \cdot$. isomers formed by electron impact ionization of phenol, phenetole, bicyclo[2.2.2]oct-2-ene-5,7-dione, and 2-phenoxyethyl chloride.[277] Figure 4-2 shows the photodissociation spectra of the m/z 94 ions derived from these molecules. The spectra indicate that two ion structures are primarily responsible for the spectra, one formed from phenol and phenetole and the other formed from phenoxyethyl chloride and bicyclo[2.2.2]oct-2-ene-5,7-dione. These structures are taken to be ionized phenol and ionized cyclohexadienone, respectively. More information is available from the kinetics of photodissociation. If only one ion structure is formed, the time dependence of the photodissociation signal at a single wavelength obeys the following relationship:

$$N(t) = N_0(t) e^{-Ikt} \qquad (4\text{-}3)$$

where $N(t)$ is the number of parent ions at time t with the light on, $N_0(t)$ is the number of parent ions at time t with the light off, I is the intensity of the light (in photons), and k is the rate constant for photodissociation (per photon). This relationship is not obeyed when a fraction of the ions in the parent ion population does not

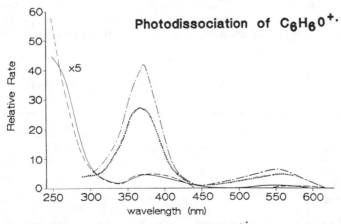

Figure 4-2 ■ Photodissociation spectra of several $C_6H_6O^+ \cdot$ isomers derived from phenol (—), phenetole (---), bicyclo(2.2.2)oct-2-ene-5,7-dione ($\cdots$), and 2-phenoxyethyl chloride (—·—). (Adapted from reference 277.)

photodissociate at the wavelength used. A likely cause for this is that some of the ions are of a different structure. In the study referred to previously, it was found that the $C_6H_6O^+$ · ions formed from phenoxyethyl chloride and bicyclo[2.2.2]oct-2-ene-5,7-dione did not strictly obey the kinetic relationship, indicating that some of the initially formed cyclohexadienone may rearrange to ionized phenol.

The relative abundances of daughter ions produced by photodissociation can provide further information in ion structure studies. An example for which both photodissociation spectra and photofragment ion spectra were obtained, using a BE spectrometer, was reported by Wagner-Redeker and Levsen.[262] $C_5H_8^+$ · ions produced by electron impact ionization of 1,2-pentadiene, 1,3-pentadiene, 2-pentyne, cyclopentene, and dissociative ionization of 3-octyne were studied as a test of the utility of photodissociation as an ion structure probe. $C_5H_8^+$ · isomers were chosen because, with the exception of ionized 1-pentyne and 2-pentyne, the CID spectra are very similar (distinguished only by the charge-stripping peaks).[367–369] Furthermore, the different isomeric ions were expected to have distinct chromophores; viz, cumulative double bonds and conjugated double bonds. All isomers could be distinguished based on the photodissociation spectra except for the ion formed from 3-octyne. A mixture of structures was deemed to be produced from this precursor molecule. The spectra of daughter ions produced by photodissociation were obtained at two wavelengths, 515 and 351 nm. The spectra are summarized in Table 4-3 (515 nm) and Table 4-4 (351 nm). The spectra acquired at 515 nm (2.4 eV) are indistinguishable, whereas the spectra acquired at 351 nm (3.5 eV) show significant differences. It was concluded that an energy diagram similar to that of Figure 4-1a applies in this case. Isomerization of the ions apparently can occur at energies lower than the thresholds for dissociation. The lower-energy photodissociation and, presumably, CID tend to sample preferentially ions with energies sufficient to isomerize, whereas the higher-energy photodissociation samples noninterconverting ions.

As discussed in Section 3.2.1.1, the kinetic energy released in a dissociation is sensitive to ion structure. This topic was discussed with regard to metastable ions for which T values are determined for relatively long-lived ions with just sufficient energy to dissociate. The measurement of T accompanying photodissociation, however, better reflects the structure of ions initially formed with low internal energies. An important advantage for the photodissociation technique is the ability to obtain T values for

Table 4-3 ■ Photoionization MS/MS Spectra for $C_5H_8^+$ · Irradiated
with 515-nm Photons

Precursor molecule	m/z	Relative abundances[a]				
		40	42	53	66	67
1,3-pentadiene		9	7	84	(21)	(2600)
1,2-pentadiene		11	12	77	(10)	(260)
Cyclopentene		9	10	81	(26)	(1950)
2-pentyne		10	11	79	(12)	(190)
3-octyne		9	10	81	(12)	(350)

[a]Abundances relative to the total daughter ion current excluding m/z 66 and m/z 67 (shown in parentheses), which result from low-energy processes.

Table 4-4 ▪ Photoionization MS/MS Spectrum for $C_5H_8^{+\cdot}$ Isomers Irradiated
with 351-nm Photons

Precursor molecule	m/z	Relative abundance[a]							
		39	40	41	42	53	65	66	67
1,3-pentadiene		1	15	2	16	61	5	(12)	(180)
1,2-pentadiene		2	7	4	7	55	26	(10)	(52)
Cyclopentene		—	9	5	10	57	19	(12)	(250)
2-pentyne		—	6	2	5	75	12	(7)	(66)
3-octyne		1	7	6	7	60	20	(10)	(80)

[a]Abundances relative to total daughter ion current excluding m/z 66 and m/z 67 (shown in parentheses), which result from low-energy processes.

several very well-defined energy inputs and, presumably, for dissociating ions of well-defined internal energies. This capability has to date been little used for ion structure studies, primarily due to the very few BE spectrometers fitted for photodissociation in the second reaction region. Muhktar et al.,[118] however, have demonstrated the usefulness of this approach with isomeric xylenes formed by 70-eV electron impact ionization. The mass spectra of these ions are identical, as are the charge-stripping spectra and the metastable ion spectra. Figure 4-3 shows the T values measured at 10% peak height for the loss of $CH_3^{\cdot}$ from the three isomers as a function of photon energy. Differences are observable in the dependence of T on photon energy, particularly at 3.47 eV. Interestingly, T values are seen to decrease with increasing photon energy over the range of 2.41 to 2.71 eV. This contrasts with what might be expected from the quasiequilibrium theory (see Section 3.2) and with what is usually observed. This may indicate nonstatistical behavior, but may also simply reflect a poor understanding of energy partitioning as a function of internal energy. The case of the isomeric xylenes appears to be another example for which an energy diagram similar to Figure 4-1a applies. The ions can apparently isomerize at energies less than that required for the lowest energy dissociation. Noninterconverting ions, however, can be sampled by photodissociation.

A final ion structural tool that uses light as the activating agent is multiphoton photodissociation using the same or different colored photons. Multiphoton irradiation studies of ions are typically performed in ion cyclotron instruments and have not generally involved MS/MS techniques.[370] With the advent of FT–ICR, however, these types of studies can now be performed in conjunction with MS/MS.[371] To date, most of these studies have not been directed toward distinguishing ion structures. Rather, they have been directed toward, for example, understanding the dynamics of multiphoton absorption of ions, relaxation studies, and other such fundamental studies. Nevertheless, the degree of control afforded by multiphoton photoexcitation should find more widespread application to ion structure studies. The number of variables in a multiphoton experiment, such as wavelengths of the photons, pulsed versus continuous radiation, time of irradiation, and background pressure, makes a description of all the possible studies beyond the scope of this discussion. A few selected experiments, therefore, are mentioned. Bomse et al. have used slow infrared (IR) multiphoton excitation of organic ions to study the energetics of their unimolecular and bimolecular

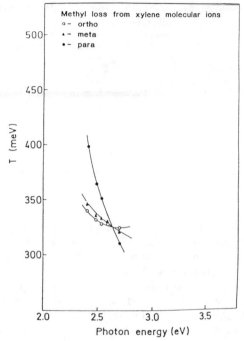

Figure 4-3 ■ Kinetic energy released, T, measured at 10% peak height, for the loss of $CH_3\cdot$ from ionized xylene isomers as a function of photon energy. (Adapted from reference 118.)

chemistry.[370] They have shown that the low-power IR irradiation of parent ions produces photodissociation and that the relative abundances of the daughter ions formed can be used to distinguish isomeric ions.[259] For example, although the low-power IR photodissociation of the symmetrical proton-bound dimers of n-propanol and isopropanol are identical, the relative abundances of the daughter ions are different.[370] However, only the products from the lowest critical energy dissociation channel are observed with this approach. Since the ions are energized slowly, this approach tends to sample those same ions sampled in a study of metastable ions. Dunbar and co-workers have observed an enhancement of photodissociation yields when single IR photon absorption is followed by irradiation with ultraviolet light.[257] These workers have also shown the utility of using two photon experiments to study the rearrangement chemistry of ionized halotoluenes.[257]

4.2.2. Bimolecular Chemistry

Isomeric ion structures have also been distinguished by their ion/molecule reactions. Several approaches have been employed, and most of these have involved the use of ion cyclotron resonance mass spectrometers. The most common approach has been to compare the charged products from the ion/molecule reactions of isomeric ions with common neutral molecules. This approach will be discussed further. Other, lesser used, approaches are the comparison of proton affinities by the proton transfer reaction with

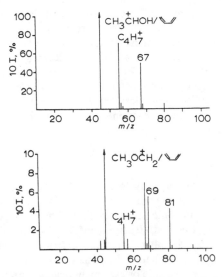

Figure 4-4 ■ MS/MS spectra of $C_2H_5O^+$ isomers in collisions with 1,3-butadiene at a collision energy of 1 eV. (Adapted from reference 377.)

a common base[372,373] and the comparison of equilibrium and/or rate constants for a particular ion/molecule reaction.[374]

The ion/molecule reactions of $C_2H_5O^+$ isomers have been studied using an ICR mass spectrometer,[375,376] a triple-quadrupole mass spectrometer,[377] and by trapping ions in the space charge of an electron beam.[378] Figure 4-4 shows MS/MS spectra obtained when CH_3CHOH^+ and $CH_3OCH_2^+$ were collided with 1,3-butadiene at a collision energy of 1 eV. Dramatic differences in the spectra are obvious. These isomers were also readily differentiated with benzene used as the target molecule. In the ICR studies, several isomers were distinguished using isopropanol as the reactant molecule, but protonated ethylene oxide could not be differentiated from protonated acetaldehyde.[375] It was later shown, however, that protonated ethylene oxide reacts with hydrogen sulfide and with phosphine, whereas protonated acetaldehyde does not.

It is presumed that if two ions have different reactivities with a common reagent molecule, their structures are different. It is less safe, however, to conclude that two ions have a common structure if their reactivities with a particular molecule are the same. The results for protonated ethylene oxide and protonated acetaldehyde described previously constitute a case for which two ions cannot be distinguished by their ion/molecule reactions with a particular reagent, but can be distinguished by their reactions with another. If the reaction proceeds through a collision complex, the charged products from the *unimolecular* dissociation of the collision complex are used to differentiate ion structures. It is reasoned that if the structures of the collision complexes are different, the structures of the ionic reactants must therefore differ. However, the energy surface of the collision complex may allow for rearrangement to a common structure or interconverting structures prior to dissociation. It is therefore important, although not easy, to select a neutral reactant that leads to distinct, noninterconverting collision complex structures.

A further complication may arise from the effect of parent ion internal energy on the extent or occurrence of an ion/molecule reaction. In the ICR instrument, the lifetime of the ion prior to reaction is usually on the order of milliseconds, so that stable ions are sampled. Nevertheless, an effect on the reactivity of the ion has been observed in several cases from changes in the ionizing electron energy.[375,379,380] In addition to internal energies of the reactants, the center-of-mass collision energy also adds to the energy of a collision complex. The collision energy can also, therefore, have a major effect on an ion/molecule reaction if it proceeds through a long-lived collision complex (see Section 3.4).

4.2.3. Neutral Structures

A major use of mass spectrometry is to provide structural information regarding a neutral sample, the identity of which is usually unknown. This information, of course, is obtained indirectly via an ionic surrogate of the neutral species and is generally restricted to constitution. Detailed structural information on a neutral molecule can usually be obtained directly via other forms of spectroscopy such as the light spectroscopies and nuclear magnetic resonance. In some cases, however, these techniques are not applicable. For example, unstable or extremely reactive species are difficult to study by the conventional spectroscopies. In some cases, unstable and/or very reactive species can be formed in a mass spectrometer and studied using methods that employ MS/MS. Two notable examples, radical trapping and neutralization–reionization mass spectrometry, are described.

Radical Trapping. The structures of radicals formed in a chemical ionization source have been studied by McEwen and Rudat using MS/MS.[381–385] The approach involves several steps: first, one of several compounds found to react rapidly with radicals is admitted into the ion source; second, the radical complex is ionized by one of several methods; and third, the ions are studied using CID and MS/MS. These workers have found three tetracyano compounds, tetracyanoethylene (TCNE), tetracyanopyrazine (TCP), and tetracyanoquinodimethane (TCNQ), to be useful as carbon-centered radical trapping reagents. Under conditions in which high radical concentrations are present, the positive ion and negative ion mass spectra are dominated by ions formed by the reaction of radicals with the trapping reagent. This is due to the fact that the rate for the radical/trapping reagent reaction is large enough for the reaction to occur within the residence time of the radical in the ion source. The sequence of reactions in the experiment for TCNQ $[(CN)_2[C_4H_4](CN)_2]$ and a tertiary radical $\cdot CR_1R_2R_3$ is:

$$TCNQ + \cdot CR_1R_2R_3 \rightarrow (TCNQCR_1R_2R_3)^{\cdot} \qquad (4\text{-}4)$$

$$(TCNQCR_1R_2R_3)^{\cdot} + e^- \rightarrow (CN)_2C-[C_4H_4]-C^{\cdot}(CN)(CR_1R_2R_3)^+ + \cdot CN + 2e^-$$
$$(4\text{-}5)$$

$$(CN)_2-[C_4H_4]-C^{\cdot}(CN)(CR_1R_2R_3)^+ \rightarrow \cdot R_1 + (CN)_2C-[C_4H_4]-C(CN){=}CR_2R_3^+$$
$$(4\text{-}6)$$

where reaction (4-6) represents the CID MS/MS experiment in which the major ionic products typically arise from losses of each of the R groups. For example, the MS/MS spectrum generated from the preceding sequence of reactions for the sec-butyl radical shows predominantly losses of $H^{\cdot}$, $CH_3^{\cdot}$, and $C_2H_5^{\cdot}$. This approach has been used to study the degree to which radicals isomerize before being trapped. The time between radical formation and trapping can be crudely controlled by the partial pressure of the radical trapping reagent. Deuterium labeling studies have been performed to provide further information regarding the isomerization of radical species.

Even-electron ion structure information can also be inferred from this approach. Radicals tend to isomerize more slowly than ions so that even-electron ions that are neutralized in the ion source and trapped by a trapping reagent may more accurately reflect the structure of the initially formed ion than the ions that reach the collision cell. This approach has been useful in studying even-electron hydrocarbon ions.

Neutralization–Reionization Mass Spectrometry. Neutralization–reionization mass spectrometry (NRMS) is discussed briefly in Section 3.5.2.3. The method relies on the production of fast neutral species via charge exchange, CID, and/or metastable ion dissociation. The fast neutral species are then reionized via collision with a target gas. Recently, McLafferty et al. and Holmes et al. have used this method to study ion structures and reactive and unstable neutrals. In several cases, ion structures have been clearly distinguishable via NRMS when they were not using CID. For example, the ions $CH_3CH_2CH_2NH_2^{+\bullet}$ and $^{\bullet}CH_2CH_2CH_2NH_3^+$ could not be distinguished via CID, presumably due to the fast isomerization of the latter structure to the former structure after collisional activation and prior to dissociation. The NRMS experiment clearly distinguished these species since the hypervalent species $^{\bullet}CH_2CH_2CH_2NH_3^{\bullet}$ formed by charge transfer dissociates without appreciable rearrangement. It was also noted that the cross section for charge transfer with Hg vapor was greater for $CH_3CH_2CH_2NH_2^{+\bullet}$ than for the distonic form.[347]

Reactive and unstable neutrals may be formed in the neutralization reaction or via fragmentation. These neutral species can be studied following reionization. For example, both Holmes et al.[386] and McLafferty et al.[13] have shown that HNC (not HCN) is formed, along with $C_5H_6^{+\bullet}$, from the dissociation of metastable aniline cation. Holmes et al. have shown that ionized methyl acetate fragments to give CH_3CO^+ with $^{\bullet}CH_2OH$ as the neutral species formed rather than $CH_3O^{\bullet}$.[387,388]

4.3. Reaction Mechanisms

A mass spectrum is a convolution of variety of simple and complex chemical reaction mechanisms controlled by both kinetic and thermodynamic factors. While this makes interpretation of mass spectra extremely difficult for the analytical chemists, those interested in exploring chemical reactions on the time scale of the mass spectrometer are presented with vast opportunities for research. In fact, the mass spectrometer has been termed a "complete chemical laboratory" by Beynon.[389]

In the study of gas-phase reactions with a mass spectrometer, the reacting species are isolated and thus there are no effects from solvents. The chemistry that occurs is the "intrinsic" chemistry of the ion, and the study of such reactions offers in some cases indirect evidence of the effects (or lack thereof) that solvents have on chemical reactions and reactivity. The following discussion of reaction mechanisms is divided into unimolecular and bimolecular studies and also between those occurring in the ion source and the reaction region.

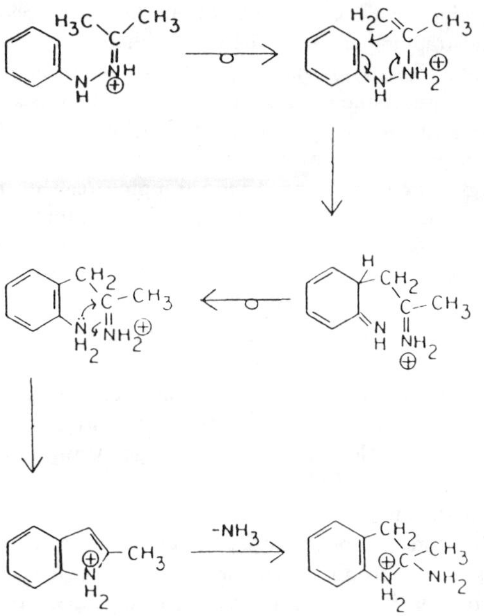

Figure 4-5 ■ Proposed solution reaction mechanism for the Fischer indole synthesis.

4.3.1. Unimolecular Reactions in the Ion Source

One of the first cases in which MS/MS was used to investigate in detail the correspondence between gas-phase and solution-phase reaction mechanisms was a comparison between the gas-phase acid-catalyzed elimination of ammonia from protonated phenylhydrazones with the Fischer indole synthesis in solution.[390] The proposed solution mechanism for the Fischer indole synthesis is shown in Figure 4-5 for acetone phenylhydrazone. As shown, this reaction is quite complex, with two hydrogen transfers and numerous bond cleavages and new bond formations, leading to the elimination of a molecule of ammonia with formation of protonated 2-methylindole. To demonstrate that this same reaction occurs in the gas phase, it is necessary to observe an ion at the mass in the chemical ionization mass spectrum corresponding to the protonated indole. For acetone phenylhydrazone, this ion would be m/z 132, which is indeed observed in its isobutane chemical ionization mass spectrum. By selecting the protonated acetone phenylhydrazone ion (m/z 149) as the parent ion in an MS/MS experiment, it was then shown that a major peak in the daughter ion MS/MS spectrum corresponded to loss of 17 daltons to give an ion at m/z 132. This result is consistent with the formation of protonated 2-methylindole, although many other reactions can be envisioned that could account for the loss of 17 daltons from the protonated acetone phenylhydrazone. Therefore, it was necessary to identify the structure of the m/z 132 ion. This was accomplished by selecting m/z 132 from the ion source and obtaining its daughter ion MS/MS spectrum. This was compared with the daughter ion MS/MS spectrum obtained from m/z 132 formed when 2-methylindole was introduced into the

UPPER - (2-methylindole+H)$^+$

LOWER - (acetonephenylhydrazone+H-NH$_3$)$^+$

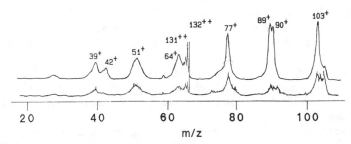

Figure 4-6 ■ Comparison of the daughter ion MS/MS spectrum of protonated 2-methylindole with the daughter ion MS/MS spectrum of the species formed in the chemical ionization source corresponding to ammonia loss from protonated acetone phenylhydrazone. The high degree of similarity provides strong evidence of formation of the Fischer product in the chemical ionization source. (Adapted from reference 390.)

chemical ionization source. As shown in Figure 4-6, the two spectra are essentially identical, indicating that the two ions have the same structure.

While this demonstrates that 2-methylindole is formed from protonated acetone phenylhydrazone, it says little about the reaction mechanism. To learn more about the reaction mechanism, labeling experiments were completed using d_6-acetone. From Figure 4-5, it can be seen that with perdeutero-methyl groups, the molecule of ammonia eliminated should contain two deuteriums. Loss of NHD$_2$ is indeed observed in the daughter ion MS/MS spectrum of protonated d_6-acetone phenylhydrazone, giving strong support to the assumption that the gas-phase and solution reaction mechanisms are similar. Further evidence for this similarity was obtained by using the ability of MS/MS to mass-select incompletely labeled ions for study. Some of the d_6-acetone phenylhydrazone underwent partial H/D exchange in the ion source, forming the d_4 and d_5 species. Daughter ion MS/MS spectra of these ions showed that they lost NHD$_2$, NH$_2$D, and NH$_3$ in a ratio consistent with the mechanism in Figure 4-5. Furthermore, the daughter ion MS/MS spectra provided proof that the label was definitely on the methyl groups.

Other protonated phenylhydrazones were also shown to form protonated indoles in the chemical ionization source via comparison of the daughter ion MS/MS spectra of the ion formed in the ion source with those from the authentic protonated indoles. This combination of MS/MS experiments provided strong evidence that the gas-phase reaction mechanism leading to a protonated indole from a protonated phenylhydrazone is the same as the solution-phase mechanism. Finally, by use of the appropriate ketone (methyl isopropyl ketone), the Fischer indole synthesis according to the mechanism in Figure 4-5 would be blocked and instead an indolenine should be formed. Comparisons of the daughter ion MS/MS spectra demonstrated that this in fact did occur.

In summary, MS/MS provided the following information in elucidating the gas-phase reaction mechanism: (1) that the protonated phenylhydrazones fragment directly to an ion corresponding to the appropriate protonated indoles; (2) that an isotopically labeled protonated phenylhydrazone loses ammonia by a mechanism consistent with that proposed; (3) that the structure of the ion formed by ammonia loss was the same as that of the appropriate protonated indole; and (4) that the reaction could be diverted to other products by appropriate variation of the reagents.

The preceding example, and most other investigations into gas-phase reaction mechanisms that have made use of MS/MS, used the most common MS/MS mode, daughter ion scans. However, there have been a few cases in which charge permutation MS/MS spectra have been used to characterize gas-phase reactions in the ion source. In particular, most of these have involved the charge inversion reaction. In general, charge inversion is used with negatively charged parent ions, which lends itself to the study of base-catalyzed reactions, in contrast to the acid-catalyzed Fischer indole synthesis described previously.

An example in which charge inversion was indispensable in the elucidation of a gas-phase reaction mechanism is the gas-phase barbituric acid synthesis.[391] In this study, the $(M - H)^-$ ion from methyl diethyl malonurate was shown to fragment via loss of 32 daltons, which could possibly be the result of the intramolecular cyclization by which barbituric acids are synthesized. The daughter ion MS/MS spectrum of this $(M - H\text{—}CH_3OH)^-$ ion contains the same ions as the daughter ion MS/MS spectrum of $(M - H)^-$ ions from the appropriate barbituric acids, but the relative abundance of one ion (m/z 42) in the spectra is substantially different. This suggests that the barbituric acid is not being formed. However, the charge inversion spectra for the $(M - H\text{—}CH_3OH)^-$ from methyl diethyl malonurate and the $(M - H)^-$ ion from the barbituric acid are identical. As discussed in Section 3.5.2.1, the charge inversion reaction is a very high-energy process, typically requiring much more energy than many CID processes. Since these higher-energy pathways are generally more informative for structure analysis, as discussed in Section 4.2, it was proposed that this was evidence that the barbituric acid was indeed being formed by an intramolecular cyclization. The difference in the relative abundances of the ions at m/z 42 in the daughter ion MS/MS spectra was assumed to be due to internal energy differences resulting from the different means of formation of the parent ions.

4.3.2. Unimolecular Reactions in Reaction Regions

This section describes examples of some of the experiments in which MS/MS can be used to probe unimolecular reaction mechanisms for ions dissociating in a reaction region. The mechanisms of unimolecular reactions occurring in reaction regions have not been demonstrated as rigorously as the ion source reactions discussed previously. This is due to the fact that structural information on the product ion of a unimolecular dissociation in a reaction region can only be obtained in an MS/MS/MS experiment, and few such studies have been performed. In general, mechanisms of reactions occurring in a reaction region have been indirectly established on the basis of kinetic energy release measurements and isotopic labeling and have not been explicitly demonstrated by the additional step of structural determination of the product ion. While in many cases, the source-generated ion of the same m/z value could be used for structure confirmation via MS/MS, the structures of the fragment ion formed in the source and the daughter ion formed in a reaction region are not necessarily the same.[392]

The first area to be discussed is the determination of competing mechanisms leading to the formation of product ions with the same empirical formulas. A classic study in this area was performed by Beynon et al.[154] In this study, it was found that *para*-substituted nitrobenzene ions undergo loss of NO·by two distinct mechanisms, one giving a rather small value for kinetic energy release and the other giving a much larger value. Since the parent ion was the same for both reactions, the differences in kinetic energy release most likely resulted from either different transition states (and therefore different critical energies) or different structures of the product ions. The second of these possibilities would also likely require a different transition state as well.

It was concluded in this study that the larger kinetic energy release was a result of a mechanism in which oxygen transfer to the benzene ring occurred through a three-member cyclic transition state to the *ipso* position, while the smaller kinetic energy release was a result of a four-member cyclic transition state that transferred the oxygen atom to the *ortho* position of the benzene ring. From this and several other studies, it has been generally postulated that in reactions that involve cyclic transition states, as the number of atoms in the "ring" decreases, the percentage of reverse critical energy partitioned into kinetic energy release increases.[393] This is rationalized by assuming that the smaller transition states are "tighter," and therefore less effective in partitioning internal energy into vibrational modes in the products. This then allows more internal energy to be partitioned into the energy of separation, the kinetic energy released.

Another aspect of the study of the loss of NO·from *para*-substituted nitrobenzene radical cations was the effect of the ring substituents. It was observed that the abundance of the ions formed by the process giving the large kinetic energy release increased compared to the products of the process giving the smaller kinetic energy release as the *para*-substituent became more electron-withdrawing. In addition, while the value of the larger kinetic energy release increased as the electron-withdrawing power of the substituent increased, the value of the smaller kinetic energy release was relatively insensitive to the substituent. This was concluded to be due to the stability of the product ions. The reaction associated with the larger value, from the three-member cyclic transition state, gives a product ion that can resonantly interact with the *para* substituent, providing greater product stability than the other mechanism, which results in a product ion that does not interact with the *para* substituent.

While most studies using the kinetic energy release to provide mechanistic information have relied on the presence of competing mechanisms with different values of kinetic energy release, this is not always necessary. Williams and Hvistendahl have used the magnitude of the kinetic energy release in cases of unimolecular H_2 loss, in conjunction with appearance potential measurements and labeling experiments, to postulate the mechanism of the hydrogen loss reaction.[394,395] For the reacting ions listed in Table 4-5, the loss of H_2 was shown to be a concerted 1,2-elimination. The kinetic energy release associated with all these reactions was quite large, especially for the latter four (Table 4-5). It was argued that this large magnitude of the kinetic energy release was due to the fact that a concerted 1,2-elimination should be orbital symmetry-forbidden. The appropriate correlation diagram is shown in Figure 4-7. It can be seen from the diagram that as the electronic reorganization takes place, the highest occupied molecular orbital becomes mutually repulsive between the carbons and hydrogens.

Williams and Hvistendahl then used these results to propose mechanisms for H_2 loss from ions for which the labeling experiments were inconclusive due to hydrogen shifts prior to dissociation. The parent ions in this study were $C_2H_4^{+\cdot}$, $C_2H_5^+$, $C_6H_7^+$,

Table 4-5 ■ Kinetic Energy Releases Resulting from the 1,2-Elimination of H_2

Reacting ion	Product ion	T (kcal/mol)
$H_3C—CH_3^{+\cdot}$	$H_2C—CH_2^{+\cdot}$	4.4
$H_2C=OH^+$	$HC\equiv O^+$	33
$H_2C=NH_2^+$	$HC\equiv NH_2^+$	20
$H_2C=SH^+$	$HC\equiv S^+$	20
$H_3C—NH_2^{+\cdot}$	$H_2C=NH^{+\cdot}$	19

$C_3H_7^+$ and $C_7H_9^+$. Hydrogen (H_2) loss from the first three of these ions occurs with a relatively small kinetic energy release, while the latter two ions dissociate with a relative large kinetic energy release. From measurement of the critical energy of the reaction and orbital symmetry considerations similar to those discussed previously, it was concluded that the ions that lose H_2 with a small kinetic energy release do so by a concerted, symmetry-allowed, 1,1-elimination. The large kinetic energy release associated with H_2 loss from the other two ions was postulated to be a result of a symmetry-forbidden 1,2- or 1,3-elimination.

Another method used to postulate reaction mechanisms has been the study of the dissociations of a series of similar or homologous ions. The use of the observed dissociations along with known thermochemical values allows a hypothetical potential energy surface to be constructed, from which a possible transition state can be deduced. A good example of this is the work of Bowen and Williams, which studies the effects of ion/dipole interactions in the dissociation of certain oxonium ions.[396] The metastable dissociations of the six members of the homologous series $C_3H_7O^+=CR_1R_2$ $(R_1, R_2 = H, CH_3)$ were observed in this study. The two possible isomers are the n-propyl and isopropyl structures, shown as I and II in Figure 4-1. For the smallest

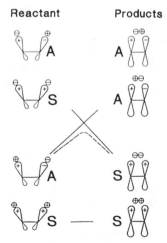

Figure 4-7 ■ Molecular orbital correlation diagram for the loss of H_2 from ethane showing the symmetry-forbidden nature of the reaction. (Adapted from reference 394.)

members of this series ($R_1, R_2 = H$), both structures show the same metastable dissociations (loss of H_2O and CH_2O) in similar abundances and with the same values of kinetic energy release. These results indicate that the two ions fragment from a common structure. These results also suggest a potential energy surface as shown in Figure 4-1a. Since the measured appearance energy for the reaction involving loss of CH_2O was less than that required to form the n-propyl cation, the product ion is assumed to be the isopropyl cation. Furthermore, there must be some stabilization in the transition state to allow the isomerization of n-propyl to isopropyl at energies below the dissociation energy. This stabilization is proposed to occur via an ion/dipole interaction in a structure in which the propyl cation is loosely bound to the CH_2O. Since the heat of formation of the two propyl cations differ by 16 kcal/mol, the ion/dipole stabilization must be greater than this value. A stabilization of 18 kcal/mol was calculated for a distance of 0.3 nm between the point charge and point dipole, which is consistent with the rest of the data.

The next set of isomeric ions in the series, n-$C_3H_7O^+$=$CHCH_3$ and iso-$C_3H_7O^+$=$CHCH_3$, both show loss of H_2O and C_3H_6, but in different relative abundances. The H_2O loss increases in importance with ion lifetime, providing evidence that it is the lower energy process. The H_2O loss is more predominant in the isopropyl isomer, which suggests that the isomerization from the n-propyl to the isopropyl isomer is rate-determining (the potential energy surface shown in Figure 4-1c). Further evidence supporting this assumption is the measurement of the kinetic energy release for loss of H_2O as 65 meV for the n-propyl isomer versus 52 meV for the isopropyl isomer. For the isomerization to be rate limiting, the energy of the transition state must fall between the heat of formation of CH_3CH=$OH^+ + C_3H_6$ (147 kcal/mol) and CH_3CH=$O + C_3H_7^+$ (152 kcal/mol). Assuming a similar transition state as in the preceding example, an ion/dipole stabilization of 21 kcal/mol is calculated, which would lower the energy of the transition state to 150 kcal/mol from that expected of an isolated n-propyl cation and acetaldehyde. This is in the range necessary for consistency with the rest of the data.

The last set of isomers in the series ($R_1 = R_2 = CH_3$) both fragment in the metastable time frame by loss of C_3H_6; however, the n-propyl isomer does so with a value of kinetic energy release about 60% greater than the isopropyl isomer. Using similar methods as in the previous examples, this observation is consistent with a rate-determining isomerization from the n-propyl form through an ion/dipole stabilized transition state.

Another method that is sometimes used to investigate unimolecular reaction mechanisms occurring in a reaction region is study of the isotope effects observed when hydrogens are replaced by deuteriums. Williams and co-workers applied the use of H/D isotope effects to reactions in which the H(D) was *not* incorporated into the neutral species lost. Since the H (or D) is not lost, the typical method of partially labeling a compound and comparing the relative abundance of the fragments containing H with those containing D cannot be used. Instead, a competitive metastable dissociation that does not involve the label must be used as a reference to compare the relative abundance of the metastable reaction involving the H(D).

An example in which an isotope effect provides insight into a reaction mechanism that does not involve the loss of the H(D) is the loss of CO from p-bromophenol.[397] In addition to loss of CO, the molecular ions also lose Br$\cdot$ in a competitive metastable dissociation. Since loss of CO must involve a hydrogen transfer at some point in the reaction, the abundance of CO loss relative to loss of Br· was compared for the undeuterated species and the OD analog. It was found that the OD species lost CO by a factor of 3 less than OH species relative to the loss of Br: This is taken as evidence that the H(D) transfer occurs as the rate-determining step in the loss of CO from the molecular ion of *p*-bromophenol.

In this same study, an isotope effect was used to provide evidence for the mechanism and structure of the $(M - C_2H_4)^{+ \cdot}$ from *p*-bromophenylethylether. The $(M - C_2H_4)^{+ \cdot}$ had been postulated to have the same structure as the ionized phenol, as discussed previously. The ratio of CO loss to loss of Br· for the unlabeled $(M - C_2H_4)^{+ \cdot}$ is the same as for the phenol ion, while the ratio for these metastable dissociations from $(M - C_2D_4)^{+ \cdot}$ (from perdeuteroethylether) is the same as for the OD phenol. This suggests that the C_2H_4 loss leads to the ionized phenol structure and occurs via a four-centered reaction and *not* a five-center path, which would give an ion that should not have an isotope effect associated with CO loss (Figure 4-8).

While all the previous examples of unimolecular dissociations in reaction regions have rationalized reaction mechanisms without specifically examining product ion structures, Beynon and co-workers have performed experiments where the product ion was examined by MS/MS/MS.[398] The determination of kinetic energy release values was the dominant method used to elucidate mechanisms. One of the examples from this work is the loss of NO· from nitroaromatic cations. As discussed earlier in this section, nitroaromatic cations lose NO· by two different mechanisms, giving product ions of different structures. A composite metastable peak shape is observed due to the different distributions of kinetic energy release associated with the distinct reactions. By selection of different parts of the metastable peak formed in the first reaction region of a BE instrument with good angular collimation, it was shown for a variety of compounds that the structure associated with the smaller kinetic energy release can subsequently lose CO in a second reaction region dissociation. The ion formed in the reaction that leads to the large kinetic energy release does *not* give a metastable dissociation in the second reaction region. However, it does lose HCO in a CID reaction, whereas the ion of the other structure does not.

The kinetic energy release for the loss of CO from ions that lost NO· in a previous reaction region was also measured, in this case via an MS/MS/MS experiment. This value was compared with the value of the kinetic energy release obtained in an MS/MS

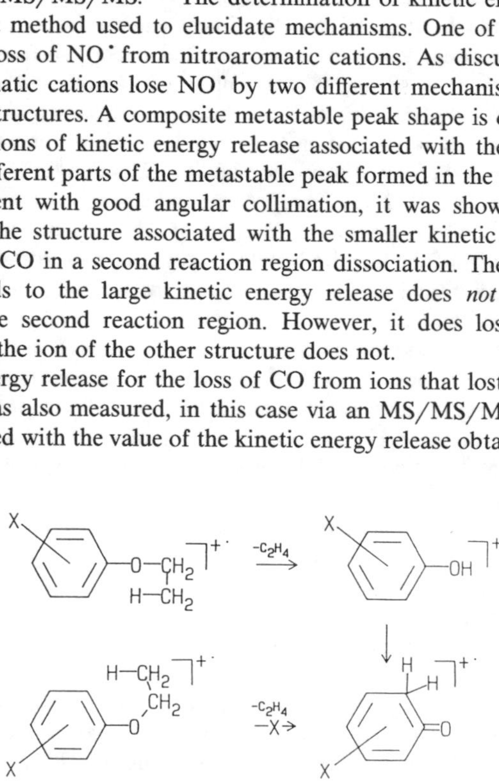

Figure 4-8 ■ Possible reaction mechanisms for the loss of C_2H_4 from substituted phenetole. The isotope effect observed for the loss of CO from the $(M - C_2H_4)$ ion indicates that the four-centered mechanism occurs, leading to the corresponding ionized phenol, rather than the five-centered mechanism. (Adapted from reference 397.)

experiment for the loss of CO from $M - NO^{\cdot}$ ions formed in the source. For the majority of the nitroaromatics studied these values were very similar, which provides good evidence that the structures of the product ions are the same. An exception to this generalization is 2-nitronaphthalene for which the measured kinetic energy releases are different for $M - NO^{\cdot}$ ions formed in the ion source and in the first reaction region. This suggests that the ions formed by loss of $NO^{\cdot}$ in different regions of the instrument have different structures.

4.3.3. Bimolecular Reactions in the Ion Source

Most mass spectra are a result of a series of competitive and consecutive unimolecular reactions. However, when high-pressure ion sources are used, such as a chemical ionization source, bimolecular ion/molecule reactions can occur. While the predominant reaction typically is proton transfer, the use of other reagent gases allows the study of more complicated association reactions.

Although ion cyclotron resonance (ICR) spectrometers are ideally suited to study many bimolecular ion/molecule reactions, high-pressure ion sources have the advantage of allowing collisional stabilization of the ion/molecule complexes. Thus, the intermediates in exothermic reactions may be observed in many cases, while only the ionic products will be detected in an ICR spectrometer. Gross' group has been one of the leaders in the use of MS/MS to study such chemistry and to elucidate ionic gas-phase reaction mechanisms.

A good example of the use of MS/MS to determine a bimolecular gas-phase reaction mechanism is the study by Miller and Gross of the reaction of alkyl iodides with the benzene cation.[399] A previous study with ICR had shown that isopropyliodide reacts with $C_6H_6^{+\cdot}$ from benzene, but not with other $C_6H_6^{+\cdot}$ species.[380] In addition, n-propyliodide and other isopropyl halides do not react. However, using the ICR, all that could be determined was that $C_9H_{13}^+$ was the product ion, and that the benzene radical cation, and not the isopropyl cation, was the reacting ionic species.

In a chemical ionization source on an EBE instrument, not only was the $C_9H_{13}^+$ product ion observed but also a $C_9H_{13}I^+$ ion. The CID daughter ion MS/MS spectrum of this latter species showed as a major peak the loss of $I^{\cdot}$; along with ions corresponding to loss of a propyl radical and benzene, and ions assumed to be benzene and propyl cations. Observation of these ions suggests that the ion/molecule complex can best be described as covalently bonded rather than an associative complex with iodide as the bridge. This conclusion is supported further by deuterium labeling experiments in which some H/D scrambling was observed.

The CID daughter ion MS/MS spectrum of the $C_9H_{13}^+$ ion generated in the ion source when benzene and 2-iodopropane were introduced as the reagents was also obtained. This MS/MS spectrum is shown in Figure 4-9, along with the CID daughter ion MS/MS spectra of several protonated alkylbenzenes. This figure shows that the CID daughter ion MS/MS spectrum of the benzene/iodopropane reaction product closely resembles that of protonated isopropyl benzene and is quite different from the spectra of the rest of the isomers.

To confirm that the $C_9H_{13}^+$ ion formed in the source is formed from the $C_9H_{13}I^+$ collision complex, an MS/MS/MS experiment was performed. CID of $C_9H_{13}I^+$ in the first reaction region of the EBE instrument produced the $C_9H_{13}^+$ ion, and this ion was then passed onto the third reaction region, from which its CID daughter ion spectrum

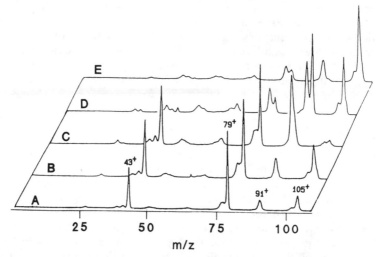

Figure 4-9 ■ Comparison of daughter ion MS/MS spectra from (*A*) $C_9H_{13}^+$ formed in the ionization source from the reaction of ionized benzene with 2-iodopropane, (*B*) protonated isopropylbenzene, (*C*) protonated *n*-propylbenzene, (*D*) protonated 1-ethyl-2-methylbenzene, and (*E*) protonated 1,3,5-trimethylbenzene. The correspondence between *A* and *B* indicates that the reaction product has the same structure as protonated isopropylbenzene. (Adapted from reference 399.)

was obtained. This spectrum proved to be identical to the CID daughter ion MS/MS spectrum of $C_9H_{13}^+$ formed in the ion source. In sum, the various MS/MS experiments defined the $C_9H_{13}^+$ ion formed in a reaction between benzene and isopropyliodide as identical to that formed from isopropylbenzene.

4.3.4. Bimolecular Reactions in a Reaction Region

Ion-trapping instruments have been used quite frequently in the study of bimolecular reactions. This is a result of the near-thermal kinetic energies of ions in trapping instruments as opposed to the eV to keV kinetic energies required to separate various mass-to-charge ratios in beam instruments. Thus, the latter type of instrument typically employs a deceleration step to reduce ion kinetic energies to less than an electron volt prior to reaction.

There are two advantages to a study that carries out bimolecular reactions in a reaction region as opposed to the ion source. The first of these is that the reactants are explicitly defined, and the second is the ability to vary the kinetic energy of the reacting ion. However, pressures in the reaction regions are typically too low to allow collisional stabilization of ion/molecule complexes, and thus the intermediates in ion/molecule reactions cannot generally be observed.

An example of the advantage of being able to define explicitly the reactants comes from the previously discussed work of Miller and Gross.[399] In this study, they showed that the $C_9H_{13}^+$ species could be formed by reaction of $C_3H_7^+$ species with benzene, in addition to being formed in the reaction between the benzene cation and iodopropane.

Thus, an MS/MS/MS experiment was required to confirm that the reaction occurring in the source was that of elimination of I˙ from the benzene/iodopropane cationic complex and not reaction of $C_3H_7^+$ from iodopropane with benzene. On the other hand, it was necessary to have the high source pressures of a chemical ionization source to observe the iodopropane/benzene cation complex in the first place.

Many early studies of bimolecular ion/molecule reactions in a reaction region investigated the reactions of hydrocarbon species. Most of these were motivated by the development of chemical ionization and the rich chemistry that was found to occur in a "high-pressure" or chemical ionization source. However, due to the variety of ions resident in a chemical ionization source, it was difficult to determine unambiguously the reaction mechanisms. Selecting a single ionic species and allowing it to react with a neutral gas in a reaction region provided a means to follow more accurately the relevant chemistry and to elucidate reaction mechanisms. A representative example of this type of experiment is the work by Bone and Futrell studying the ion/molecule reactions occurring in propane.[400] When propane is ionized by electron ionization, a number of ions are formed that can then react with neutral propane. Since many of these reactions lead to the same ions that are formed during the initial ionization process, Bone and Futrell mass-selected specific ions and reacted these with propane in the reaction region of a multisector instrument.[50]

An advantage of the ability to mass-select the reactant ion is the concurrent ability to observe the reactions of labeled compounds to determine whether the charge was transferred from the ion to the neutral and/or whether atoms were transferred between the ion and neutral. A case study involves reactions of the molecular ion with the neutral molecule. At kinetic energies of less than 1 eV, the only reaction of the molecular ion of propane with propane is charge exchange. At higher energies (2 and 4 eV), ions at m/z 32 and 34 were observed when $C_3D_8^+$ was the ion and C_3H_8 the neutral. It was shown that these ions were the result of CID by using Kr in the reaction region.[400]

Very complex chemistry of propane and propane ions and of propane with ions from other hydrocarbons was delineated in the work of Bone and Futrell.[400] In addition to the normal complement of experiments, the daughter ion MS/MS spectra were recorded as a function of collision energy of the ion/molecule pair. In summary, the most intense ion formed in the electron impact ionization of propane is $C_3H_7^+$; this ion does not react with propane molecules at an appreciable rate. Previous studies had suggested that the $C_3H_7^+$ ion was formed by hydride abstraction from propane by a variety of ions. Bone and Futrell confirmed that hydride abstraction is essentially the only low energy reaction of CH_3^+, $C_2H_5^+$, and $C_3H_5^+$ with propane molecules. Ethylene and propene ions also react with propane via hydride abstraction, but both species have other reaction paths as well. For instance, both of these latter ions also produce a $(M - 2H)^+$ ion from propane and the alkane corresponding to the reactant ion. For the propene ion with near-thermal kinetic energy, this is the major pathway for reaction, with a rate of about 12 times that of hydride abstraction. For ethylene ions, hydride abstraction proceeds at twice the rate of $(M - 2H)^+$ formation.

The reactions of ethylene ion and propene ion with propane were both shown to be sensitive to the kinetic energy of the reactant ion. For propene ions, this effect was manifested as a change in the relative rates of reaction, with hydride transfer becoming more competitive at higher kinetic energies. However, for ethylene ions, a new reaction path appears at higher kinetic energies. This is transfer of a hydrogen atom from the

Table 4-6 ■ Ion/Molecule Reaction Products for the Reaction
of $C_3H_3^+$ with C_3D_8

m/z	Ion	Relative intensity[a]
43	$C_3H_3D_2^+$	1.1 (1.5)
44	$C_3H_2D_3^+$	1.0 (0.9)
45	$C_3HD_4^+$	0.8 (0.4)
50	$C_3D_7^+$	10.0 (10.0)
59	$C_4H_3D_4^+$	0.06 (0.2)
60	$C_4H_2D_5^+$	1.0 (0.1)
61	$C_4HD_6^+$	0.06 (—)
62	$C_4D_7^+$	0.01 (—)

[a]At quasithermal ion kinetic energy and at 2 eV (in parentheses).

propane neutral to the ethylene ion to form the ethyl ion. This is an example where the kinetic energy of the ion drives the reaction, as this reaction is about 6 kcal/mol endothermic. Several other ionic products were also observed at higher kinetic energies but these were shown to be the result of either CID or dissociative charge exchange.

Two other ions were investigated in this study, $C_2H_3^+$ and $C_3H_3^+$. The most striking aspect of these ions in their reaction with propane is their formation of "long-lived" complexes in which substantial exchange of atoms occurs. This was determined by reacting an unlabeled ion with a labeled neutral. Complex formation is also evidenced in the $C_3H_3^+$ system by the observation of ions at masses greater than either the reactant ion or neutral. The product ion distribution for the reaction of $C_3H_3^+$ with C_3D_8 is shown in Table 4-6. Since a reaction complex is formed, it is difficult to determine an empirical reaction mechanism, i.e., to determine which carbon skeleton ends up with the charge. However, by increasing the kinetic energy, the interaction time for atom exchange is reduced, and reaction mechanisms were postulated by the authors. The $C_2H_3^+$ ion also showed evidence of undergoing a reaction different from all those previously discussed. At higher kinetic energies, instead of the formation of a reaction complex, it was postulated that proton transfer occurred from the ion to form a transient $C_3H_9^+$ species, which subsequently dissociated to $C_2H_5^+$.

A summary of all the reactions discussed previously is shown in Figure 4-10. It can be seen that even an apparently simple system such as propane has a very complex and rich gas-phase ion/molecule chemistry. This complexity in such a basic system demonstrates the utility of being able to study the reactions of an isolated species in a reaction region.

While ICR instruments have historically been used to study ion/molecule reactions, it was not until the acceptance of the FT–ICR technique that trapping instruments were widely used to isolate an ion of specific mass-to-charge ratio and then to react this ion with a neutral species. This occurred in part because methodology developed for the FT–ICR experiment also facilitated the MS/MS experiment, and also because of newer features incorporated into the FT–ICR instrument (such as pulsed valves) that facilitated the experiments.

One of the major areas of ion/molecule chemistry that has been explored using the FT–ICR technique is the gas-phase reactions of transition metal ions. Freiser's group has been one of the leaders in this area. Their general experimental scheme in these studies involves a multistep sequence, taking advantage of the ability to do multiple

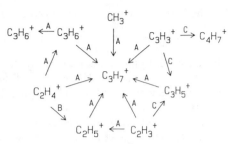

Figure 4-10 ■ Ion/molecule reaction pathways for the reactions of fragment ions from propane with neutral propane. The neutral reactant in all cases is propane. (*A*) reactions in which the product ion is from the neutral propane, (*B*) reactions in which the product ion is form the reactant ion, and (*C*) reactions in which atom exchange occurs between the neutral and reactant ion, leading to a product that is not totally from one or the other.

stages of MS, since the selection, dissociation, and analysis occur sequentially in time, not space. The first step is often laser ionization followed by isolation of the reactant ion via ejection of the other ions from the ICR cell. Then the selected ion is allowed to react with the neutral molecule, which can either be continually present or pulsed into the system. The desired ion/molecule reaction product is isolated via ejection of the rest of the ions from the cell. CID is then performed with this isolated ion/molecule reaction product and the daughter ions analyzed. A nice example of this type of experiment is the formation of $NiC_4H_8^+$ isomers by the reaction of Ni^+ with various neutral molecules.[401] Four reactions leading to the $NiC_4H_8^+$ isomers are shown in equations (4-7) through (4-10).

$$Ni^+ + n\text{-}C_4H_{10} \rightarrow Ni(C_2H_4)_2^+ + H_2 \qquad (4\text{-}7)$$

$$Ni^+ + n\text{-}C_6H_{14} \rightarrow NiC_2H_5CH{=}CH_2^+ + C_2H_6 \qquad (4\text{-}8)$$

$$Ni^+ + \underset{CH_2CH_2}{\overset{CH_2CH_2}{\diagdown}}CO \rightarrow \underset{CH_2CH_2}{\overset{CH_2CH_2}{\diagdown}}Ni^+ + CO \qquad (4\text{-}9)$$

$$Ni^+ + (CH_3)_4C \rightarrow Ni(CH_3)_2C{=}CH_2^+ + CH_4 \qquad (4\text{-}10)$$

All the preceding product ions show loss of C_4H_8, giving Ni^+, in the CID daughter ion MS/MS spectrum. For reaction (4-10), this is the only dissociation observed. The product ion from reaction (4-7) also shows loss of C_2H_4, while that from reaction (4-8) does not, but rather dissociates by loss of H_2. Both these fragmentations are observed in the CID daughter ion MS/MS spectrum of the reaction product in reaction (4-9). Thus, at least four different isomers of $NiC_4H_8^+$ are stable. Of note is that all of these isomers are formed via insertion into a carbon–carbon bond. This is in contrast to the reactions of nickel and Ni-complexes in solution, which proceed via insertion of the metal into carbon–hydrogen bonds.

Recently, the reaction intermediate scan has been developed to investigate organic reaction pathways.[402] The reaction intermediate scan is an MS/MS/MS experiment that can be considered as a more general version of the consecutive metastable ion dissociation reactions discussed in Section 4.3.2. The reaction intermediate scan involves consecutive MS/MS experiments—a daughter ion scan followed by a parent ion scan. Equation (4-11) is the general equation for an MS/MS/MS experiment:

$$m_p^+ \rightarrow m_d^+ \rightarrow m_{gd}^+ \tag{4-11}$$

where m_{gd}^+ is a granddaughter ion. In the reaction intermediate scan, m_p^+ and m_{gd}^+ are fixed. The instrument is scanned such that the data identify all the m_d^+ species that are intermediates in the two-step fragmentation of the selected m_p^+ to the selected m_{gd}^+. As a simple example of the utility of the reaction intermediate scan, O'Lear et al. studied fragmentations of 2-ethylphenol. A reaction intermediate scan with m_p^+ set to be equal to the molecular ion (m/z 122) and m_{gd}^+ set as m/z 79 showed, as expected, that the ion at m/z 122 fragmented to the product ion at m/z 79 by consecutive loss of $CH_3^{\cdot}$ and CO (Figure 4-11). Since the ion at m/z 53 is a major ion in the mass spectrum and loss of C_2H_2 is a common fragmentation route of highly unsaturated ions, it was

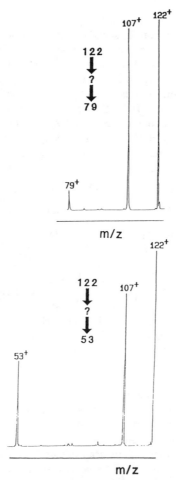

Figure 4-11 ■ Reaction intermediate spectra showing the dissociation pathway of the molecular ion of 2-ethylphenol (m/z 122) dissociating to m/z 79 and m/z 53. Loss of a methyl radical is an intermediate step in both fragmentation pathways. (Adapted from reference 402.)

postulated that m/z 53 was formed from the ion at m/z 79. However, a reaction intermediate scan with $m_p^+ = 122$ and $m_{gd}^+ = 53$ showed that the intermediate ion in this reaction sequence is not that at m/z 79 but instead m/z 107 (Figure 4-11). From these data, it can be said that the most favored pathway for formation of m/z 53 *from* m/z 122 is via m/z 107. The data do *not* exclude m/z 53 being formed from m/z 79 in another reaction pathway, however. The strength of the reaction intermediate scan, therefore, is its ability to establish directly intermediate steps in fragmentation pathways.

4.4. Thermochemistry

Much of the fundamental chemical research using MS/MS has been directed toward obtaining thermochemical data on ionic and neutral species. Most of these studies rely on the measurement of kinetic energy released in a dissociation, gain the kinetic energy loss associated with a collision-induced reaction, or the relative abundances of product ions. Examples of thermochemical determinations will be given for each of these types of measurements.

4.4.1. Energy Gain/Loss Measurements

Section 3.3.1.1 describes the basis for energy gain/loss measurements. Briefly, under suitable conditions, the endothermicity, q, of a collision between a high-velocity ion and an essentially stationary neutral target is reflected in the translational energy loss of the ion. This measurement, therefore, requires that the initial translational energy of the ion is well defined and that its final translational energy must be measurable. Virtually all such measurements performed with an MS/MS instrument have been made using BE geometry mass spectrometers. Energy gain/loss measurements have been made for several types of collision reactions. Some of these have involved charge stripping (Section 3.5.2.2), charge inversion (Section 3.5.2.1), atomic projectiles colliding with atoms or molecules, and polyatomic ions colliding with atomic or molecular targets. Energy loss measurements have recently been used in conjunction with field ionization to study neutrals formed by charge exchange (Section 3.5.1.1). An example is described in the following discussion for each of these types of collision.

Energy loss measurements have been used extensively to measure second ionization potentials via the charge stripping reaction (see Section 3.5.2.2).[185,328,403-406]

$$m_p^{+\cdot} + N \rightarrow m_p^{2+} + N + e^- \qquad (4\text{-}12)$$

This approach to measurement of the second ionization potential avoids many of the problems associated with the conventional electron impact ionization efficiency measurement. Principal among these problems are the low signals generated by doubly charged ions and the quadratic threshold law for their formation.[407,408] In the energy loss measurement, the second ionization potential is generally assigned as equal to the minimum energy loss, obtained by extrapolation of the high-energy side of the charge-stripping peak to the baseline. The method is widely applicable in that it can be used on molecular ions, fragment ions, protonated molecules, or virtually any ion that can be formed in an ion source, provided that it forms a doubly charged ion stable on the order of microseconds. When CID reactions do not give products that interfere with the charge-stripping peak, experimental accuracies in the measurement of the second ionization potential of between 0.1 and 0.3 eV are attainable. In some cases, a doubly charged ion can be observed via the charge-stripping reaction when it is not observed to be formed in electron ionization. For example, CH_4^{2+} has been formed by charge stripping of $CH_4^{+\cdot}$,[409] while it is not observed in the electron ionization mass spectrum of methane. This is probably due to the unfavorable Franck–Condon factors for the formation of a stable dication in a vertical transition from CH_4. A stable CH_4^{2+} is accessible from $CH_4^{+\cdot}$, however. Theoretical calculations indicate that the measured energy loss value for the reaction $CH_4^{+\cdot} \rightarrow CH_4^{2+}$ is closer to the adiabatic ionization potential of $CH_4^{+\cdot}$ than to the vertical ionization potential.[410-412] In other cases, fair agreement has been observed for the energy loss measurement and the theoretically calculated vertical ionization potential.[329]

A related translational energy change experiment has been described for the reaction involving the single-electron capture from a target atom or molecule to a fast doubly charged ion[124]:

$$m_p^{2+} + N \rightarrow m_p^{+\cdot} + N^{+\cdot} \tag{4-13}$$

This reaction is described in Section 3.5.1.3 and can be considered as the opposite of the charge-stripping reaction. Just as with the charge-stripping reaction, under the proper conditions, the change in the kinetic energy of the projectile can reflect the amount of kinetic energy converted to internal energy or vice versa. For ground state m_p^{2+} in a reaction with ground state N, in which ground state $m_p^{+\cdot}$ and $N^{+\cdot}$ are formed, the kinetic energy change of the projectile reflects the difference in the ionization potentials of $m_p^{+\cdot}$ and N:

$$\Delta E_{\text{LAB}}\left(m_p^{2+} \rightarrow m_p^{+\cdot}\right) = \text{IP}\left(m_p^{+\cdot}\right) - \text{IP}(N) \tag{4-14}$$

Unlike the charge-stripping reaction, which is a collisional ionization reaction, charge transfer reactions are often exothermic so that an energy gain may be observed for the projectile. This reaction provides an alternative means for measuring the second ionization potential of m_p. If the reaction is accompanied with internal excitation of $m_p^{+\cdot}$ and/or $N^{+\cdot}$, the kinetic energy of the projectile is decreased according to the kinetic energy transferred to internal energy. Fragmentation of $m_p^{+\cdot}$ often occurs after charge transfer, and the dissociations of $m_p^{+\cdot}$ can be used to characterize the structure of m_p^{2+}.

Energy loss measurements have been used to obtain electron affinities from the charge inversion reaction:[185,316,317]

$$m_p^{-\cdot} + N \rightarrow m_p^{+\cdot} + N + 2e^- \tag{4-15}$$

This reaction, discussed in Section 3.5.2.1, involves double-electron detachment from $m_p^{-\cdot}$, which requires that the electron affinity and ionization potential of m_p be overcome. Provided all reactants and products are formed in their respective ground states and that the kinetic energies of the liberated electrons are very small, the kinetic energy change of the projectile closely approximates the sum of the ionization potential and the electron affinity of m_p:

$$\Delta E_{\text{LAB}} = \text{IP}(m_p) + \text{EA}(m_p) \tag{4-16}$$

This type of measurement is most useful for atomic ions. The Franck–Condon factors for double-electron detachment from polyatomic anions usually results in the formation of a highly excited $m_p^{+\cdot}$ ion that subsequently dissociates. A peak for $m_p^{+\cdot}$ is therefore often not observed.

Energy loss measurements have recently been made, in conjunction with field ionization in a reaction region, to study high Rydberg state atoms formed by charge transfer in the reaction[413]:

$$m_p^{+\cdot} + N \rightarrow m_p + N^{+\cdot} \qquad (4\text{-}17)$$

This reaction, described in Section 3.5.1.1, is used as the first step in neutralization–reionization mass spectrometry. Ground-state atoms and molecules can be ionized by field ionization only in very high electric fields (10^8 V/cm). Lower electric fields (10^3 V/cm) are effective in ionizing atoms and molecules in high Rydberg states. An electrode structure that provides field strengths on the order of 10^3 V/cm has been incorporated into the second reaction region of a BE spectrometer after a collision cell. This allows Rydberg atoms formed by charge transfer in the collision cell to be reionized before they enter the electric sector. Two mechanisms for the formation of argon ions in high Rydberg states in collisions with xenon could be distinguished from the energy gain/loss spectra. The mechanisms are (i) the formation of atoms in high Rydberg states from the collision of metastable argon atoms formed in the ion source with xenon in the reaction region, and (ii) the formation of argon ions in high Rydberg states from the collision of ground-state argon ions with xenon. These reactions can be written, respectively:

$$Ar^{+\cdot*} + Xe \rightarrow Ar^* + Xe^{+\cdot} \qquad (4\text{-}18)$$

$$Ar^{+\cdot} + Xe \rightarrow Ar^* + Xe^{+\cdot} \qquad (4\text{-}19)$$

where $Ar^{+\cdot*}$ represents argon ions in metastable excited electronic states and Ar^* represents argon in high Rydberg states. Field ionized Ar^* formed from reaction (4-19) are observed at about 20-eV lower kinetic energy than when they are formed from reaction (4-18). The results also indicate, from peak intensities, that the cross section for reaction (4-18) is roughly 200 times greater than that for reaction (4-19).

Energy loss measurements have also been reported for the neutralization–reionization mass spectrometry experiment.[413] The measured energy loss of a projectile ion that undergoes neutralization and subsequent reionization contains contributions from kinetic energy changes in each step. Under the proper experimental conditions, namely, conditions in which the change in kinetic energy of the fast collision partner largely reflects the internal energy changes involved in the reactions, the kinetic energy change of the projectile is largely determined by the energetics of the neutralization reaction and by the subsequent ionization reaction. For ground-state reactants and products, the internal energy change in the former reaction is determined by the difference between the ionization potential of the target gas and the recombination energy of the ion. The minimum energy change associated with the collisional ionization reaction is the ionization potential of m_p with the liberated electron having negligible kinetic energy (in the center-of-mass frame of reference). Holmes et al. have reported a calibration method for energy loss measurements in NRMS based on the collisional ionization of krypton atoms in high Rydberg states.[413]

Translational energy gain/loss measurements have been used extensively to study collisions of atomic and diatomic ions with atomic and molecular targets.[170] Electronic and vibronic transitions between low-lying states can be observed in either the projectile ion, the target, or both. When ions are the subject of study, helium is often used as the target to avoid transitions in the target interfering with those in the ions.

Conversely, when the target is of interest, protons, helium ions, or lithium ions are often used as the projectile. An example of the former type of study is provided by Illies et al.,[414] who studied keV energy collisions of $Kr^{+\cdot}$ and $N_2^{+\cdot}$ with helium using a BE geometry mass spectrometer. The kinetic energy spectrum of krypton ions following collisions with helium at a collision energy of 8000 eV shows both an energy loss and an energy gain peak each at about 0.7 eV from the nominal parent ion kinetic energy. Transitions between two angular momentum states of the electronic ground state are responsible for these peaks, with reaction (4-20) representing an energy gain and reaction (4-21) an energy loss:

$$Kr^{+\cdot}\left(2P_{1/2}\right) \rightarrow Kr^{+\cdot}\left(2P_{3/2}\right) \tag{4-20}$$

$$Kr^{+\cdot}\left(2P_{3/2}\right) \rightarrow Kr^{+\cdot}\left(2P_{1/2}\right) \tag{4-21}$$

At an ionizing electron energy of 70 eV, the ratio of $Kr^{+\cdot}(2P_{3/2})$ to $Kr^{+\cdot}(2P_{1/2})$ formed in the ion source is approximately 2 to 1. The lifetimes of these states are long compared to the flight times through the instrument so that a similar ratio of states should characterize the ions that enter the collision cell. From the intensities of the peaks in the spectrum, the authors concluded that the transition probabilities of reactions (4-20) and (4-21) are essentially equal. The relative intensities of the energy gain and energy loss peaks were used to study the ion/molecule reaction chemistry of the two krypton ion states. Independent data suggest that the $Kr^{+\cdot}(2P_{3/2})$ undergoes charge transfer with N_2O roughly 27 times faster than does $Kr^{+\cdot}(2P_{1/2})$. This difference in reaction rates was qualitatively confirmed by measuring the ratio of the energy gain/energy loss peak intensities as a function of the N_2O pressure in the ion source. The ratio increases with increasing N_2O pressure, indicating that $Kr^{+\cdot}(2P_{3/2})$ reacts faster with N_2O, and is therefore depleted more quickly than $Kr^{+\cdot}(2P_{1/2})$.

The energy gain/loss spectrum of 8-keV $N_2^{+\cdot}$ incident on He shows two major energy loss peaks corresponding to transitions to the two lowest electronic states of the ion. Structure on the peaks is consistent with vibrational energy spacings, indicating vibronic transitions. The relative probabilities of the vibronic transitions indicated in the spectrum match the calculated Franck–Condon factors for a vertical transition. This is expected for high collision velocities in which the ions are collected with a small angular acceptance centered around 0°. Energy gain peaks were also observed that arise from transitions from vibronically excited states of the ion formed in the ion source with lifetimes sufficient to reach the collision cell.

A number of energy gain/loss studies have involved collisions of atomic projectiles with diatomic molecules.[206,415–417] An early example was provided by Fernandez et al.[216] with experiments involving collisions of argon ions and helium ions with molecular nitrogen. Spectra were obtained at several incident beam energies between 1.0 and 3.0 keV and at several scattering angles from 0° to 3.0°. Figure 4-12 shows a series of energy loss spectra obtained for collisions with $Ar^{+\cdot}$ on N_2 at several scattering angles at a collision energy of 2.0 keV. Peak *A* corresponds, at low scattering angles, to elastically scattered argon ions. Peak *B* corresponds to a transition from the ground state of N_2 to an excited electronic state, and a shoulder on the low-energy side of Peak *B*, labeled *C*, is interpreted as due to the ionization of N_2. As the scattering angle increases, the observed energy loss increases for all peaks. When the energy loss is corrected for energy loss due to elastic scattering, the collision inelasticity, *q*, is observed to increase with scattering angle for both peaks *A* and *B*. Figure 4-13 shows *q*

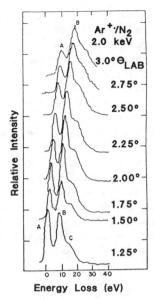

Figure 4-12 ■ Series of energy loss spectra of ionized argon in collisions with N_2 obtained at several scattering angles and a collision energy (E_{LAB}) of 2.0 keV. (Adapted from reference 216.)

plotted as a function of scattering angle for collision energies of 1.0 and 2.0 keV. This observation is interpreted as due to vibrational excitation via momentum transfer collision. For peak *B*, this is thought to occur after a vertical electronic excitation of the ion. Similar data were observed with experiments that involved the collision of helium ions with nitrogen.

Energy gain/loss measurements involving polyatomic species, either as the projectile or the target, are generally more difficult to make and are usually less informative

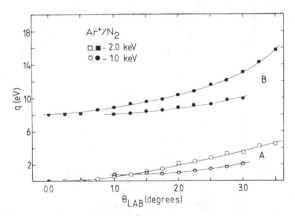

Figure 4-13 ■ Collision inelasticity for $Ar^{+\cdot}/N_2$ plotted as a function of scattering angle for collision energies of 1.0 and 2.0 keV. (Adapted from reference 216.)

than energy gain/loss measurements of collisions involving atoms or diatoms. The energy level spacings are much smaller for polyatomics than for atoms and diatoms, and the energy resolution of the measurement is generally inadequate for a differentiation between states. Furthermore, the broad distribution of energies transferred in keV collisional activation (see Section 3.3.1.3) is also reflected in a range of energy losses. The most probable energy loss is typically on the order of an electron volt. Most of the intact projectile ions that undergo collision, therefore, are not shifted much in kinetic energy from the ions that do not undergo collision. The main beam signal is a major interference in energy gain/loss measurements of the intact projectile ions. Energy loss measurements can be made on daughter ions, thereby avoiding interference from the main beam. Kinetic energy release in the dissociation, however, broadens the kinetic energy distribution of the daughter ions, making difficult the assignment of the minimum energy loss. For these reasons, fewer energy gain/loss measurements have been made for collisions of polyatomic species than for collisions involving atomic species. Nevertheless, informative studies have been based on energy gain/loss measurements for both polyatomic ion/atomic target and atomic ion/polyatomic target collisions. An example of each will be discussed.

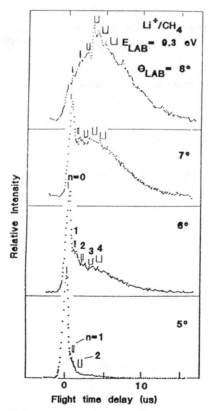

Figure 4-14 ■ Time-of-flight spectra of Li^+ following collisions with CH_4 at a collision energy of 9.3 eV at various scattering angles. (Adapted from reference 418.)

Toennies et al. have studied the energy losses associated with low-energy collisions (1–10 eV) involving atomic ion projectiles (H^+ and Li^+) and polyatomic targets (such as CH_4 and SF_6) at a variety of scattering angles.[245,418] Interference from the main beam is avoided by measuring energy losses through a time-of-flight analysis at nonzero scattering angles. Typical results are shown in Figure 4-14, which shows a series of Li^+ time-of-flight spectra following 9.3-eV collisions with CH_4. A signal at zero flight time delay corresponds to elastic scattering (no energy loss), and greater time delays correspond to greater inelasticity associated with the collision. Collision endothermicity shows a strong dependence on scattering angle. The energy losses correspond to excitation of vibrational levels of methane. This approach is obviously inferior to light spectroscopies in the measurement of energy level information for methane, but these studies are providing new insights into the mechanisms of energy transfer in collisions involving polyatomics.

Annis et al. have reported energy loss measurements for the polyatomic ions $UF_6^{-\bullet}$ and $MoF_6^{-\bullet}$ in collisions with a variety of targets at collision energies of tens to hundreds of electron volts and at several scattering angles.[419–421] The measurements were carried out with an apparatus consisting of an ion source, a collision chamber, and an electric sector. The masses of the ions that make up the peaks observed in the kinetic energy spectra were determined by time-of-flight analysis. Laboratory scattering angles of up to 5° were accessible with an angular resolution of about 0.5°. Figure 4-15 shows a series of kinetic energy spectra obtained for collisions of 198.2-eV $UF_6^{-\bullet}$ with

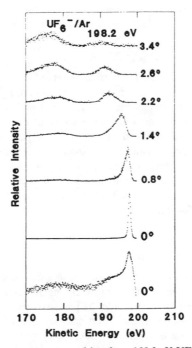

Figure 4-15 ■ Ion kinetic energy spectra resulting from 198.2-eV $UF_6^{-\bullet}$ projectile ions impinging on argon obtained at several scattering angles. (Adapted from reference 419.)

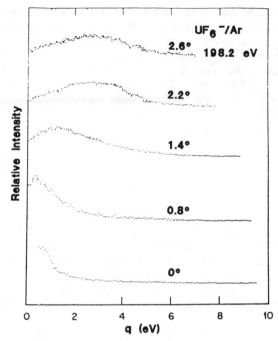

Figure 4-16 ■ Plot of the collision endothermicity, q, as a function of scattering angle for the inelastic $UF_6^{-\cdot}$ peak in the data of Figure 4-15. (Adapted from reference 419.)

argon at several scattering angles. The signals observed at kinetic energies of 170 to 180 eV are due to UF_5^- daughter ions arising from CID of $UF_6^{-\cdot}$. The data show a trend toward greater energy losses with increasing scattering angle and a greater relative degree of UF_5^- formation. Figure 4-16 shows a plot of the collision endothermicity, q, as a function of scattering angle for the inelastic $UF_6^{-\cdot}$ peak in the data of Figure 4-15. These plots reflect only the inelastic energy loss and therefore clearly show an increase in translational to internal energy conversion with increases in scattering angle. Similar trends were observed when Xe, SF_6, and UF_6 were used as the targets. Larger values of q were observed, however, for the molecular targets. This observation was interpreted as due to excitation of the target molecule.

4.4.2. Kinetic Energy Release Measurements

While kinetic energy release measurements have often been used to determine ion structures and reaction mechanisms, thermochemical determinations have not often been made from such measurements. This is primarily due to the poor understanding of energy partitioning between internal and translational modes. A useful application of kinetic energy release measurements in thermochemistry, however, is in the correction of heats of formation derived from appearance energy (AE) measurements. This application is most accurate when metastable ions are studied.

If there is an energy barrier to the reaction:

$$m_d^+ + m_n^{\cdot} \rightarrow m_p^{+\cdot} \tag{4-22}$$

i.e., if there is a reverse critical energy to the dissociation of $m_p^{+\cdot}$ to m_d^+ and $m_n^{\cdot}$, the appearance energy measurement will lead to an overestimate of the heat of formation of m_d^+. If metastable $m_p^{+\cdot}$ ions fragment to give m_d^+, the peak shape associated with m_d^+ can give an indication for the presence of a significant kinetic shift and/or a reverse critical energy. If the T value measured at half-height is less than a few tens of millielectron volts,[332] it is safe to conclude that neither a significant kinetic shift nor a reverse critical energy are involved. If the T value is large, the measured AE value leads to an upper limit for the heat of formation. Unfortunately, with the present poor understanding of energy partitioning from internal modes to translation in fragmentation processes, it is not possible to correct reliably the heat of formation using the measured value of T. If m_d^+ is not formed from metastable $m_p^{+\cdot}$, the AE measurement is likely to lead to a reasonably accurate heat of formation. This follows since the absence of a metastable peak indicates a steeply sloped $k(\epsilon)$ versus ϵ curve. Very small kinetic shifts are associated with reactions that increase in rate rapidly with internal energy, and reactions with reverse critical energies are most frequently characterized by slowly rising $k(\epsilon)$ versus ϵ curves.

Holmes et al. have shown that in some cases the measurement of the AE of a metastable ion can be useful, particularly if the specialized instrumentation required for conventional ionization energy and appearance energy measurements is unavailable.[366,422-425] This method also allows measurements to be taken on daughter ions that are the products of several dissociation routes. For example, Burgers et al.[423] have measured the appearance energies of HCO^+ from ionized methanol formed via two different channels:

$$HCOH^{+\cdot} \rightarrow HCO^+ + H^{\cdot} \tag{4-23}$$

$$H_2COH^+ \rightarrow HCO^+ + H_2 \tag{4-24}$$

The energetics of these processes are not distinctly accessible from AE measurements in the ion source since both processes contribute to the formation of HCO^+.

4.4.3. Relative Product Ion Abundances

The relative abundances of ions in a mass spectrum or MS/MS spectrum are important pieces of information that are used in addition to the mass-to-charge ratios of the ions themselves. In organic mass spectrometry, relative peak intensities are most often used as the basis to distinguish isomeric ions. In MS/MS, this comparison is the basis of the abundance ratio test described in Section 4.2. Relative product ion abundances have also often been used in MS/MS, however, as measures of energy rather than structure. Two types of measurements are described in this section to illustrate the study of relative peak abundances in MS/MS to obtain information on energetics. The first involves the determination of relative gas-phase thermochemical quantities via a kinetic method applied to ion-bound dimers. The second involves the characterization of parent ion internal energies.

4.4.3.1. Thermochemical Information from Ion-Bound Dimers.

Mass spectrometric techniques have provided chemists with gas-phase thermochemical values for ionic and neutral species. Many of these values have been based on ionization energy and appearance energy measurements. Newer techniques, however, have provided new types of information and have made many previously unknown ionic heats of formation accessible through the use of thermochemical cycles. Hundreds of measurements have been based on the establishment of an equilibrium between ionic and neutral reactants in either a high-pressure ion source[426,427] or an ion cyclotron mass spectrometer.[428,429] In some cases, the forward and reverse reaction rates involved in an equilibrium are measured in turn.[428,429] The equilibrium measurement approach has provided many new insights into the intrinsic chemistry of gas-phase species. Thermochemical measurements based on the establishment of an equilibrium can be made in instruments that can perform MS/MS experiments, such as the FT–ICR, but they do not require MS/MS. This section describes a kinetic method that relies on the formation and subsequent dissociation of an ion-bound dimer in an MS/MS instrument to obtain thermochemical values. The description of the approach begins with a general discussion that compares and contrasts the kinetic method with the equilibrium approach.

A widely measured quantity is gas-phase basicity, defined as the negative of the free-energy change occurring upon protonation of a base:

$$B + H^+ \to BH^+ \qquad GB(B) = -\Delta G_{rxn} \qquad (4\text{-}25)$$

where GB(B) denotes the gas-phase basicity of B. A related quantity is gas-phase acidity, defined as the free-energy change associated with the heterolytic A—H bond cleavage:

$$A{-}H \to A^- + H^+ \qquad GA(AH) = -\Delta G_{rxn} \qquad (4\text{-}26)$$

Gas-phase acidities are equivalent to the gas-phase basicity of the conjugate base:

$$GA(AH) = GB(A^-) \qquad (4\text{-}27)$$

Reaction enthalpies can be derived from reaction free energies by correction of the entropy contribution $(T\Delta S)$ to the free energy.[430-432] The negative enthalpy of reaction (4-26) is commonly referred to as the proton affinity of base B:

$$B + H^+ \to BH^+ \qquad PA(B) = -\Delta H_{rxn} \qquad (4\text{-}28)$$

where PA(B) denotes the proton affinity of B. The negative enthalpy of reaction (4-26) is the heterolytic bond dissociation energy of A—H and is symbolized as $D(A^-{-}H^+)$.

Quantitative measures of gas-phase thermochemistry are often made by measurement of the equilibrium constant, or the forward and reverse reaction rates associated with, proton transfer between two bases:

$$B_1H^+ + B_2 \underset{k_{-1}}{\overset{k_1}{\rightleftarrows}} B_1 + B_2H^+ \qquad (4\text{-}29)$$

where k_1 and k_{-1} are the forward and reverse reaction rate constants, respectively, and where k_1/k_{-1} is equal to the equilibrium constant K. The free-energy change in this reaction is related to K through:

$$\Delta G_{rxn} = -RT^0 \ln K \qquad (4\text{-}30)$$

where T^0 is the temperature and ΔG_{rxn} is the difference in gas-phase basicities of B_1 and B_2:

$$\Delta G_{rxn} = GB(B_1) - GB(B_2) \qquad (4\text{-}31)$$

The equilibrium method yields relative thermochemical quantities. A given measurement yields the difference in gas-phase basicities, for example, of the bases involved in the proton transfer reaction, but not the absolute GB value for either base. A few absolute values have been established[433] that serve as anchor points in the assignments of GB and PA values to bases that have been studied via the equilibrium technique.

The equilibrium approach to the measurement of relative thermochemical quantities has provided an extensive list of proton affinities, gas-phase basicities, gas-phase acidities, and other quantities that have been useful in many applications. It stands as the most accurate and reliable method for this purpose. However, there are several difficulties associated with these measurements that limit its range of applicability. For example, the method relies on the establishment of an equilibrium that may be difficult to achieve due to competing reactions or because the rate of approach to equilibrium is inherently slow.[434] The approach requires the accurate measurement of the neutral bases in addition to the measurement of the abundances of the ions. Therefore, the approach is limited to pure species with an appreciable vapor pressure, and, furthermore, these species must undergo a single, well-defined reaction. It is these considerations that have led, in part, to an alternative method for the measurement of relative thermochemical quantities based in the kinetics of dissociation of ion-bound dimers. The kinetic approach does not require the establishment of equilibrium, the measurement of ion/molecule reaction rates, high-purity reagents, the measurement of sample pressures, or that the sample molecules be volatile.

The underlying criterion for use of the kinetic method is the formation of a loosely bound ion consisting of a central ion associated with two ligands. Cluster ions such as these generally produce very simple daughter ion MS/MS spectra. Abundant daughter ions arise from loss of either of the ligands; the relevant reaction for the dissociation of a proton-bound dimer consisting of bases B_1 and B_2 is:

$$B_1 - H^+ - B_2 \begin{array}{c} \xrightarrow{k_1} B_1H^+ + B_2 \\ \\ \xrightarrow{k_2} B_1 + B_2H^+ \end{array} \qquad (4\text{-}32)$$

where k_1 and k_2 are the rate constants for the losses of B_1 and B_2, respectively. If B_1 and B_2 are similar species, the relative rates of fragmentation (which are proportional to the rate constants) should have very similar frequency factors. Since the dissociations are simple cleavages, these reactions should have very small reverse critical energies.[12] Under these conditions, the rates of the dissociations are controlled by the relative critical energies of each reaction channel, the difference in which is equivalent to the difference in gas-phase basicities. In the absence of consecutive fragmentations, the relative abundances of the daughter ions are determined by the dissociation rates. The relative abundances of the daughter ions therefore reflect the relative gas-phase basicities of B_1 and B_2.

Figure 4-17 shows the daughter ion MS/MS spectrum of the proton-bound dimer consisting of s-butylamine and n-propylamine as bases, obtained using a BE geometry mass spectrometer. It is clear from the spectrum that the gas-phase basicity of s-butylamine is greater than that of n-propylamine. If the dimer ion internal energy is characterized by a Boltzmann distribution, the difference in gas-phase basicities is given by:

$$\Delta GB = RT^0(\ln k_1/k_2) \qquad (4\text{-}33)$$

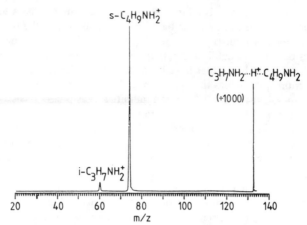

Figure 4-17 ■ Daughter ion MS/MS spectrum of the proton-bound dimer ion consisting of s-butylamine and n-propylamine as obtained using a BE geometry mass spectrometer.

where T^0 is the temperature or, when peak abundances reflect the dissociation rates:

$$\Delta GB = RT^0 \left(\ln \left[B_1 H^+ \right] / \left[B_2 H^+ \right] \right) \tag{4-34}$$

where $[B_1 H^+]$ and $[B_2 H^+]$ are the relative abundances of these ions. This relationship has also been presented in the context of the quasiequilibrium theory,[249,435] which does not assume a Boltzmann distribution of internal energies.

The distribution of the dimer ion internal energies is not Boltzmann, and is in fact generally unknown. Without a knowledge of the internal energy distribution, it is not possible to calculate accurately the difference in gas-phase basicities from a single MS/MS spectrum. It has been demonstrated that $\ln([B_x H^+]/[B_0 H^+]$, where $[B_x H^+]$ represents the abundance of one of a series of chemically similar protonated bases, and $[B_0 H^+]$ represents the abundance of a protonated reference base, is linearly related to ΔGB (or ΔPA when entropy effects cancel). Such a plot can be constructed from the daughter ion MS/MS spectra of a number of proton-bound dimer ions consisting of chemically similar bases. Provided that GB or PA values for some of the bases are known, a line can be determined, and values can be assigned to bases of unknown GB or PA. Figure 4-18 shows an example of such a plot constructed with data from the daughter ion MS/MS spectra of a number of proton-bound dimers consisting of substituted pyridines. The proton affinity of quinoline was assigned from this plot. From these and other data, it has been found that thermochemical values can be assigned with accuracies of ± 0.5 kcal/mol.[436]

The correlation between ion abundances and critical energies is not restricted to proton-bound dimer ions. Bursey et al. have also shown such a correlation for covalently bound species.[435] Other central ions have also been studied.[436-439] For example, Figure 4-19 shows the daughter ion MS/MS spectrum obtained for the [109]Ag$^+$ ion solvated by propylene and isobutylene. The spectrum indicates that the isobutylene binds silver cation more strongly than the propylene.[439] Burinsky et al.

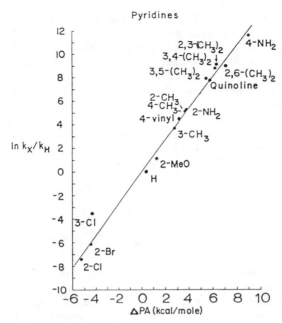

Figure 4-18 ■ Plot of the natural logarithm of the abundances of protonated substituted pyridines relative to that of protonated pyridine obtained from daughter ion spectra of pyridine proton-bound dimer ions as a function of proton affinity, PA. (Adapted from reference 436.)

have measured electron affinities using the kinetic method.[438] Other central ions that have been studied include Cu^+, Na^+, and Al^+.

4.4.3.2. Ion Internal Energies from Ion Abundances. Another application of the measurement of daughter ion abundances in MS/MS has been in the determination of ion internal energies. Although not a measurement of a thermochemical quantity per se, the characterization of ion internal energy in MS/MS has been particularly valuable in the study of activation reactions. As discussed in Section 3.2, the daughter ion

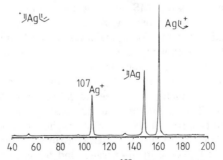

Figure 4-19 ■ Daughter ion MS/MS spectrum of $^{109}Ag^+$ solvated by propylene and isobutylene. (Adapted from reference 439.)

MS/MS spectrum is determined by the time scale of the experiment, by the consecutive and competitive dissociation pathways of the ion and by the internal energy distribution of the ion. When the ion structure is known and when the internal-energy-dependent dissociation behavior is known, the relative abundances of the daughter ions in the MS/MS spectrum can be used to follow changes in the internal energy of the parent ions. In ion structure differentiation, on the other hand, it is necessary to remove or to minimize ion internal energy as a variable in the experiment.

A number of ions have been used as parent ions in MS/MS experiments designed to measure ion internal energies. The parent ion most widely used for this purpose has been the *n*-butylbenzene molecular ion. The molecular ion of *n*-butylbenzene undergoes two competitive dissociations to give daughter ions at *m/z* 91 and *m/z* 92:

$$(C_6H_5)CH_2CH_2CH_2CH_3^{+\cdot} \left< \begin{array}{l} C_7H_8^{+\cdot} + C_3H_6 \\ \\ C_7H_7^+ + C_3H_7^\cdot \end{array} \right. \tag{4-35}$$

The ratio of the daughter ion abundances is very sensitive to the parent ion internal energy. The rearrangement reaction to give $C_7H_8^{+\cdot}$ dominates at low parent ion internal energies, and the simple cleavage reaction to give $C_7H_7^+$ dominates at high parent ion internal energies. This case is a classic example of a competitive rearrangement and simple cleavage reaction. The rearrangement reaction has the lower critical energy, and the rate of the simple cleavage reaction increases with internal energy at a much faster rate than does the rate of the rearrangement. Beynon et al. first used this ion to characterize the internal energy deposition in high-collision-energy CID and in photodissociation.[222] Since then, charge exchange experiments have been performed to determine experimentally the $91^+/92^+$ ion ratio.[244] This ion ratio has been used to characterize the mechanisms of photodissociation,[440,441] low-collision-energy CID,[43,240,241] and angle-resolved mass spectrometry.[442] The use of the $91^+/92^+$ ratio to calibrate internal energy deposition has been discussed in detail.[443]

The $91^+/92^+$ ratio for fragment ions formed from ionized *n*-butylbenzene, like most other relative daughter ion abundances used to measure the parent ion internal energy, provides only an average internal energy. A more or less broad distribution of internal energies characterizes parent ions in MS/MS. The shape of the distribution is expected to vary significantly with the method of parent ion formation and particularly with the activation method used in MS/MS. The distribution of internal energies in the parent ion is usually unknown. However, Kenttämaa and Cooks have used a simple method, based on relative daughter ion abundances, to approximate the internal energy distribution of parent ions following various types of activation reactions.[246]

The method is based on a thermochemical approach that assumes that once the critical energy for a dissociation is reached, the reaction goes to completion. Implicit in this assumption is that the kinetic factors (entropic factors) for the dissociations of the parent ion are the same. This assumption can rigorously be made for a limited number of ions, and the critical energies for the dissociations must be known. The triethylphosphate molecular ion appears to fit the criteria because it fragments via a series of consecutive reactions of known critical energies, and the entropy requirements of the dissociations appear to be similar. The parent ion gives four daughter ions through the reaction scheme:

$$m_p^{+\cdot} \xrightarrow{\epsilon_{O_1}} m_{d_1}^+ \xrightarrow{\epsilon_{O_2}} m_{d_2}^+ \xrightarrow{\epsilon_{O_3}} m_{d_3}^+ \xrightarrow{\epsilon_{O_4}} m_{d_4}^+ \qquad (4\text{-}36)$$

where ϵ_{O_n} represents the respective critical energies. The internal energy distribution of the parent ions that dissociate can be roughly determined from the relative abundances of the daughter ions by assuming that all $m_{d_n}^+$ arise from the parent ions with an internal energy between ϵ_{O_n} and $\epsilon_{O_{n+1}}$. In practice, the relative abundance of $m_{d_n}^+$ is plotted as the unweighted average of ϵ_{O_n} and $\epsilon_{O_{n+1}}$. This method has been used to study the differences in the parent ion internal energy distributions following keV collision energy CID, eV collision energy CID, surface-induced dissociation, and 70-eV electron ionization.[246-248]

Characteristics of MS/MS for Analytical Applications

5.1. Sample Considerations

5.1.1. Sample Collection

Extraordinary care is usually given to the instrumental and experimental aspects of data acquisition in MS/MS, and equal emphasis must be placed on the sampling process. For any analysis, "an understanding of the principles of sampling is important to the analyst because 1) the greatest care taken in performance of the analysis may be in vain if the sample has been obtained carelessly, and 2) the proper method of treating the sample or samples depends on the purpose of the sampling. These points are all the more important because the analyst is usually not directly responsible for obtaining the samples."[444]

The care in sampling that precedes any chemical analysis[445, 446] is required also in MS/MS, even more so than in GC/MS. Reduction in sample preparation for MS/MS does not imply a tolerance of careless sampling. The requirement of sample homogeneity is especially stringent for MS/MS as compared to techniques, like GC/MS, that require extensive sample preparation procedures before analysis. For instance, analysis of a complex mixture for a targeted component is an analytical problem to which MS/MS is often applied. If only a small fraction of the originally collected sample is used for analysis, care must be taken to ensure that the sample is representative of the complex mixture as a whole. Although several grams of sample may be collected or submitted for analysis, in MS/MS only micrograms of the total sample (at most) are introduced into the ion source to minimize source contamination and sample carryover. Correct sampling becomes even more significant because the need for rapid analysis (that may have initially dictated the use of MS/MS) often precludes a thorough statistical comparison with data derived from other fractions from the same sample. Since the tendency for minimal sample pretreatment in MS/MS analyses can preserve both original and induced variations in the sample, at least a rudimentary homogenization step, such as grinding, mixing, or dissolution, should always be performed as a minimum effort to ensure a homogeneous sample and to minimize these effects on the quality of the MS/MS data obtained. Conversely, if proper attention is paid to sampling, rapid analysis by MS/MS may eventually make it possible to characterize a large and diverse sample population with greater rigor than possible with more time-consuming methods of analysis.

Discussions of MS/MS often mention its ability to discriminate sample signals from *chemical noise*. As will be seen in the following chapter on analytical applications of MS/MS, this sometimes leads to a decrease in the limit of detection for targeted compounds. However, results of MS/MS analyses have sometimes been found to be less precise than those of GC/MS on the same samples. The instrumental parameters in MS/MS can be controlled with the same precision as in GC/MS. There are differences in sampling requirements, and *sampling noise* may contribute to the reported variations in quantitative MS/MS analyses. Sampling noise is the variation encountered in an analysis when different portions of a large sample are analyzed and is due to incomplete homogenization of the sample in terms of the analyte of interest.

In that MS/MS often deals with complex mixture analysis of samples for which pretreatment has been minimized, the *concentration* step often inherent in sample treatment methods is not available. As Figure 5-1 illustrates, the absolute amounts of the sample components introduced into the instrument are often lower in MS/MS than in GC/MS. The absolute amount of sample may therefore inadvertently fall below the limit of detection of the instrument simply through the lack of any concentration step. When chemical noise is not a limit to the sensitivity, GC/MS may provide a higher sensitivity than MS/MS for identification of unknowns by virtue of this sample concentration. Again, if time is not a consideration in the analysis, sample concentration can be completed prior to MS/MS in exactly the same manner as with any other method of analysis. Accordingly, many excellent analytical results with GC/MS/MS have been reported.

Choice of the form of the sample (whole blood vs. blood plasma vs. blood serum, for example) should also reflect the analytical strengths and weaknesses of an MS/MS analysis. In certain applications, such as forensic chemistry, the analyst may exert no

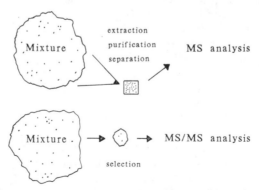

Figure 5-1 ■ Contrast in sample concentration for samples prepared for mass spectrometry in which most of the matrix components are removed and the sample concentrated (top) vs. the direct sampling of a portion of a larger sample (bottom), in which the sample concentration is not increased, as in many MS/MS experiments.

control over the sampling process. Whenever possible, the form of the sample should be chosen to minimize contamination of the ion source and to reduce the possibility of a matrix ionization effect. As a general rule, minimize the complexity of the sample matrix in order to avoid the possibility of matrix effects (see Section 5.2.3). A carefully considered balance must be drawn between the time and resources required for sample workup and the possibility of interferences and artifacts related to the complexity of the sample matrix.

5.1.2. Sample Contamination

Common contaminants can be introduced during sample storage, such as phthalates that dissolve from bottle-cap plasticizers, ions that leach from glassware, and contaminants introduced during sample handling (e.g., derivatization reactants, side-products, and sample preservatives). Ende and Spiteller have reviewed the effects of contaminants and artifacts in mass spectrometric analysis, with an emphasis on GC/MS analysis.[447] The effect of contaminants in GC/MS can be compared with their effect in MS/MS. In GC/MS, components of the sample are separated in time as they elute from the column. A phthalate contaminant, for instance, will elute with its characteristic retention time. Quantitation of components that elute at other retention times is unaffected unless there is a carrier effect. In the simplest model of sample introduction in MS/MS, all of the sample is introduced into the source, with simultaneous ionization of sample and contaminants. For trace component analysis, the contaminant may exert a significant matrix effect (see the next section) by changing the ionization efficiency for the compound of interest. Contamination of samples can thus be of greater significance for MS/MS than for GC/MS. For qualitative analysis, even the presence of matrix effects of this sort may be acceptable. In quantitative analysis, if an isotopically labeled internal standard is used, both the sample and the internal standard will experience an equivalent matrix effect in the efficiency of the ionization process.

In many MS/MS experiments, the direct insertion probe is used to introduce the sample into the source of the mass spectrometer, and fractional distillation during a programmed temperature increase can serve to separate the evaporation of the matrix components from the sample of interest. The ionization of the sample can therefore take place at a separate time from most, but probably not all, of the matrix components. Again, a balance must be drawn between the desorption profile of the sample molecules and the persistence of the ion beam from the source. It is of interest to note that this fractional distillation of sample from the direct insertion probe is in effect a crude separation in analogy to simple GC/MS/MS experiments.

Finally, the storage of MS/MS samples must preserve the homogeneity of the sample, since it is likely that only a small fraction will be taken for MS/MS analysis (vide infra). Prior to selection of an aliquot for analysis, any temperature, density, and size gradients in the sample should be eliminated. Of course, in screening analyses, the large numbers of samples analyzed by MS/MS must all be treated identically to ensure that valid comparisons result.

5.1.3. Sample Derivatization

For GC/MS, sample derivatization typically increases the volatility of the sample so that it can be separated by gas chromatography.[448] Derivatization can also be used to

increase the efficiency of the ionization process, as shown by the addition of elec-
trophoric groups (those that enhance the ability of the molecule to capture a thermal
electron) to compounds analyzed by negative ion mass spectrometry.[449] Derivatization
has also been used to lower the ionization energy of compounds of interest and to
direct fragmentation processes.[450]

In MS/MS, derivatization is often used to emphasize specific dissociation reactions
of a sample compound so that it can be more accurately identified in a complex
mixture. For example, derivatization might be designed to produce a product with a
distinctive pattern of fragmentation. A characteristic daughter ion or an expected
neutral loss can then be monitored in the MS/MS experiment. Parent ion and neutral
loss MS/MS spectra are both widely used to identify the members of functional group
sets in a complex mixture; characteristic dissociations can be enhanced when a
functional-group-specific reaction is incorporated into the sample preparation procedure. Figure 5-2 illustrates this concept for the analysis of derivatized steroids; the
addition of the derivatization label provides a characteristic MS/MS process that
pinpoints all of the reactive species. Sample preparation reactions in MS/MS, as in
GC/MS, raise concerns about the completeness of the derivatization reaction, the
possible formation of side products of the reaction, the reproducibility of the reaction
itself, and the increased sample contamination that may arise from additional handling
of the sample. Sample derivatization increases the complexity of the matrix and the
probability of matrix effects.

Zakett and Cooks[451] have described the use of a derivatization reaction with specific
application to MS/MS. The phenols in a complex coal liquid sample were treated with
acetyl chloride to provide the methyl ester derivative. The daughter ion MS/MS
spectra of these derivatives contain ions corresponding to regeneration of the phenolic
ions by loss of neutral ketene (42 daltons). A neutral loss scan for 42 daltons of the

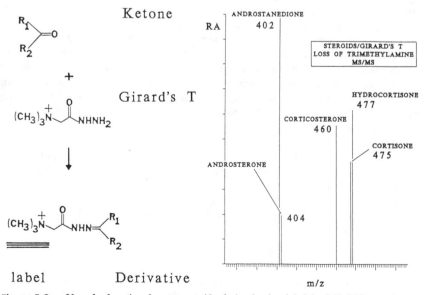

Figure 5-2 ■ Use of a functional-group-specific derivatization label for MS/MS experiments.

derivatized mixture provides the distribution of phenolic compounds in the coal mixture directly. Hunt et al.[65] have adapted a number of derivatization reactions into a comprehensive scheme for the analysis of environmental pollutants. The hallmarks of these methods are the speed and completeness of the reaction and the characteristic MS/MS behavior of the derivatized products.

The increasing use of FAB and SIMS ionization techniques is catalyzing the development of new derivatization reactions for MS/MS. These ionization techniques are especially suited for the sputtering of preformed ionic species from solution.[452] Derivatization reactions that produce these ionic products have consequently been developed.[453-458] Since many of them involve the addition of charged subgroups to the original molecule, the daughter ion MS/MS spectra often contain ions characteristic of the added groups. A parent ion or a neutral loss MS/MS spectrum pinpoints those molecules in a mixture that have reacted with the derivatization reagent.[459,460]

5.2. Choice of Ionization Method

5.2.1. Review of Ionization Methods

Almost every method of molecular ionization has been used in MS/MS experiments. Ions formed by any ionization method can be studied by MS/MS with suitable instrumentation. As might be expected, the ionization methods in widest general use (electron ionization, chemical ionization, and fast atom bombardment) are most often used as sources of ions for MS/MS experiments, often in complex mixture analyses. MS/MS experiments that provide information about ion structures and ion energetics are providing fundamental mechanistic information for these methods and also for the newer ionization methods. Excellent reviews of the various ionization methods are available.[461,462] A brief listing and description complete this section, and a discussion of the characteristics relevant to MS/MS experiments is included in the next section.

Electron Ionization. Organic molecules in the gas phase are bombarded by energetic electrons. An energy of 70 eV is usually chosen to create reproducible spectra that contain odd-electron ions formed by ejection of an electron from the molecule.[463,464] Subsequent fragmentation of the excited molecular ion produces the spectrum of fragment ions that is interpreted to deduce the structure of the original molecule. About 20% of electron ionization spectra of small organic molecules do not contain a significant relative abundance of the molecular ion, thus complicating the assignment of the molecular mass of the compound. Negative ions are not formed with high abundance in an electron ionization source operated in the usual manner.

Chemical Ionization. In chemical ionization, molecules in the gas phase are ionized in an ion/molecule reaction with a positively or negatively charged reagent ion that may itself be formed as a result of electron ionization or chemical ionization reactions.[465,466] The protonated molecule or the deprotonated molecule is usually a prominent ion in the chemical ionization mass spectrum. Fragmentation occurs to a degree influenced by the exothermicity of the ion/molecule reaction, i.e., the amount of internal energy resident in the ion initially formed, and can be controlled by judicious choice of the chemical ionization reagent gas. Atmospheric pressure ionization[467] is a form of chemical ionization.

Desorption Ionization. The term desorption ionization embraces several newer ionization methods designed to produce ions from nonvolatile or thermally fragile organic molecules in the condensed phase. These include methods such as fast atom bombardment,[468,469] in which the bombarding particle is an energetic (keV) neutral particle, secondary ion mass spectrometry, in which the bombarding particle is an energetic (keV) ion,[470,471] and plasma desorption, in which the bombarding particle is a very energetic (MeV) ion derived from spontaneous fission of a radionuclide or extracted from a particle accelerator.[472,473] Laser desorption may also be included in this category to the extent that phenomena resembling sputtering phenomena may occur in organic samples bombarded by an intense laser beam.[474] The processes through which ions are formed and ejected into the vacuum are thought to be nonequilibrium in nature.[475] The molecular ions formed are most often even-electron ions such as the protonated and deprotonated molecules described previously. The degree of fragmentation is, under most conditions, an increasing function of the energy density of the primary ion beam and, since the ions are ejected from a condensed phase (liquid or solid), also of the chemical and physical nature of the sample and its support. A number of novel ions, such as those formed by the cationization of an organic molecule with a metal cation,[476] may be found in desorption ionization mass spectra.

Nebulization Ionization. Several ionization methods are based on the extraction of ions from a liquid dispersed into fine droplets and evaporated into a vacuum. In thermospray,[477,478] ions formed statistically as the result of the nebulization process are sampled by the mass spectrometer. In electrospray[479,480] and electrohydrodynamic ionization,[481,482] the creation of ions is aided by an external induction current. Several workers have described a liquid ionization mechanism in which both processes occur.[483,484] In general, these ionization methods produce even-electron molecular ions of high relative abundance, with a minimum amount of fragmentation.

Photon Ionization. A highly focused light source, usually a laser, is used to ionize a molecule by photon-induced ejection of an electron.[485,486] Odd-electron ions can be formed with various internal energies based on the energy of the laser photons used and the number of photons absorbed. The degree of ion fragmentation, and the selectivity of the ionization process itself, is variable.

Field Ionization. A potential gradient of 10^7 to 10^8 V/cm is maintained between a sharp edge such as a razor blade and the ion entrance slit of the mass spectrometer. Volatile molecules that diffuse into this gradient are ionized by a process in which an electron is removed by tunneling.[487,488] A minimum of energy is left within the molecular ion, and only a small degree of fragmentation is observed for the ions normally observed in the mass spectrum. The technique provides molecular ions in many instances for which electron ionization does not, but the sensitivity is usually 10 to 100 times less than that of electron ionization.

Field Desorption. In field desorption, nonvolatile samples loaded onto an electrode emitter are subjected to a very high potential field gradient (10^8 V/cm). Tunneling of an electron to a positive electrode produces an odd-electron molecular ion of very low internal energy.[489-491] At the same time, dissolved metal salts in the viscous film can attach to the organic sample molecules, and cationized molecules such as $(M + Na)^+$ are also formed and extracted into the vacuum. The resulting spectra contain dominant molecular ions, although additional fragmentation of the ions can be induced by heating the emitter to a higher temperature than that needed for creation of the molecular ions.

5.2.2. Analytical Requirements for Sample Ionization

The emphasis in this section will be on those ionization methods that have most often been used with MS/MS in targeted compound analysis or mixture analyses, rather than those used in detailed studies of ion structure. In this latter work, MS/MS can be used, for example, to study isomeric ions formed by different ionization methods, under different conditions, with different lifetimes, and perhaps existing as the end result of a different series of chemical reactions. Choices of ionization method are then dictated by the need to compare the MS/MS behavior of ions formed with these diverse backgrounds.

The value of MS/MS depends explicitly on the *reproducibility* of the ionization method. The method of ionization should produce from the same sample the same flux of ions of the same characteristic structure, or mixtures thereof, each time the sample is analyzed. MS/MS spectral interpretation often relies on a comparison of current data with those obtained from previous samples or standards. Any uncharacterized or uncontrolled changes in the number or reactivity of ions produced compromises the integrity of that comparison. This requisite reproducibility is often at odds with the analytical problem at hand, viz the identification and quantitation of a targeted compound in an otherwise uncharacterized mixture. Again, the need for speed of analysis, which initially established the analytical demand for MS/MS, can preclude a careful evaluation of reproducibility.

5.2.2.1. Molecular and Ionic Structural Correspondence.
The complexity of mixture analysis by mass spectrometry has prompted the predominant use of "soft" ionization methods that by definition produce only a few species of ions, and ideally only one ion indicative of molecular mass, for each component of the mixture present in the source. The utility of such ionization methods extends to MS/MS as well. In the separation stage of MS/MS, ions are selected by their mass-to-charge ratios in the first mass analyzer. The ions formed in the source are assumed to retain the structures of the neutral molecules from which they are formed. Ideally, each component in the mixture produces one ion that is representative of that particular component, and the masses of all of the ions so formed are at different masses. The efficiency of ionization for each component should therefore be high and constant. The structures of the ions formed should be directly related to the structures of the neutral components in the mixture. Ideally, an ion should neither isomerize nor rearrange as a result of ionization, but if it does, the change must be understood in detail, or be at least constant in effect. For closely related mixture components, preservation of minor structural differences is paramount.

Electron ionization mass spectrometry has provided numerous examples of rearrangements and isomerizations of odd-electron organic ions due to the relatively large amount of internal energy deposited during ionization and the low activation energy for rearrangements. Other ionization methods that form odd-electron species suffer from similar disadvantages. Even-electron ions are generally formed by ionization methods such as chemical ionization (CI), field desorption (FD), and the three desorption ionization (DI) techniques—fast atom bombardment (FAB), laser desorption (LD), and secondary ion mass spectrometry (SIMS). The nebulization ionization (NI) techniques—liquid ionization, electrohydrodynamic ionization, electrospray ionization, and thermospray—also produce even-electron ions. Although such ions are less prone to undergo complicated rearrangements, no ionization method is completely immune from processes that do not preserve the parallels between molecules and ions. Chemical ionization is a case in point. Many compounds provide multiple basic sites at which the proton from the reagent gas can be added, and thus one compound can form several isomeric ions. Furthermore, protonated molecules can be generated using various reagent gases. Variation in the exothermicity of the proton transfer reaction may result in changes in the site or sites at which the proton is bound or to variations in the amount of internal energy of the protonated molecule.[492,493] The protonated molecule formed by different reagent gases or under different conditions of source temperature and pressure may vary in the site of protonation and subsequent chemical behavior elucidated by MS/MS.[494] Similar considerations apply to the site of attachment of other groups, as in the adduct ions with $C_2H_5^+$ and NH_4^+ formed in chemical ionization experiments with methane and ammonia as the reagent gas, respectively.[495]

Many of the newer DI and NI ionization methods create novel ions of unknown structures. It is often assumed and seldom demonstrated that the structure of the ion resembles that of the parent molecule. As these methods are used more extensively for the analysis of complex molecules, the extent of their own inherent rearrangements and isomerizations will in time also be established. As in chemical ionization, the even-electron ions are created in DI and NI by addition of a charged group, or by the direct transfer into the gas phase of a preformed even-electron ion. In FAB and SIMS, for instance, $(M + Na)^+$ is a commonly encountered ion. The sodium affinities of neutral molecules are not accurately known, and although the actual site of attachment is often assumed to be that of the analogous proton, this has not been explicitly demonstrated.[496,497] Daughter ion MS/MS spectra of these ions may therefore be found to vary with the method of formation of the cationized molecule and with the cationizing species, and such experiments themselves may provide detailed information about ion structure and ionization mechanism.

5.2.2.2. Ion Flux. A practical requirement for the ionization method chosen in MS/MS is a reasonable magnitude and persistence of the ion signal generated. For instance, although a daughter ion MS/MS spectrum can be obtained in a few seconds, the optimum experimental parameters can require a few minutes to establish. Data system experiment control has been described[498,499] in which the largest peaks in the mass spectrum sampled every few seconds are automatically identified, and the daughter ion MS/MS spectra of those selected ions obtained. Even this automated experiment requires a persistence of ion signal for several tens of seconds. Input from the operator at the instrument console is a rate-determining step in the completion of most MS/MS experiments.

The practical limit of ion flux in MS/MS experiments is related to the specificity of the experiment. If sufficient specificity can be assured, the limit of detection for MS/MS, as with single-stage mass spectrometric analysis, drops into the regime in which a signal intensity of a few ions detected per second can be considered analytically significant. It is useful to consider the numbers relating to ion transmission in an MS/MS instrument and the dynamic range expected between the lowest practical signal level (1 ion/s) and that expected under three sets of conditions (a, b, and c) (Figure 5-3). The three sets of conditions refer to initial ion currents from the source of about 10^9, 10^8, and 10^7 ions/s, respectively. At constant conditions of 10% transmission through the first mass analyzer, 10% collision efficiency (as in a CAD process), and 10% transmission through the second mass analyzer, the number of ions at the detector

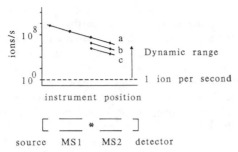

Figure 5-3 ■ Ion transmission through the components of an MS/MS instrument.

ranges from 1 to 10^6 ions/s (a) at best. In each stage of ion analysis or processing, a considerable fraction of the total ion current is lost. It is the concomitant gain in selectivity and the high gain of modern detection systems that make MS/MS experiments possible at all. Increasing resolution in parent ion selection is always accompanied by a rapid drop-off in the absolute magnitude of the ion current. Since this loss occurs at a point early in the transmission of the ions through an MS/MS instrument, it is all the more significant in the final analysis. True dynamic range is smaller than that illustrated in Figure 5-3, since practical daughter ion MS/MS spectrum measurement cannot be completed at ion counts of 1 or even 10 ions/s in most instruments. At some point, the number of parent ions will be sufficiently low that special scanning or data integration routines are necessary for measurement of the complete daughter ion MS/MS spectrum. Weak parent ion fluxes mandate slower scan speeds for the second analyzer and the application of multichannel analyzers and signal averaging techniques. For the measurement of full daughter ion MS/MS spectra, a sufficient number of parent ions must be reacted to determine not only the masses of the daughter ions, but also to determine accurately their relative abundances.

   Most ionization methods can provide a long-term signal as long as the sample flux into the source remains constant. The need for a signal to persist at a constant level throughout an MS/MS experiment then places constraints upon the operation of the sample inlet device, be it a reservoir, a direct insertion probe, or a chromatographic column. For instance, fractional distillation with changing temperatures of the direct insertion probe can provide a crude separation of the components in a complex mixture,[500,501] but, alternatively, may restrict the availability of the signal from the component of interest to a relatively short part of that time. Changing concentration profiles across an eluting GC peak may also affect the quality of the MS/MS data, especially when the MS/MS scan cycle times are similar to the width of the eluting peak. The current widespread use of FAB ionization with MS/MS reflects the general persistence and stability of the ion beams generated by FAB, but even this method of ionization is not without the occurrence of time-dependent phenomena.[502,503]

5.2.3. Matrix Effects

It is not surprising that reports of significant matrix effects in MS/MS analyses have appeared, but unsettling that they have so seldom been discussed. Voyksner et al.[504] have shown that the signal for selected reaction monitoring of the m/z 320 to m/z 257 transition for a tetrachlorodibenzodioxin is reduced by one-third when 100 pg of a standard of the dioxin is mixed with 1 μg of methyl stearate in the direct insertion probe. This effect reflects a change in the ionization conditions, as the ionization of dioxin is suppressed by the addition of diluents. Several types of matrix effects are known in MS/MS. The first is that just described, in which a change in the composition of the mixture introduced to the source decreases the ionization efficiency for the compound of interest. With fewer ions available for MS/MS analysis, the signal measured in the MS/MS experiment will perforce decrease. The second form of matrix effect is more subtle. In these cases, the presence of the matrix alters the MS/MS data by changing the structure or reactivity of the ion selected for MS/MS analysis. To return to a familiar example, the site of protonation in a molecule can be evident in the daughter ion MS/MS spectrum. Change in the composition of a mixture in the source may change the extent of protonation at one site or the other, altering the quantitative results that might be obtained. Practically, changes in the absolute amount of material vaporized into an ionization source may change the instantaneous pressure.[505] Higher source pressures can decrease the reactivity of the ion through collisional stabilization.

Finally, a third type of matrix effect should be noted. A discussion of the concentration step in sample preparation and its implications for GC/MS and MS/MS have been provided in the preceding section. There are a few examples in which the concentration step is detrimental to the analysis. Such a situation arises when components of interest in a mixture are intermolecularly reactive. When dispersed in the inert matrix, they remain isolated and unreactive and can be identified by MS/MS. Attempts to concentrate the material of interest increases the relative concentration of these components, the intermolecular reaction proceeds more rapidly, and the compound of interest may be lost.

5.3. Interpretation of MS/MS Spectra

The interpretation of mass spectra is based on a correlation of dissociation processes with the structure of the molecule introduced into the source of the instrument. Simple bookkeeping methods were first developed to tabulate the losses, and "identification" of the sample structure ultimately depended upon a comparison of the spectrum to that of the authentic compound. As the spectra of authentic compounds began to accumulate, rules of fragmentation were developed, and the mass spectral fragmentation behavior of important classes of compounds became apparent. An appreciation of the underlying chemistry of gas-phase unimolecular fragmentation led to conceptual advances such as the *ortho* effect, Stevenson's rule, and the McLafferty rearrangement. The mass spectral database grows more rapidly than the number of compounds available for analysis, because several ionization methods are likely to be applied to each compound. Standard operating procedures for electron ionization mass spectrometry, at least, were developed, and databases of electron ionization mass spectra were created. Computers now search these files to match the current spectrum against those in the library file and to suggest a compound identification. Several automated systems have been developed in an attempt to codify and to extend the interpretive ability of the mass spectrometrist. However, innovations in computer hardware and relatively simple data manipulation software far outpaced developments in the programming field. Much routine mass spectral interpretation, for electron ionization mass spectrometry, at least, begins with a computer search of the mass spectral library.

Interpretation of MS/MS spectra has developed along similar lines, with the added complication that there are noticeably more experimental parameters that exert a drastic effect on the MS/MS spectrum. The development of MS/MS has included a progression from visual spectral comparison, to standardization and library creation, library searching, and to the application of automated search systems.

As an example, Figure 5-4 revives an early example of MS/MS in which papaverine was identified in raw opium on the basis of the similarity in the daughter ion MS/MS spectra.[506] Since papaverine is known from other experiments to be present, the agreement in the spectra is certainly not fortuitous. Nevertheless, simple visual comparisons, even when augmented by rationalizations of the fragmentation processes, are fundamentally unsatisfying. The need is apparent for a standard set of conditions under which MS/MS spectra might be obtained, compiled, and searched with the computer systems that had been developed for mass spectral interpretation. The following discussion summarizes the situation for high- and low-collision-energy MS/MS spectra.

5.3.1. High-Energy MS/MS Spectra

MS/MS data obtained in high-energy collisions have for the most part been obtained with BE instruments, with the process of collision-activated dissociation occurring in a collision cell (a reaction region) located between the sectors. Linked scans now make it possible to probe reactions occurring in other field-free regions of the instrument (see Chapter 2), but for the purposes of MS/MS spectral interpretation, this discussion will be limited to the type of spectra shown in Figure 5-4. These MS/MS spectra are obtained with instruments of BE geometry and are the result of processes of dissociation occurring in the second reaction region of the instrument.

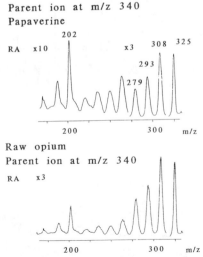

Figure 5-4 ■ Identification of papaverine in raw opium based on a comparison of daughter ion MS/MS spectra obtained on a BE instrument.

As described in Chapters 2 and 3, the (amplified) kinetic energy release characteristic of a fragmentation process is measured by the kinetic energy analyzer (E) in the BE instrument. Information about the energetics of the ion dissociation process is of fundamental interest. However, the width of the peaks observed in the kinetic energy spectrum so obtained sometimes makes it difficult to assign masses and relative abundances accurately in the daughter ion MS/MS spectrum, especially if ions are formed at adjacent masses.

The energy resolution of a typical electric sector is perhaps a few tenths of an electron volt. An MS/MS spectrum is obtained by setting the voltages on the electric sector plates to pass the mass-selected parent ion. These voltages are then decreased to transmit the daughter ions in order of decreasing kinetic energy, which is also in order of decreasing mass. The scanned axis is not mass per se, but a voltage that is proportional to daughter ion masses scaled from 0% to 100%, for which 100% is the electric sector voltage at which the parent ion is passed. The number of resolution elements remains constant irrespective of the mass of the parent ion selected. Given that there may be 1000 resolution elements in the range of 0 to 250 V and given a peak with a typical kinetic energy release value, the mass of the daughter ion can usually be established to within a value of 0.3 daltons for a parent ion of 500 daltons. Mass assignment accuracy drops to 0.6 daltons for a parent ion of 1000 daltons, and less than 1 dalton accuracy at 1600. Furthermore, nominal mass assignment assumes that there is no kinetic energy lost during the collisional activation process, which is not strictly the case for CID[221] (see Chapter 3). Figure 5-5 illustrates this effect for two daughter ion MS/MS spectra obtained with the same instrument, but for parent ions of different masses. MS/MS analysis of higher mass ions requires that special proce-

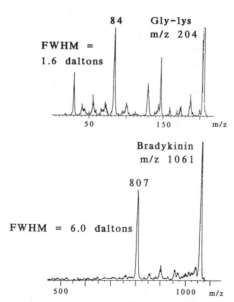

Figure 5-5 ■ Accuracy of mass assignment in daughter ion MS/MS spectra obtained on a BE instrument.

dures be used to increase the effective mass assignment accuracy for the daughter ions, such as floating the collision cell at voltages different from the accelerating voltage of the source.[58,507]

When daughter ions of adjacent masses are formed, the kinetic energy analysis of daughter ion masses can result in overlapping signals for the ions. When the daughter ions are of near equal abundance, the overall peak shape can be deconvoluted to provide the individual peak shapes.[508] When the difference in relative abundances of the daughter ions becomes large, the presence of the ion of lower abundance may not be apparent, since the signal is a small component on one side of a much larger signal.

In most MS/MS spectra obtained with BE instruments, the relative abundances of the daughter ions are not reported with high precision. An accurate assessment of the relative abundances of the daughter ions requires an area rather than a peak height calculation, which presumes a knowledge of the peak shape. Several peak shapes for daughter ions can be observed in the metastable or daughter ion MS/MS spectrum. The daughter ion MS/MS spectra of nitrogen- and phosphorus-containing parent ions often contain signals for ions produced by a charge-stripping process, i.e., the oxidation of the singly charged mass-selected parent ion to the doubly charged parent ion (see Section 3.5.2.2). The mass is the same, but since the charge has doubled, the kinetic energy-to-charge ratio is halved, and the signal for this ion appears at $0.5E$, where E is the electric sector voltage required to pass the parent ion beam through the sector.[337] This signal is typically very narrow in width, as the oxidation reaction does not involve a kinetic energy release. The daughter ion MS/MS spectrum given in Figure 5-6 shows that the relative abundance of the charge-stripping ion appears to be about 10% if peak heights are compared. When calculated using the peak area, the actual relative abundance is far less than 1%, reflecting the relatively low cross section for the charge-stripping reaction.

In a BE instrument, the daughter ions formed as the result of a collision-activated dissociation are not of equal kinetic energy, with the kinetic energy distributed among

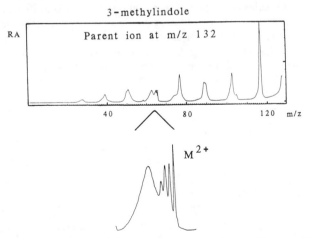

Figure 5-6 ■ Relative abundances of charge-stripping peaks in daughter ion MS/MS spectra obtained on a BE instrument.

the products of the dissociation reaction in direct proportion to mass. Although all ions are accelerated into the first dynode of the electron multiplier typically used as a detector (provided that the potential of the first dynode is of opposite polarity), their impact velocities will differ, and the response of the electron multiplier varies with the kinetic energy of the impacting ion.[509] A scaling factor based on the relative kinetic energy of the daughter ions has been applied to generate relative abundances of ions used in isotope-resolved MS/MS experiments.[510] However, since the detection efficiency is a function not only of the kinetic energy of the ion, but also its structure,[509] scaling is not routinely applied, and reported relative abundances of daughter ions in the daughter ion MS/MS spectrum can be highly instrument dependent. The recent use of conversion dynode electron multiplier systems[511] and postacceleration detectors[512] have further complicated the assignment of intrinsic abundances to daughter ions, both for BE instruments and for the multiquadrupole instruments.

Rumpf et al.[513] have recently discussed these factors in detail for daughter ion MS/MS spectra measured with a BE instrument. The collection efficiencies of daughter ions of different mass formed from the dissociations of parent ions in the second reaction region were calculated based on ion trajectories through a series of defined slits and apertures. Several situations arise in which a rigorous calculation of ion abundances corrected for collection efficiencies are required. The first is in comparisons of theoretical with experimental data. Failure to consider the mass discrimination factors in the daughter ion collection efficiencies tends to underestimate the abundances of the lower mass daughter ions. A second situation in which the corrected ion abundances must be used is in a consideration of isotope effects in the mechanisms of the collision-activated dissociation. The mass discrimination effects are in many cases larger in magnitude than the isotope effects, and serious misinterpretations of the chemistry may result from the use of the uncorrected data. Finally, the relative paucity of lower mass daughter ions formed by dissociations of high mass parent ions (peptides, for example) can be attributed in part to a mass discrimination in the transmission of lower mass ions into the electric sector of a BE instrument.

McLafferty et al.[166, 514] have emphasized that the relative abundances of the daughter ions formed by collision-activated dissociation in an MS/MS spectrum should be considered separately from those that are formed from metastable ions. The usual method for the determination of the metastable ion abundances is a pressure-resolved study (Figure 5-7) that assigns metastable ion abundances by extrapolation to zero-collision gas pressure in the cell. Other methods are based on a study of peak shapes[515] or

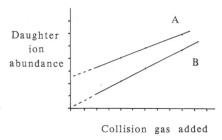

Figure 5-7 ■ Data from a pressure-resolved plot determines the metastable abundances of daughter ions.

on the shift in ion signal with a potential applied to the collision cell.[516] Such differences in ion origins are seldom pursued in the MS/MS analysis of mixtures, in accordance with the simple comparison of spectra that is often the end use of the data.

Despite difficulties in assessing the masses and relative abundances of daughter ions in an MS/MS spectrum acquired with a BE instrument, a library of such spectra has been collected.[517] Most of the entries are for lower mass ions of interest in ion structural characterizations and for molecular structure assembly.[518] In contrast to the bar-type daughter ion MS/MS spectra produced by multiquadrupole instruments, which can easily be formatted for computer storage, these spectra are more difficult to store accurately. The noise level in the spectrum is typically higher, which means that for practical reasons, there will be some signal at each mass. Furthermore, the retention of accurate peak shapes requires that the value acquired in each resolution element (the 1000 channels of energy resolution described previously, or more, if available) be retained. The only high-energy MS/MS library assembled to date[517] is not so configured. The spectral data stored in this library is a listing of peak masses and abundances, with a few selected instrument parameters and the literature reference stored with each entry. Advances in computer storage and retrieval may eventually make it possible to store the data in a more sophisticated manner.

5.3.2. Low-Energy MS/MS Spectra

Most low-energy daughter ion MS/MS spectra are obtained with multiquadrupole instruments operating in the ion kinetic energy range of 5 to 150 eV, and the discussion here will be so directed. Interpretive procedures and libraries developed for MS/MS data obtained with such instruments should be applicable to hybrid instruments fitted with a collision quadrupole and a final quadrupole mass analyzer (see Chapter 2). It should not be assumed that conclusions drawn from experiments on such instruments extend to others that vary substantially in dissociation efficiency, ion transmission, and ion collection parameters.

Commercial multiquadrupole instruments are interfaced to sophisticated data collection and processing computers. Libraries of standard electron ionization mass spectral data are usually included, and provision made for the addition to the library file of mass spectra generated by the user. Since MS/MS data generated by the multiquadrupole instruments are in the same format as the electron ionization mass spectral data, i.e., a two-dimensional array of relative abundance versus mass-to-charge ratio, the generation of a daughter ion MS/MS spectral library is straightforward. The same search routines used for the regular mass spectral library can be used to access data in the MS/MS library, since the routines do not depend upon an a priori identification of the molecular ion in a mass spectrum or the mass-selected parent ion in an MS/MS spectrum.

Almost immediately after the first in-house MS/MS libraries were created, discussions were held at the annual meetings of the American Society for Mass Spectrometry to establish standard operating conditions so that library data could be shared among the growing numbers of MS/MS users. A 1982 workshop[519] organized by Dawson addressed this issue. Basic difficulties arise because low-energy daughter ion MS/MS spectra are a strong function of the instrument operating conditions, including collision energy and target thickness, defined to be the product of the collision gas pressure and the length of the collision cell (see Sections 3.3.1.4 and 3.3.1.5). Details of the operation of the first and third quadrupoles vary with the manufacturer and can also affect the daughter ion MS/MS spectrum recorded. In a situation parallel to that developed for the searching of electron ionization mass spectral libraries,[520] an argument was put forward that the absence or presence of a daughter ion at a particular mass-to-charge ratio provided sufficient information in a library search, especially considering the variable relative abundances reported under nominally the same conditions of instrument operation. The consensus was to retain the relative abundance information; most matching algorithms place a secondary importance upon it in any case.

The 1982 workshop provided the impetus for a round-robin test for spectral comparison of a few selected compounds to assess the reproducibility and compatibility of spectra generated on different multiquadrupole instruments. The results were summarized at the 1983 meeting of the American Society for Mass Spectrometry[521] and also published.[522] Seven laboratories responded to the call for data, representing two brands of commercial instruments. The two parent ions investigated were n-butylbenzene ($C_{10}H_{14}^+$, m/z 134) and protonated dimethylphthalate ($C_{10}H_{11}O_4^+$, m/z 175). Butylbenzene had been previously used as a model system to compare the effects of internal energy on the dissociation to daughter ions at m/z 91 and m/z 92.[193,222,240,443,523] For low-energy collisions, it had been established that the $91^+/92^+$ ratio in the daughter ion MS/MS spectrum of n-butylbenzene was sensitive to the collision energy, but was not a strong direct function of the target thickness. Other ion ratios in the daughter ion MS/MS spectrum, notably $65^+/91^+$, are a strong function of the target thickness.

Standard operating conditions were established as follows. A sufficient pressure of either argon or nitrogen was added to the collision region to attenuate the parent ion beam intensity to first one-third and then one-tenth of its original value. Then, for each case, the daughter ion MS/MS spectra of n-butylbenzene and protonated dimethylphthalate were recorded as a function of collision energy. Figure 5-8 shows the data recorded from the various laboratories for the logarithm of the $91^+/92^+$ ratio at the two chosen target thicknesses. The agreement is moderately good between the various sets of data and suggests that with careful control of the instrumental parameters, a standard library of daughter ion MS/MS spectra could be generated.

These initial results suggested, however, that the instrument parameters as recorded from the instrument readouts are not a sufficiently reproducible means of specifying library-quality operating conditions for low-energy MS/MS analysis. A change was suggested[521] that resulted in standard definitions of collision energy and target gas thickness to attain a specified ratio of $91^+/92^+$ from the molecular ion of n-butylbenzene. Ideally, a parent ion must provide (1) two daughter ions similar in mass for which the relative abundances are a strong function of the collision energy and not the target thickness, and, conversely, (2) two daughter ions similar in mass for which the relative abundances are a strong function of the target gas thickness and not of the collision energy. To date, no candidates have been suggested. Indeed, none may exist that fulfill the criteria as stated, which may not be separable. The results of a proposed second round robin of analysis using the new definition of standard operating procedure have not been reported, and the availability of a general MS/MS library remains uncertain.

Martinez and Dheandhanoo[524,525] have emphasized that the creation of MS/MS libraries is contingent upon the development of instrument-independent operating conditions for the collection of data. Such a requirement is especially stringent in the case of the triple-quadrupole instrument, in which parameters such as the axial kinetic

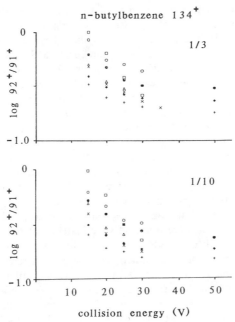

Figure 5-8 ■ Data comparison for low-collision energy daughter ion MS/MS spectra obtained in a round-robin study.

energy of the ions, the target gas thickness, and the amplitudes of the rf voltages applied to the rf-only quadrupole can have significant effects on the abundances of the ions observed in daughter ion MS/MS spectra.[64,125] Martinez and Dheandhanoo suggest that a series of ion dissociation reactions of known cross sections be used to develop conditions of instrument operation such that the measured MS/MS data can be correlated without regard to instrument. First results suggest that the resonant charge exchange reaction of argon ions colliding with argon collision gas can serve to standardize target gas thickness in the triple-quadrupole mass spectrometer. A round-robin evaluation of instrument results is underway.[526] The results of this comparison may lead to better lab-to-lab reproducibility in MS/MS data. However, the large number of commercial instruments already in the field are installed and operated with various preselected parameters not likely to be changed. A large and diverse MS/MS library, especially with data from triple-quadrupole instruments, is unlikely to appear in the near future.

5.3.3 Automated Systems

An automated search system has been described that attempts to match an unknown daughter ion MS/MS spectrum against MS/MS spectral data stored in a library or against the data in a standard library of electron ionization mass spectral data.[527,528] Most of the candidate compounds in the library are eliminated from consideration in a prescreening step. The remaining files are analyzed to produce match factors that gauge the degree of correlation between the library spectrum and the comparison spectrum.

The automated search system has been used to estimate the structural information contained in daughter ion MS/MS spectra by an empirical identification of matches between characteristic ions and the structures that they represent. This use of the system is predicated on a library of low-collision-energy MS/MS data of sufficient size to highlight patterns of ion/structure correlation. As the system develops, it will include access to a structure generator that will allow the computer to suggest molecular substructures based on the patterns observed in the MS/MS spectra. At present, this determination remains the province of the chemist.

The application of automated computer programs to the interpretation of daughter ion MS/MS spectra, in particular those of mixtures, has been described by Cross and Enke.[529] If the parent ion selected in an MS/MS experiment is a mixture of two or more isobaric ions, then the daughter ion MS/MS spectrum measured will be a combination of the discrete spectra. Recognition of the presence of isobaric overlaps in complex mixture analysis is requisite to accurate quantitation. A presearch of the spectrum eliminates many of the candidates from the library, and the spectra of a few likely components are then more intensively evaluated. Relative contributions of the individual spectra are scaled and added to match the observed spectrum as closely as possible. A residual spectrum is calculated, and the process repeated until all components in the parent ion mixture have been resolved, or until the residual spectrum cannot be effectively matched against the data in the MS/MS library. This automated search system has been further refined to include not only the information available in daughter ion MS/MS spectra, but that in parent ion and neutral loss MS/MS spectra as well.[530] The algorithms developed to identify the patterns of ion/structure correlation are applied to a training set of MS/MS data, and then multiple clues to specific molecular substructures are identified by cross correlation. A comprehensive review of this work has recently become available.[531]

Giblin et al.[532] have applied factor analysis to the correlation of daughter ion MS/MS with structure. In composite daughter ion MS/MS spectra of isomeric or isobaric parent ions, there may exist no daughter ions uniquely characteristic of each component in the parent ion mixture. The algorithm normalizes the peak intensities within a daughter ion MS/MS spectrum and then performs a multiple linear least-squares regression of the mixture data against similar data stored in a library file. The program determines the relative mole fractions along with the standard errors of the determination. The overall error of calculated mole fractions was about 15% for 1 : 1 mixtures of ions, but increased to about 60% for the smaller component in 30 : 1 mixtures. Examination of the spectral data indicates that variation with time and with collision gas pressure provides sufficient variation to account for these errors. The programs described are noteworthy in that they deal with daughter ion MS/MS spectra measured with a BE instrument.

It should be noted that there is some controversy over the validity of the assumption that the daughter ion MS/MS spectra of isomeric/isobaric parent ions can be considered as a linear combination of the individual daughter ion MS/MS spectra.[515,533] In general, the experimental errors of measurement are at least as large as errors that arise because the system is assumed to be linear when in fact it is not. Errors may also arise if the CAD cross sections of the parent ions are substantially different, as they are likely to be in most situations.[534]

The extensive application of artificial intelligence systems to MS/MS data can provide new capabilities. Palmer and Enke[535] have recently described an algorithm that can determine the empirical formula of a parent ion by examination of the daughter ions formed in the MS/MS spectra of the ^{13}C isotopes of the parent ion. The number of carbon atoms in the parent ion is then established by comparison of the intensities of the isotopic and nonisotopic daughter ions. The carbon number information is passed on to a second program that generates self-consistent empirical formulas. Although described to date only for determination of the number of carbon atoms, extension to other isotopic atoms is straightforward in concept, with limitations imposed by the dynamic range of the MS/MS measurement. Using this variant of the isotope-resolved MS/MS experiment, the number of candidate compounds in a library search routine can be decreased accordingly. The use of isotope-resolved MS/MS data to generate empirical formulas for parent ions was described earlier by Bozorgzadeh et al.[536]

More recently, Palmer et al.[537] have described a method for analyzing patterns in mass spectra and MS/MS data. A pattern recognition/artificial intelligence program is used to predict the presence or absence of defined substructures in unknown compounds. The system is predicated on the development of rules that correlate the features in the mass spectrum (specific losses, characteristic ions) with the presence of substructures identified in a training set of MS or MS/MS data.

Brand et al.[538] have described a knowledge-based tuning program for triple-quadrupole MS/MS instruments. The system algorithms were originally developed to aid in the acquisition of multivariable data from an MS/MS instrument and are extended to aid in the tuning of the instrument for daughter ion, parent ion, and neutral loss MS/MS spectra. Wong et al. suggest that increases in sensitivity of 2- to 30-fold are possible using the expert tuning system. Such algorithms are already incorporated into some commercial instrumentation.[539] More sophisticated computer control of the instrument allows real-time variation of the ion path parameters as a quadrupole mass filter is scanned, and the system can be optimized for any of the various MS/MS experiments.

Weber at al.[540] have described the use of pattern recognition techniques for analysis of daughter ion MS/MS spectra of eight isomeric C_9H_{12} isomeric alkylbenzenes. Using standard chemometric software, the pattern recognition algorithms generated a differentiation between four different types of daughter ion MS/MS spectra corresponding to n-propyl-, isopropyl-, ethylmethyl-, and trimethylbenzene molecular structures. A clear differentiation of the type of alkyl substitution on the ring could thus be obtained, but the position of the substituent on the ring could not be determined. It is noteworthy that the pattern recognition algorithms were applied to the daughter ion MS/MS data of parent ions created by electron ionization, that the electron ionization mass spectra themselves are very similar, and that the high-collision-energy daughter ion MS/MS spectra also appear similar. Application to data sets derived from other ionization methods, or other MS/MS experiments, is expected to become an exciting area of research in the next few years as computational capabilities no longer become the limiting step.

6

Analytical Applications

Applications of MS/MS to fundamental chemical studies, including those of small organic ion structures, reaction mechanisms, and ion thermodynamics, have been addressed in Chapter 4. Applications to problems of targeted component analysis, complex mixture analysis, and general structural analysis are covered in this chapter, which is divided into nine sections that reflect the uses of MS/MS in environmental problems, bioorganic structural analysis, the analysis of industrial chemical products, foods and flavors applications, forensic chemistry problems, application to fuel and petroleum chemistry, geochemical applications, agricultural research, and, finally, process control applications. This chapter is not intended to be a complete compilation of reported applications of MS/MS in these areas. It concentrates on a few selected examples in each area to provide a useful description of the capabilities and limitations of the method. Both sector and quadrupole instruments have been widely used, and in most cases, data of comparable value are obtained. Certain instances require higher resolution of the parent ion or unit resolution of the daughter ion; these will be specifically noted.

6.1. Environmental Applications

Analyses of samples of environmental origin are of two general types. Determinations of the first type are concerned with members of a list of compounds of interest established by decree or regulation. The MS/MS experiment is expected to provide information about the quantitative distribution of these targeted components in any particular sample. An example of this type of analysis is the monitoring of chemicals on the priority pollutant list of the U.S. Environmental Protection Agency. A determination of the second type focuses more specifically on one particular class of compounds, or perhaps one specific compound, the quantitative determination of which must occur in the presence of varying amounts of closely related compounds, including isomers. An example of this second type of analysis is the determination of tetrachlorodibenzodioxins as a class, and/or the much publicized toxic 2,3,7,8-tetrachlorodibenzodioxin in particular, both in the presence of polychlorinated biphenyls or other halogenated interferences.

6.1.1. Priority Pollutant Analysis

In a rigorous study and excellent example of the analytical power of MS/MS, Hunt et al.[541-543] described a master scheme for the direct analysis of organic compounds (the priority pollutant list compiled by the EPA) in environmental samples. The scheme uses chemical ionization and daughter ion, parent ion, and neutral loss MS/MS experiments with a triple-quadrupole mass spectrometer, taking full advantage of the sophisticated computer control of that instrument. Most of the chemical reaction and chromatographic separation steps of the competing GC/MS method are eliminated, although some derivatization steps remain. The protocol provides for detection of both knowns and unknowns by molecular weight and functional group. A total analysis time of under 30 min/sample is quoted. The derivatization reactions are used for phenols, amines, and polynuclear aromatic hydrocarbons and promote dissociation chemistry specific to these functional groups. Since no separation step is involved in the MS/MS analysis, the results are functional-group-specific, but often not isomer-specific. Such isomer differentiation still requires the use of chromatography/mass spectrometry.

Figure 6-1 is an example of the data generated in this comprehensive analytical scheme. Collision-induced dissociations of $(M + H)^+$ ions formed from phthalates include formation of a daughter ion at m/z 149, the protonated phthalic anhydride. This daughter ion is characteristic of the phthalate structure (except for dimethylphthalate), and a parent ion MS/MS spectrum of this selected daughter ion should indicate

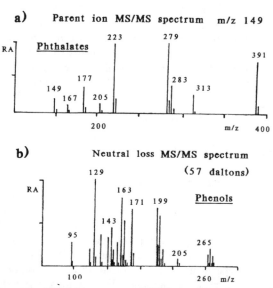

Figure 6-1 ■ (*a*) Parent ion MS/MS spectrum of the characteristic ion at m/z 149, highlighting all phthalates in a complex mixture. (*b*) Neutral loss (57 daltons) MS/MS spectrum to characterize carbamate derivatives of phenol in a complex mixture. Data were obtained with a triple-quadrupole MS/MS instrument.

the masses of the protonated molecules of phthalates present as mixture components. Figure 6-1a is the parent ion MS/MS spectrum of m/z 149 in a sample of otherwise uncharacterized industrial sludge. The ions at m/z 223, 279, 313, and 391 are the protonated molecules of diethyl-, dibutyl-, butylbenzyl-, and either di-n-octylphthalate or di-2-ethylhexylphthalate, respectively. The inability of MS/MS to differentiate between the last two possibilities is an example of the need for a chromatographic separation for unambiguous identification. Additional ions observed in the parent ion MS/MS spectrum at m/z 167, 177, and 205 correspond to phthalate fragment ions that also react to form the ion at m/z 149. Each of the indicated compounds was present at levels of 1 to 5 ppm in the sludge.

A neutral loss (57 daltons) MS/MS spectrum is given in Figure 6-1b. Phenols in a lyophilized sample are treated with methyl isocyanate to form carbamates. In low-energy CID, carbamates are generally characterized by loss of 57 daltons (methyl isocyanate, CH_3NCO) and regeneration of the phenol. A neutral loss (57 daltons) MS/MS spectrum therefore directly identifies phenols in the complex mixture, which in this example is the same industrial sludge analyzed for phthalates (Figure 6-1a), now spiked with seven different phenols at the 1-ppm level. The signals observed in the MS/MS spectrum correspond to the protonated molecules of phenol (m/z 95), dimethylphenol (m/z 123), chlorophenol (m/z 129), chloromethylphenol (m/z 143), dichlorophenol (m/z 163), trichlorophenol (m/z 197), and pentachlorophenol (m/z 265). Note that the chlorine isotope ratios are observed in the neutral loss MS/MS spectrum, and these can be interpreted directly to indicate the number of chlorines in the parent ion, since the loss of 57 daltons does not involve the chlorine atoms. Note that, in general, the intensities of the peaks in a parent ion MS/MS spectrum do not accurately reflect the relative concentrations of the components in the mixture, since each component will have a different cross section for the monitored dissociation.

6.1.2. Polyhalogenated Compounds

Polyhalogenated compounds are well-known environmental pollutants. They are distributed throughout the environment not only as a result of their previous widespread use in chemical and electrical industries, but also because they can be formed as the result of incineration of municipal refuse containing chlorinated organic material such as polyvinyl chloride. Since members of some of the compound classes and some particular isomers within those classes are acutely toxic, much analytical effort has been expended in developing rapid, sensitive, and specific protocols for the analysis of large numbers of samples that may contain these compounds. The potential of MS/MS in such work, especially as a screening method prior to a more definitive analysis by GC/MS or LC/MS, is great, and most of the work has been directed to that end.

Early research focused on the use of MS/MS to differentiate isomers of the polyhalogenated compounds, since the toxicity was known to vary widely with the positions of halogen substitution. In the case of dioxins, such work predated the development of the gas chromatographic columns and the conditions necessary to separate completely all of the tetrachlorodibenzodioxin isomers.

The propensity of many aromatic hydrocarbons to undergo hydrogen/deuterium equilibration after ionization prompted several workers to investigate the preservation of halogen positional identity through the processes of ionization and unimolecular and collisional dissociation. Of particular interest was the preservation of structure in dissociation reactions occurring in the field-free region of a mass spectrometer. Early work of Safe et al.[544] indicated a randomization of chlorine among the substituent positions in small aromatic polychlorinated molecules. Further work[545] demonstrated isomer differentiation in metastable ion dissociations and suggested that MS/MS might be able to provide some degree of isomer specificity in the analysis of mixtures of closely related polyhalogenated compounds.

The expectation that MS/MS could be used to differentiate isomeric ion structures for polyhalogenated compounds was also supported by daughter ion MS/MS spectra of parent ions from polychlorinated 2-phenoxyphenols (high-energy collisions on a BE geometry instrument).[546] These compounds undergo thermal and photochemical ring closure to the toxic polychlorinated dioxins, with a yield as high as 5%. Three isomeric tetrachloro- and three isomeric pentachloro-2-phenoxyphenols could be readily distinguished by comparison of the daughter ion MS/MS spectra of the parent $(M - H)^-$ ions formed by negative ion chemical ionization. Interestingly, the daughter ion MS/MS spectra of the positive ions of the same parent ions formed by charge inversion were identical. The daughter ions formed as a result of the dissociation of the charge-inverted ion are not isomer-specific (see Section 3.5.2.1).

One early study probed the mechanisms of CID by careful study of the kinetic energy release (high-energy collisions, BE instrument).[547] Although the presence of *ortho* and *meta* chlorines in polychlorinated biphenyls could be inferred from kinetic energy release data, a complete differentiation of all the structural isomers was not possible. Since these early reports, chromatographic methods for the separation of many of the positional isomers of the polyhalogenated compounds have been reported, and little recent MS/MS work for the sole purpose of isomer differentiation has been completed.

Much work has been directed to the analysis of dioxin residues in environmental samples. Chess and Gross[548] reported a method based on monitoring the dissociations of the metastable parent ions of dioxins ionized by electron ionization. The transitions monitored were $320^+ \rightarrow 257^+$ and $322^+ \rightarrow 259^+$, representing loss of COCl from the two isotopic forms of the molecular ion of a tetrachlorodibenzodioxin. Soil samples were obtained from the vicinity of a chemical industry landfill. After extensive sample cleanup, the detection limit for tetrachlorodibenzodioxin (no isomer specificity) was determined to be 20 pg at a signal-to-noise ratio of 2.5; this value corresponds to 5 ppt in the soil samples.

Harvan et al.[549] used a BE instrument and daughter ion MS/MS spectra to analyze 2,3,7,8-tetrachlorodibenzodioxin in extracts from air filter samples. Capillary gas chromatography was used to separate the ion signal of this particular isomer from all others. The specificity of the experiment is high, determined by the requirement that the signal appear in the proper retention time window, and that the masses of the reactants and products correspond to the selected reaction of $320^+ \rightarrow 257^+$ and $320^+ \rightarrow 285^+$, the losses of COCl and Cl, respectively. The response of the MS/MS experiment was linear over the investigated range of 4 to 100 pg of standard sample. The relative standard deviation was 9.5% for 100 pg of standard ($n = 3$) and 19% for 10 pg ($n = 4$). A signal-to-noise ratio of 8 is reported for the analysis of 5 pg of the standard dioxin sample in this MS/MS experiment. Levels of tetrachlorodibenzodioxins in the air filter samples ranged from 50 to 500 pg per filter; however, the gas chromatographic separation showed that the dioxins did *not* include the 2,3,7,8-isomer.

Daughter ion MS/MS spectra obtained on a BE instrument were used to detect a hexachlorobiphenyl and a tetrachlorodibenzofuran in blood and fat samples,[550] as part of an evaluation of methods that might be applied directly to samples subjected to little or no cleanup. Alternative methods to which a comparison was made were direct insertion probe high-resolution mass spectrometry, and gas-chromatography/high-resolution mass spectrometry. The authors conclude that MS/MS is not reliable for these analyses because of consistently and unpredictably high results relative to the values obtained with the other methods of analysis. The analytical agreement between methods was restored when gas chromatography was used in concert with the daughter ion MS/MS experiment. The problem to which these methods were applied is the identification of a targeted compound in a complex mixture of closely related compounds. The first separation step by mass selection *can* be carried out with high resolution, but this criterion was relaxed in this particular MS/MS analysis. The authors also note that the poor daughter ion resolution (due to the analysis of daughter ion masses by kinetic energy analysis) contributed to the susceptibility of the MS/MS method to interferences. More recent work with the analysis of polyhalogenated compounds by mass spectrometry[551,552] confirms the ubiquity of closely related compound interferences that require high resolution in both chromatography and mass spectrometry for complete removal.

Slayback and Taylor[505] have summarized the use of GC/MS/MS for the analysis of 2,3,7,8-tetrachlorodibenzodioxin and 2,3,7,8-tetrachlorodibenzofuran in environmental matrices. Sensitivity sufficient for analysis at the 1-ppb level in soil (10–20-g sample) and 3-ppt level in water (1-L sample) was established. Despite the development of extensive cleanup procedures for environmental samples, GC/MS of such "dirty" matrices does not provide the required specificity at these levels of analysis. Gas chromatography coupled with high-resolution mass spectrometry can provide these analytical levels of performance, but is not cost-effective for a large number of samples because of the time investment required. The advantage of GC/MS/MS in dioxin analysis is its ability to screen and to confirm, in a cost-effective manner and at a specified level, the absence or presence of 2,3,7,8-tetrachlorodibenzodioxin in environmental matrices. Sample cleanup is still required, but its extent is reduced. Specificity in the analysis is achieved by monitoring the abundances of several daughter ions in a selected reaction monitoring experiment. Report of a positive result requires that these daughter ion ratios fall within precise windows that correspond to the chemical behavior of the authentic dioxin. Confirmation and quantitation can be carried out at the 1.5-ppb level, with a percent standard deviation of 10% at that level of analysis.

Shushan et al.[553] have reported the application of MS/MS to a problem of environmental "process control" involving the monitoring of polyhalogenated compounds. A mobile triple-quadrupole mass spectrometer equipped with atmospheric pressure ionization was brought on site to a municipal incinerator. An MS/MS experiment was designed to screen for polychlorinated dioxins and furans formed as the result of incineration of chlorine-containing wastes. With a minimum of chromatographic cleanup, the use of isotopically labeled internal standards for quantitation, and a total instrumental time of 7.5 min/analysis, stack gases and fly ash were analyzed for these targeted compounds. With a 1.0-mg sample of fly ash, 1 pg of polychlorinated dioxin or furan could be reliably detected with a reproducibility of 10%. The experiment provides a screening analysis rather than a rigorous isomeric differentiation of these compounds. The on-site analysis was essential to the goal of adjusting the operating parameters of the incinerator to minimize the production of the polychlorinated dioxins and furans. It was concluded that the instrumental procedures were sufficiently sensitive and selective for a successful completion of this analysis, but that problems with sample selection, concentration, and pretreatment are the primary contributors to the stated level of reproducibility. A report[554] on the use of the mobile MS/MS instrument quotes a sample turn-around time of 20 min for on-site extraction and analysis of soil samples for 2,3,7,8-tetrachlorodibenzodioxin. The detection limits for GC/MS/MS were 0.3 ppb in soil, corresponding to 1 pg of the dioxin isomer injected onto the GC column.

6.1.3. Atmospheric Pollutants

There have been a few reports of the use of MS/MS to identify the oxidation products of organic pollutants released to the atmosphere. To date, these studies have been limited to laboratory systems and have not yet evolved to the analysis of genuine atmospheric samples. Mahle et al.[555] used daughter ion MS/MS spectra (high-energy collisions, BE instrument) to characterize the products of the hydroxylation reaction of benzene, toluene, xylene, and naphthalene under atmospheric pressure ionization conditions. This study suggested that such experiments might also be used to study the general reactions of aromatic hydrocarbons in low-temperature plasmas.

A variety of MS/MS experiments carried out on a triple-quadrupole mass spectrometer have been used to characterize the atmospheric degradation products of toluene.[556] Of particular interest in this study was the use of neutral loss and parent ion MS/MS spectra to subdivide the complex mixture into functional group classes. Isotopically labeled starting materials were used in conjunction with an appropriate shift in the neutral loss or mass of the daughter ion to differentiate true process signals from background signals and to separate by mass isomeric compounds that differentially incorporate the isotopic label. Toluene and nitrogen oxides (1–10 ppm) were irradiated for 24–30 h; the reaction vessels were rinsed with solvent, which was then reduced to a total volume of 20 μL, sufficient for 10 MS/MS experiments of various types. The *total* amount of reaction product in each run was 5 μg. In addition to confirmation of the known degradation products of toluene, several new products were identified, including hydroxy-4-oxo-2-pentanal, 5-oxo-1,3-hexadiene, and 4-oxo-2-pentenal. An intriguing experiment was briefly described that eliminated the product collection and concentration step. Using atmospheric pressure ionization, a few of the more abundant products could be identified directly in an irradiated air/toluene mixture. Sensitivity is the limiting factor in this experiment. Attempts to bring the levels of minor products in the reaction mixture above the limit of detection are self-defeating, as the higher concentrations of the compounds suppress the propagation steps in several radical-based reactions, thus changing the system under study.

Dumdei et al.[557] have described a direct air-sampling system using MS/MS for the analysis of trace volatile organic vapors in which ambient air was introduced directly into a chemical ionization source without prior entrapment or concentration. The limiting factor in these experiments is the instrument sensitivity. The pumping capacity of the source on the mass spectrometer determines the total volume of air that can be drawn into the source without raising the pressure in the mass spectrometer itself. In atmospheric pressure ionization, the ionization method itself provides some degree of sample concentration, since the ions are transported efficiently into the mass spectrometer and the bulk of the gas load is diverted. The mobile MS/MS instrument described earlier[554] has also been used for on-site analysis of air samples using atmospheric pressure ionization. Real-time detection limits for a broad range of chemical classes are in the low-ppt-to-low-ppb range (v/v) in ambient air.

6.1.4. Water Pollutants

GC/MS has been the method of choice in the analysis of organic pollutants in water sources. A few applications of MS/MS in this area have appeared. Boon et al.[558] have used MS/MS with a pyrolysis ionization source to characterize the particulate organic matter from several European river waters. The structural studies were aided by multivariant analysis of the patterns of organic structures, providing some insights into the origin of the organic matter and the extent of its degradation. Lindstrom et al.[559] have used GC/MS/MS and direct insertion probe MS/MS to determine the levels of 1,1-dichlorodimethyl sulfone in fish and mussels that resided in waters that received the effluent of a pulp mill bleach plant. A triple-quadrupole instrument was used with methane positive ion chemical ionization to assay the target compound in fish liver and mussel fat extracts submitted to a single stage of lipid cleanup. Reported levels were 1 mg/kg of fish fat and 0.5 mg/kg of mussel fat. Isotope-resolved MS/MS was used to confirm the presence and the levels of these compounds in these extracts.

Surfactants in river waters have been extracted and analyzed by MS/MS.[560] Liquid chromatography was used to separate classes of surfactants extracted from river and drinking water, and FAB ionization of the summed fraction was used to produce $(M + Na)^+$ ions selected as the parent ions in a daughter ion MS/MS experiment (BE instrument). The positions at which the long alkyl chains of the surfactant branched were apparent from the spectrum (see Section 6.3.3), and the identities of the compounds were established by comparison with the daughter ion MS/MS spectra of the authentic compounds. Schneider and Levsen[561] have used field desorption, direct exposure probe chemical ionization, and fast atom bombardment ionization methods, each in conjunction with daughter ion MS/MS experiments, to study surfactants in river surface waters, and the rates of their degradation. With the use of FD/MS/MS, no separation of cationic, nonionic, and anionic surfactants was necessary prior to the analysis, and quantitation was accomplished directly.

Rivera et al.[562] used activated carbon filters to trap organic material from a commercial water supply and fast atom bombardment and daughter ion MS/MS to identify nonionic and anionic surfactants, and polyglycols present. Rivera et al.[563] also used FAB/MS/MS to study dyes present in a municipal water supply drawn from a river that also served to dispose of wastes from a local dye and textile industry. Detection limits varied widely, depending on the structure of the dye. In general, about 100 to 500 ng was required for an unambiguous determination, providing a relatively poor sensitivity. Analysis was greatly complicated by the presence of surfactants and dispersants (also used by the dye/textile industry) as compounds in the matrix.

6.1.5. Indoor Air Pollution

In response to energy conservation measures, the rate of indoor/outdoor air exchange in modern buildings has been decreased to the point at which indoor air pollution is now of concern. Concentrating samplers have been used in conjunction with GC/MS to monitor levels of indoor air pollution. In such analyses, samples are collected on site, stored, and then transported to the analytical laboratory at which the analysis takes place some time later. MS/MS has also been used in a study of indoor air pollution; here the speed of the analysis was such that the concentrations of targeted components could be established in real time.[564] The experiment could then be modified to identify the source of the contamination by a search for related compounds. The instrument used was a commercial triple-quadrupole mass spectrometer equipped with a low-pressure chemical ionization source. The entire instrument is portable and so is brought directly into the house. The sampling device draws 90 L/min of room air, resulting in a 10-mL/min flow of air into the source of the mass spectrometer. Selected reaction monitoring experiments were used to monitor concentrations of toluene, methyl ethyl ketone, ethyl benzene and xylenes, 1,1-dichloroethane, and pentane and other C_5 hydrocarbons in room air.

The ambient air of eight single-family homes located near a hazardous waste site was monitored for the target contaminants. Two minutes of monitoring at each chosen site within the house, or each chosen site within the room, provided sensitivities such that these volatile organic compounds could be monitored in the low-ppb range and response factors of about ±45% relative standard deviation. Predictably, the relative concentrations of volatile solvents such as toluene and hydrocarbons increased as the sample was taken from near paint cans stored in the basement or from near other logical sources of volatile compound emission.

6.2. Natural Products Applications

The analysis of natural products deals perforce with the separation of complex mixtures and the combined techniques of GC/MS and LC/MS have traditionally been used by chemists in this field. The contributions of MS/MS in the study of natural products have been in three major areas. The first is in the determination of the structures of compounds found in natural products and in the differentiation of the structural isomers of those compounds. The second area of application is in the analysis of a selected compound or small set of compound in the complex mixture, i.e., a targeted compound analysis, often still combined with at least a rudimentary form of chromatography. The third area of contribution has been in the use of MS/MS to derive compound class information from a mixture of natural products, providing for example, the distribution of related alkaloids in plant extracts. This information has been used in chemotaxonomic classifications. The trend in chemotaxonomy is to rely on MS/MS for the preliminary sample workup and to use the more time-intensive methods of GC/MS and LC/MS for unambiguous confirmation. This section emphasizes MS/MS studies of alkaloids and toxins in natural products and ends with a summary of work in MS/MS-based compound and isomer identification. Games[565, 566] has provided an overview of general applications of mass spectrometry to natural products analysis.

6.2.1. Alkaloids, Lipids, and Other Naturally Occurring Compounds

A substantial body of work describes the use of MS/MS in the analysis of alkaloids found in natural products, both in the confirmation or refutation of previous work and in the extension of chemotaxonomy via mass spectrometry. Cooks has been particularly active in promoting the use of MS/MS in natural product studies. The first of his work dates to 1978, representing some of the first applications of MS/MS to problems other than ion structure. The notable advantages of MS/MS in such work are the speed and sensitivity of the analysis. The rate at which characteristic compounds can be traced in mixtures is increased by several orders of magnitude over previous methods, which typically involved several stages of sample extraction and separation before a sample of sufficient purity for structural analysis was available. Structural similarities in classes of related compounds found in natural products make the use of functional-group-specific experiments, i.e., the parent ion or neutral loss MS/MS scans, particularly attractive. Isomers of the same molecular formula can be distinguished by careful interpretation of the daughter ion MS/MS spectra, but also by the separation provided as the mixture is fractionally distilled from the direct insertion probe into the source of the mass spectrometer. The significance of the MS/MS data has extended beyond chemotaxonomy to touch upon the mechanisms of biosynthesis and the isolation of useful products from natural biological sources. Discussions of a few representative examples of this work follows.

Kruger et al.[567] described the first use of MS/MS (BE instrument) for the identification of alkaloids in cactus. Identification of ubine and mescaline in lipid-free extracts of the freeze-dried plant material was possible. Electron ionization and chemical ionization were used to form molecular ions and protonated molecules, respectively, for which the daughter ion MS/MS spectra were measured and compared to synthetic standards. The presence of mescaline and other trace alkaloids in cactus plant tissues was similarly determined by Pardanani et al.[568] with daughter ion MS/MS experiments. Of interest at the time these experiments were completed was the fact that the levels of these psychoactive drugs found in previously unstudied plant materials were as high as those found in plant materials under legal control as potential materials of drug abuse.

In 1978, Kondrat and Cooks[569] described further application of MS/MS to the direct analysis of alkaloids in plant tissues. Results were presented for the analysis of papaverine and morphine in raw opium, coniine in hemlock, and cocaine in coca leaves. Sensitivity was described as sufficient to detect and identify 1 to 10 ng of the alkaloid. The only sample preparation involved was a grinding of the plant tissue or raw sample under liquid nitrogen and then introduction of the mixture into the source of the mass spectrometer with the direct insertion probe. The amount of sample introduced into the source with each analysis was on the order of milligrams, and source contamination using this procedure was severe. Nevertheless, the information obtained from direct analysis of plant materials was thought to be of sufficient value to justify the frequent cleaning of the source. McClusky et al.[570] used the direct analysis daughter ion MS/MS experiment to analyze for the toxic compound gyromitrin and related compounds in mushroom tissues. Although the amount of sample loaded onto the direct insertion probe was 1 to 40 mg, spectra could be obtained from this single loading for several hours.

A classic example of the use of MS/MS in the analysis of plant materials, cited in several reviews, is the quantitative determination of cocaine directly in plant material.[571] Detection limits below 1 ng were demonstrated, with a quantitative accuracy of 30% at the nanogram level. With multiple-reaction monitoring, a total of 6 min was required for each analysis, using a temperature vaporization profile to evaporate the cocaine from the direct insertion probe. The detection of benzoylecognine, the major metabolite of cocaine, in urine at the 1-ng level was also demonstrated. Further studies by Youseffi et al.[572, 573] demonstrated that daughter ion MS/MS analysis could be used to map the distribution of cocaine in plant tissue with a spatial resolution as small as 1 mm^3. No sample preparation or prefractionation was needed, and the total analysis time was again only a few minutes per sample using multiple-reaction monitoring. Cinnamoylcocaine was also monitored in these experiments, and the ratio of the two compounds established for such diverse parts of the plant as the stems, the leaves, and the berries. The leaves contained the highest relative concentrations of cocaine, and the older plant tissues (the stems and the berries) contain the highest relative concentrations of the cinnamoylcocaine. Of interest in this experiment is the fact that plant tissues collected from two geographic sites were analyzed by MS/MS and that the ratio of the two compounds were characteristic of each site. The authors emphasized that the MS/MS experiment requires 1 mg of sample and that alternative procedures of extraction and concentration require much larger initial sample sizes.

Other psychoactive drugs found in plant materials have been analyzed by MS/MS. The presence of psilocin (the decomposition product of psilocybin) in mushroom tissues was demonstrated.[574] The plant steroid ergosterol and several other related compounds were identified by MS/MS analysis of a simple ethanolic extract of the mushroom tissue. Eckers et al.[575] used MS/MS to determine the presence of ergot alkaloids in fermentation broths. Several isomers of the alkaloids that do not provide distinctive electron or chemical ionization spectra were differentiated on the basis of their daughter ion MS/MS spectra. Overall, the most consistent results were obtained when liquid chromatography was used to clean up the sample mixture (as from the fermentation broth) and to separate the major components and when MS/MS was used for the identification of the components in a given LC fraction.

Extensive use has been made of MS/MS in the study of trace alkaloids in various species of cactus. The first example of this work was the use of a daughter ion MS/MS experiment to screen nine related Mexican columnar cacti for the presence of a series of tetrahydroisoquinoline alkaloids.[576] The alkaloids were identified in the plant material directly and also in progressively less complex fractions of a chromatographic cleanup. Significant improvement in the quality of the daughter ion MS/MS spectra was apparent as the sample preparation became more extensive; extensive sample preparation was necessary to identify alkaloids present in lower relative concentrations. However, the extraction and cleanup procedures used did result in the formation of an artifact from the isomerization of N-methyl- to 1-methyltetrahydroisoquinoline. Since the ratio of these compounds is an important descriptive value, the discovery of this isomerization, and the fact that it is avoided in the direct MS/MS analysis, was of particular note. Further MS/MS work[577] focused on the cactus species *Carnegiea gigantea*, known to contain many different alkaloids in both phenolic and nonphenolic fractions of a workup of the cactus extract. Daughter ion MS/MS experiments (BE instrument) using chemical ionization provided confirmation of the presence of alkaloids such as carnegine, arizonine, and dopamine. Further study was prompted by the identification of putative molecular ions for previously unidentified alkaloids. Directed by the MS/MS results, new alkaloids such as salsolidine and 1,2-dehydrosalsolidine were identified with careful chromatographic workup of the fractions. It was initially postulated that the dehydrosalsolidine may have been formed as an artifact during the extraction process. MS/MS spectra obtained directly from the plant material provided

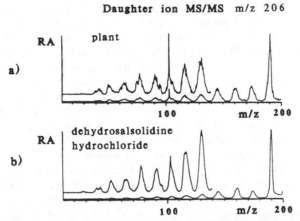

Daughter ion MS/MS m/z 206

Figure 6-2 ■ (a) Daughter ion MS/MS spectrum of m/z 206 found in the spectrum obtained directly from plant material. (b) The same spectrum obtained from m/z 206 of the standard sample (dehydrosalsolidine) synthesized as the hydrochloride salt.

evidence that such was not the case. Figure 6-2 illustrates the quality of the data from which these structural comparisons were drawn. Figure 6-2a is the daughter ion MS/MS spectrum of m/z 206 obtained directly from the plant powder, and Figure 6-2b is the daughter ion MS/MS spectrum of the standard material synthesized as the hydrochloride salt. Unit mass resolution of daughter ions on the multiquadrupole and hybrid instruments (vide infra) has made possible more rigorous spectral matches, provided that the instrumental conditions are held constant.

Twenty alkaloids, of which 13 were new, were identified in *Pachcereus weberi*, a columnar cactus.[578] The identities of the compounds were confirmed by synthesis and thin-layer chromatography. Several isomeric alkaloids were discovered. Although the MS/MS spectra of these isomers recorded on the BE instrument were difficult to distinguish, fractional distillation from the direct insertion probe provided a characteristic desorption profile of the material into the source of the mass spectrometer. In this analysis, 1 kg of plant material was extracted, the lipids were removed, and a fraction collected from a simple chromatographic cleanup that contained a mixture of all the alkaloids originally present in the plant sample. A total of 20 mg of alkaloid-containing fraction was analyzed by mass spectrometry. Chemical ionization with isobutane as the reagent gas was used to create protonated molecules for each of the various alkaloids present in the mixture. Significantly, the MS/MS data also indicated that compounds previously reported to be present in this plant material may have been mistakenly identified. The detection limits for the various new and confirmed alkaloids are estimated at 10 to 30 ng. Since the initial sample workup used 1.0 kg of material, this analysis represents the identification of new natural products at the sub-ppb level. An MS/MS study of the alkaloids in *Backbergia militaris* identified a number of new alkaloids, including fully aromatic oxygenated isoquinolines, their di- and tetrahydro analogs, and various isomeric phenethylamines.[579] The total sample consumption for the project, which involved thin-layer chromatographic separation and confirmation of the identifications made by MS/MS, was 10 g of plant material. The strategies used in the MS/MS approach have been summarized in a recent review.[580]

Many compounds that occur in natural products are nonvolatile or degrade upon heating. The electron and chemical ionization methods used in the examples cited previously both require that the sample be evaporated in the ion source to produce the parent ion for the MS/MS analysis. This volatility limitation was illustrated in the study of the mushroom constituents[575] in which psilocyben could not be identified directly, but only through its degradation product psilocin. Newer ionization methods such as laser desorption, fast atom bombardment, and secondary ion mass spectrometry produce ions from nonvolatile samples directly, without the need for sample evaporation (see Section 5.2.1). Several examples of the applications of these desorption ionization methods in natural products studies with MS/MS have been reported. Daughter ion MS/MS spectra (BE instrument) were used to identify the quaternary alkaloid candicine chloride in crude extracts of cactus tissue at the 1-ppt level.[581] The intact cation of the ammonium salt was formed by laser desorption and selected as the parent ion. The use of MS/MS analysis avoided the tedious procedures usually required for the isolation of these alkaloids, and the speed of the analysis made possible a survey of the cactus tissue extracts from various species for candicine chloride and structurally related alkaloids. A total of 3 g of plant material was used for the MS/MS analysis. Sample preparation consisted of grinding the dried plant sample, extracting with ethanol, and then partitioning the sample between chloroform and water. The alkaloid residues were analyzed from the water fraction.

Secondary ion mass spectrometry also provides ions from compounds such as the ammonium salts that can degrade upon heating. The enhanced secondary ion yield for preformed ionic compounds in the presence of a matrix of uncharged material makes possible a direct analysis of the plant extracts to identify the intact cations of compounds present as ionic ammonium salts. Compounds present in this form can also be examined by laser desorption MS/MS in a complementary experiment. Methylated analogs of candicine, identified in several cactus species, may exist in several isomeric forms. Daughter ion MS/MS experiments with both laser desorption[581] and secondary ion mass spectrometry[582,583] demonstrate that the isomeric differentiation is possible in some cases.

As Figure 6-2 illustrates, the daughter ion resolution obtained with a BE MS/MS instrument is low. Multiquadrupole instruments operate in a lower collision energy regime and can provide fragment ions different from those formed in high-energy collisions. In most cases, however, a general similarity in the pathways of dissociation is noted, and the MS/MS spectra prove to be equally valuable. However, multiquadrupole instruments, unlike the BE instrument, also provide unit mass resolution of the daughter ions.

Both high-collision-energy (BE) and low-collision-energy (QQQ) MS/MS instruments were used in a study of the presence of xanthones in *Psorospermum febrifugum*.[584] Several compounds with antitumor properties had previously been identified in the extracts of this plant material, but the presence of at least one of them was suggested to be an artifact of the isolation and concentration procedures used in the analysis. Daughter ion MS/MS experiments confirmed the presence of all previously identified compounds in the plant material and also provided evidence for the presence of closely related impurities that may not have been separately identified in previous analyses. Figures 6-3 and 6-4 contrast the daughter ion MS/MS spectra obtained from the mass-selected parent ion at m/z 377 with a BE instrument (Figure 6-3) and a multiquadrupole instrument (Figure 6-4). In each case, the daughter ion MS/MS

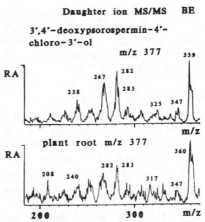

Figure 6-3 ■ Daughter ion MS/MS spectrum of the protonated molecule at m/z 377 of authentic 3',4'-deoxypsorospermin-4'-chloro-3'-ol compared with the daughter ion MS/MS spectrum of the ion at m/z 377 obtained directly from the plant root material. Data were measured on a BE instrument at high collision energy.

spectrum obtained from the plant root is compared with that of the authentic sample (3',4'-deoxypsorospermin-4'-chloro-3'-ol). The presence of an interfering compound is more obvious in the comparison of the low-collision-energy data.

Widespread availability of the multiquadrupole MS/MS instruments is likely to encourage their future use in the analysis of natural products, although only a limited

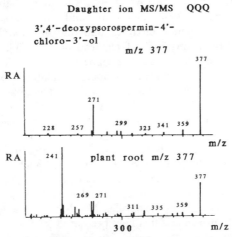

Figure 6-4 ■ Daughter ion MS/MS spectrum of the protonated molecule at m/z 377 of authentic 3',4'-deoxypsorospermin-4'-chloro-3'-ol compared with the daughter ion MS/MS spectrum of the ion at m/z 377 obtained directly from the plant root material. Data were measured on a QQQ instrument at low collision energy.

number of examples can be cited at this time. Sector instruments (BE) have been commercially available for a longer period, and now that the methodology for mixture analysis has been established, new areas of applications in natural products are being reported. An example is the use of daughter ion MS/MS to establish the presence of alpha-hederin in crude ethanolic extracts of *Hedera helix* (ivy) leaves.[585,586] A detection limit of 1 ng in the leaf extract was quoted. Daughter ion MS/MS spectra of the fragment ion at m/z 248 found in the electron ionization mass spectra of both alpha-hederin and the related compound hederacoside were used to quantitate these compounds both in extracts of the leaves and in commercial cosmetic formulations that include these compounds. Daughter ion MS/MS spectra have been used by Plattner[587] to identify brassinolide (a plant-growth stimulant) and castatsterone (its biogenetic precursor) in partially refined extracts of *Brassica napus* (rape) and *Alnus glutinosa* (alder). The antitumor compound sesbanimide was similarly quantitated at the ppm level in simple ethanolic extracts of *Sesbania drummondii*.

Several other applications to the analysis of natural products do not involve the analysis of alkaloids. Ohashi and Nagai[588] have used FAB/MS/MS to study glycosphingolipids and phosphonosphingolipids (here considered to be natural products) in plant materials. For certain classes of sphingolipids, a characteristic daughter ion characteristic of the fatty acid side chain was observed. Parent ion MS/MS scans were therefore used to identify this particular structural class in the plant extract mixture. Manning et al.[589] used FAB/MS/MS to identify a glucoronidase inhibitor isolated from the seeds of *Baphia racemosa*. The compound 2(S)-carboxy-3(R),4(R),5(S)-trihydroxypiperidine, based on an imino sugar, is thought to offer some protection from herbivores. Masucci et al.[590] have studied the daughter ion MS/MS spectra of bisbenzylisoquinoline alkaloids, which can be isolated from plants of several different families. The diverse pharmacological effects noted for these alkaloids include hypotensive, antimicrobial, muscle relaxant, and antitumor properties, underscoring the interest in the discovery and structural elucidation of such natural products.

Fraisse and Becchi[591] have used FAB/MS/MS to study C-glycosidic flavanoids in purified extracts of plants. These polar compounds are characterized by the wide variety of isomers in the natural mixtures, and daughter ion MS/MS allowed the particular isomers to be distinguished. Greathead and Jennings[592] have used linked scans on an EB instrument to characterize gibberelins in simple ethanolic extracts of grass samples. The gibberelins are plant hormones, regulating the growth and development of the plant tissues. The metabolic pathways of the gibberelins as a compound class are of interest. The daughter ion MS/MS spectra (linked scan) of the ions formed from gibberelins were sufficiently different for identification of each of nine gibberelins studied, even to the extent of differentiating geometrical isomers.

The use of MS/MS experiments other than the daughter ion MS/MS scan has not been widespread despite the presence of characteristic daughter ions in the MS/MS spectra of the alkaloids, for example, which can be used in a parent ion MS/MS scan to differentiate classes of alkaloids such as the 1-methyl or *N*-methyl tetrahydroisoquinoline compounds. The lack of use of other MS/MS scan modes is due in great part to the fact that the BE instruments that have been used for most studies to date are not equipped with the capabilities for a parent ion MS/MS or neutral loss MS/MS scan. As the use of multiquadrupole instruments in this research becomes more common, additional applications of parent ion and neutral loss MS/MS experiments can be expected.

6.2.2. Toxic Natural Compounds

Many of the alkaloids described in the previous section are quite toxic. The emphasis in this section will be on the targeted analysis of toxic compounds in natural products matrices when the presence or absence of particular compounds is of concern.

Haddon and Molyneux[593] have used MS/MS to study the toxic pyrrolizine alkaloids, constituents of *Seneccio* plant species. These plants are native to the arid grazing lands in the western parts of the United States. Ingestion of the plants in large amounts presents a health hazard to cattle, horses, and sheep that forage on these lands. The analysis is designed to screen plant samples for the content of this particular class of toxic alkaloids. Chemical ionization was used to form the protonated molecules of each component in a simple extract from the plant. Protonated molecules of the alkaloid dissociate to a characteristic daughter ion at m/z 120, corresponding to the skeletal framework for this class of alkaloids. A parent ion scan (linked scan at constant B^2/E) indicates members of this compound class present in the plant extract. Some chemical pretreatment was necessary to ensure that all of the alkaloid was in one chemical form for the analysis. With the use of an internal standard, a linear calibration range from 100 to 2000 ng of the parent compound was established. Dreifuss et al.[594] studied a number of pyrrolizidine alkaloids by negative ion chemical ionization mass spectrometry to establish a basis for structural interpretation. Daughter ion and parent ion MS/MS data were used in conjunction with exact mass measurement to establish the patterns of dissociation. Many of the $(M - H)^-$ ions from pyrrolizidine alkaloids dissociate to a characteristic daughter ion at m/z 154, and Dreifuss et al.[594] suggest that this common reaction could be used as the basis of a screening procedure for these compounds in complex mixtures.

Jones et al.[595] have expanded on the original work of Haddon and Molyneux in their use of MS/MS to differentiate isomeric compounds of the pyrrolizidine alkaloids by daughter ion MS/MS with both fast atom bombardment and chemical ionization mass spectrometry. Use of MS/MS was suggested by the fact that satisfactory gas chromatographic separations of these compounds have not yet been developed, and that sample handling increases the susceptibility of the target compounds to hydrolysis. A complicating issue in the ionization of these compounds is that several of the alkaloids produced $(M - H)^+$ ions in both fast atom bombardment and chemical ionization, and that these ions are isomeric with related alkaloids of interest. Jones et al. found that it is not always possible to differentiate between the isomers, but in several cases it *was* possible to do so based on the relative abundances of the daughter ions in the high-collision-energy daughter ion MS/MS spectrum. Furthermore, assignment of the ions in the daughter ion MS/MS spectra suggested that hydride abstraction for several of these compounds occurs on the pyrrolizidine ring rather than elsewhere in the molecular structure.

Jones et al.[596] have also used MS/MS to quantitate quinolizidine alkaloids in milk and blood samples of goats deliberately fed toxic plant materials from the lupine family. The interest is that the threat of toxic effects extends past the animal that grazes directly on plants of these species and has been traced to fetal abnormalities due to maternal ingestion of milk from affected goats. Acid/base fractionation of the biological sample was followed by extraction of the sample alkaloid into an organic solvent. One microliter of the extract was analyzed by direct exposure probe chemical ionization and high-collision-energy CID with selected reaction monitoring to increase the sensitivity of the assay. The limits of detection were about 5 ng for the target alkaloids anagyrine and lupanine. Total analysis time for 10 samples in duplicate, with generation of the calibration curve, was 3 h.

Plattner et al.[597] have recorded the daughter ion MS/MS spectra of the ergot cyclol alkaloids, with the protonated molecules formed by chemical ionization, and then chosen as the parent ion for low-energy collisions on a multiquadrupole instrument. These compounds are the products of fungus that grows on grain; when the grain is consumed by livestock, an illness called ergotism results. The similarity in structures and properties of the members of this class of alkaloids has made their separation and quantitation difficult, and indeed, even the electron and chemical ionization mass spectra are quite similar. The daughter ion MS/MS spectra are distinctive. Analysis of a crude extract of contaminated barley revealed the presence of 10 peptide alkaloids, with a total amount of 1 μg of the crude alkaloid material. The best conventional separation and identification (preparative thin-layer chromatography followed by sample elution and electron ionization mass spectrometry) revealed only the five most concentrated alkaloids, with the minor constituents suspected, but unconfirmed due to the small amount of sample.

Similar daughter ion MS/MS experiments have been used to detect and quantitate *Fusarium* mycotoxins in contaminated grains.[598] The *Fusarium* molds produce compounds such as vomitoxin, which leads to various animal disorders, and the estrogen zearalenone, which can also have undesirable effects. These compounds can be identified by GC/MS after sample cleanup and derivatization to increase the volatility of the targeted compounds, but this analysis becomes a time-intensive procedure. Plattner et al.[598] reported that daughter ion MS/MS can detect both vomitoxin and zearalenone at the 1-ppm concentration level in a simple organic solvent extract of the grain. The extraction required 30 min, and the MS/MS analysis for both of these targeted components was completed within 10 min; 50 g of grain sample was used for the analysis. At levels below 1 ppm, matrix effects in the crude extracts limited the confidence in the quantitative results measured; variance in the values at these levels could be reduced with more extensive sample cleanup.

A combination of chemical ionization, gas chromatographic separation, and multiple-reaction monitoring MS/MS has been used by Lau et al.[599] for the determination of eight underivatized trichothecene mycotoxins found in contaminated grains. Figure 6-5 illustrates the strategy for the method. The combination of retention time and selection of both parent ion and daughter ion mass provides a very selective identification of these compounds in extracts from the grain. Limits of detection for these eight compounds ranged from about 10 to 300 pg at a signal-to-noise ratio of 3.

Plattner et al.[600] have also used daughter ion MS/MS spectra to identify aflatoxins (toxic compounds from the fungi *Aspergillus flavus* and *A. parasiticus*). Negative ion chemical ionization was used to produce the radical molecular ion of the aflatoxin, and selected reaction monitoring was used to detect the dissociation of the parent molecular ion to a characteristic daughter ion. Matrix effects were observed when large amounts (equivalent to the material extracted from 25 mg of corn) of crude extracts were introduced directly into the source of the mass spectrometer directly; responses were suppressed to 25% of the values expected on the basis of a linear calibration curve established with matrix-free standards. No matrix effects were noted when smaller amounts (less than 1-mg extract equivalent) or more highly refined samples were introduced into the source. Table 6-1 summarizes the daughter ion MS/MS spectra of

Multiple selected reaction
monitoring MS/MS   **Mycotoxin analysis**

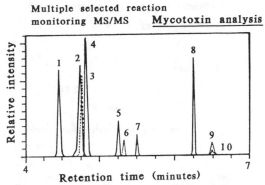

Figure 6-5 ■ Strategy for the GC/MS/MS analysis of mycotoxins in grain extracts. The parent ion/daughter ion pair used in the selected reaction monitoring experiment changes as a function of retention time of the mycotoxin in the gas chromatographic separation. For the peaks labeled 1 through 10, the target compound and parent ion/daughter ion masses are given: 1—deoxynivalenol (296 → 248); 2—monoacetoxyscirpenol (324 → 264); 3—deoxynivalenol-3-acetate (338 → 296); 4—diacetoxyscirpenol (366 → 153); 5—fusarenon-X (354 → 175); 6—nivalenol (312 = 312, selected ion monitoring); 7—neosolaniol (382 → 122); 8—zearalenone (318 → 250); 9—T-2 toxin (466 → 121); 10—HT-2 toxin (424 → 86).

various aflatoxins. The advantage of the unit mass resolution of the daughter ions provided by the multiquadrupole instrument is apparent in differentiation of the MS/MS spectra of the various isomers at m/z 312 or m/z 330. Note that since the parent ions are of the same empirical formula (compare aflatoxins G_2 and M_2; both have an empirical formula of $C_{17}H_{14}O_7$, and high-resolution mass selection of the parent ions will not differentiate these isomers.) However, with low-collision-energy daughter ion MS/MS analysis, the experimental parameters must be carefully controlled for reproducible results. Figures 6-6 and 6-7 illustrate the variation in the

Table 6-1 ■ Daughter Ion MS/MS Spectra Measured on the M⁻·. Ions of Aflatoxins Formed by Resonance Electron Capture

Compound	Molecular weight	Parent (RA)	Daughter ions m/z (RA)
Aflatoxin B_1	312	(1)	297(100), 269(4), 253(5), 209(1)
Aflatoxin B_2	314	(0.5)	299(100), 271(6), 243(8)
Aflatoxin G_1	326	(0.0)	313(33), 269(100), 241(16)
Aflatoxin G_2	330	(0.0)	315(28), 271(100), 243(11), 215(3)
Aflatoxin M_1	328	(0.5)	313(100), 270(1), 269(1)
Aflatoxin M_2	330	(0.5)	315(100), 287(1), 270(1, 259(1)
Aflatoxicol	314	(0.5)	299(100), 281(11), 271(2), 253(3), 243(3), 237(6)

Collision energy in the triple-quadrupole mass spectrometer is set at 20 V and the collision gas pressure at 2.0 mTorr.
Data from R. D. Plattner, G. A. Bennett and R. D. Stubblefield, *J. Assoc. Off. Anal. Chem.*, 1984, **67**, 734.

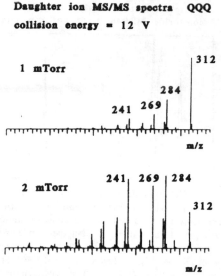

Figure 6-6 ■ Variation in the daughter ion MS/MS spectrum of an aflatoxin recorded at 12-V collision energy at collision gas pressures of 1 and 2 mTorr. Data were measured on a QQQ MS/MS instrument.

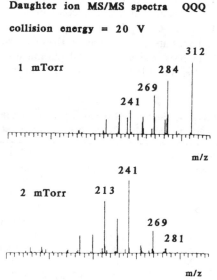

Figure 6-7 ■ Variation in the daughter ion MS/MS spectrum of an aflatoxin recorded at 20-V collision energy at collision gas pressures of 1 and 2 mTorr. Data were measured on a QQQ MS/MS instrument.

daughter ion MS/MS spectra of aflatoxin B_1 recorded at different collision gas pressures and collision energies.

The clinical symptoms of a cattle malady called "fescue foot" are very similar to those of ergotism, a response to toxic alkaloids found in ergot sclerotica. Several studies indicated a fungus contamination of fescue grass as the cause of this problem. Yates et al.[601] used daughter ion MS/MS to confirm the presence of the toxic alkaloids in infected grasses. Negative ion chemical ionization was used to produce a characteristic daughter ion A^- from each of 12 different ergopeptide alkaloids. The daughter ion MS/MS spectra of this fragment ion were unique for each of the alkaloids. A contaminated grass sample was ground and defatted to produce 10 g of grass meal, then extracted with an organic solvent to give a total of 10 mL of solution for analysis. One microliter of that extract was used for the MS/MS analysis. For the alkaloid ergotamine, levels as low as 0.01 ppm could be analyzed, corresponding to 10 pg of alkaloid. The linear range of analysis extended from about 1 pg to 60 ng; beyond this level, signal saturation and matrix effects in the negative ion chemical ionization mode were noted.

Daughter ion MS/MS spectra have been used to confirm the results of gas chromatography with an electron capture detector in a quantitative study of the levels of deoxynivalenol (a mycotoxin) in fungus-contaminated wheat.[602] Such a strategy is the reverse of that normally used, in which a rapid screening analysis by MS/MS is confirmed by a more selective and time-intensive confirmation by GC or GC/MS in samples that screen positive. In this case, the selectivity offered by the MS/MS analysis provided the required level of confirmation.

Plattner and Powell[603] have used daughter ion MS/MS spectra to study the structures of a series of maytansinoids. Both positive and negative ion chemical ionization were used to create the parent ions for these studies. Structurally similar isomers could be distinguished on the basis of the daughter ion MS/MS spectra. MS/MS experiments were used to identify three new maytansinoids present as minor components in what were originally thought to be pure reference compounds. Parent ion MS/MS spectra for structurally specific daughter ions were used to identify maytansinoids in simple extracts of the fermentation broths of their known biological source *Actinosynnema pretiosum* with no sample cleanup. Plattner has written a general review of the use of MS/MS in the analysis of toxic natural products.[604]

Krishnamurthy and Sarver[605] have used positive ion chemical ionization to create $(M + NH_4)^+$ ions from macrocyclic trichothecenes isolated from Brazilian *Baccharis* plants. The daughter ion MS/MS spectra of these adduct ions were characteristic of the macrocyclic ester bridges in the original structure and also contained several common neutral losses specific to this compound class. Selected reaction monitoring was used to quantitate the levels of these toxins in a plant extract, with detection limits of 10 to 100 pg depending on the toxin.

6.3. Industrial Products Applications

The analytical advantages of MS/MS complement the widespread use of GC/MS in the analysis of samples from the chemical industry. This section reviews the applications of MS/MS in a few areas of interest and concludes with a description of an initial application of MS/MS to process control. In such applications, the speed and specificity of an MS/MS analysis are paramount. Unfortunately, many of the most intriguing applications of MS/MS in the solution of problems of industrial interest probably remain unpublished. As with any other analytical technique, a company invests resources of personnel, instrumentation, and methods development in MS/MS. There is no advantage in the release of a successful strategy to potential competitors. With apologies to those laboratories in which the work on MS/MS goes far beyond that described here, this section describes some applications of MS/MS in the analysis of dyes, surfactants, and polymers.

6.3.1. Dyes

Industrial dyes are produced in large quantities, and the synthetic residues as well as the dyes themselves are often discharged in waste streams. Both the dyes and the synthetic by-products that accompany their production may be toxic. In addition to the identification of dyes in waste waters or synthetic mixtures, MS/MS has been of use in the characterization of the commercial products. Most dyes are sold in technical purities, with labels that define the specific dye content as anywhere between 80% and 100%. The identity of the other components in the product is often of interest.

Several studies have been described in which daughter ion MS/MS spectra are used to provide a rapid, sensitive, and selective means of analysis of these materials in complex samples. Betowski and Ballard[606] have described the daughter ion MS/MS spectra (triple-quadrupole mass spectrometer) of several dyes with prior separation and purification by liquid chromatography. Since the peaks eluting from the liquid chromatograph were 20 to 30 s wide, ample time for measurement of the daughter ion MS/MS spectra was available. Several new isomers of the commercially available dyes were identified on the basis of the MS/MS data.

Gale et al.[607] have used daughter ion MS/MS experiments (triple-quadrupole mass spectrometer) to characterize the products of the reduction of organic dyes ionized by liquid secondary ion mass spectrometry. Cationic organic dyes of low reduction potentials dissolved in glycerol produce positive ion SIMS spectra that contain not only the intact cation, but also abundant signals corresponding to the addition of one and two hydrogens to the intact cation. For the dye Janus Green, these species were mass-selected and their daughter ion MS/MS spectra recorded. Reduction of the cation significantly changes the pattern of fragment ions produced on collision-induced dissociation. The extent of reduction of dye cation is a function of the ionization method and the source conditions, and variations in these parameters was put forward as an explanation for differences in the mass spectra of these dyes obtained by diverse particle-induced desorption ionization techniques. Some caution in the interpretation of these results is warranted since some of the dyes studied are also reduced in electron and chemical ionization experiments.[608]

The capabilities of newer ionization techniques often complement MS/MS experiments. Bruins[609] has used liquid ionization to analyze a series of sulfonated azo dyes. Pneumatic nebulization of the effluent from a small LC column is followed by ion formation, catalyzed with an induction electrode, and passage of the ions through a small orifice into a triple-quadrupole mass spectrometer. In the negative ion mass spectra of the sulfonated dyes, the deprotonated $(M - H)^-$ or the doubly deprotonated $(M - 2H)^{2-}$ molecules are abundant. As is often the case with the soft ionization methods, the degree of fragmentation is minimal, and determination of the molecular mass is facilitated at the cost of an inability to interpret the mass spectrum in terms of the structure of the molecule. Daughter ion MS/MS experiments were used to generate structurally informative daughter ions from these selected parent ions. Of particular interest was the observation that daughter ion MS/MS spectra of the sulfonated azo dyes typically contain an abundant ion at m/z 80 corresponding to SO_3^-, formed by cleavage of one of the more labile bonds in the deprotonated parent molecule. A parent ion MS/MS spectrum of m/z 80 then provides an identification of all azo dyes eluting from the LC column, including indications for previously unresolved components of low relative concentrations.

6.3.2. Surfactants

In one of the earliest applications of MS/MS to an industrially related problem, Weber et al.[610] used field desorption ionization to create parent ions $(M + H)^+$, $(M + Na)^+$, or intact cations for daughter ion MS/MS analysis for the study of cationic, anionic, and nonionic surfactants. Interpretation of these spectra, obtained on a multisector instrument, allowed the determination of the length and branching of the alkyl chains found in the surfactant structures. The patterns observed were sufficiently regular that deviations in the expected spacings and abundances of the daughter ions were interpreted to indicate the presence of isobaric precursors in the beam of mass-selected parent ions. A complete daughter ion MS/MS spectrum could be obtained with as little as 5 pg of sample, and a moderate separation of surfactant types could be obtained by varying the heater current through the field desorption emitter. The emitter emission order was nonionic (6–18 nA), cationic (20–40 nA), and anionic (45–60 nA) surfactants.

In further work by Schneider at al.,[611] fast atom bombardment was used to produce both molecular weight and fragment ions for the cationic surfactants. Since surfactants are often produced as complex technical mixtures, it is difficult to distinguish between quasimolecular and fragment ions. The intact cations from a mixture of methyltrialklyammonium salts were selected as parent ions, and the daughter ion MS/MS spectra recorded (BE instrument). The MS/MS spectra are dominated by alkane loss and alkyl loss with subsequent alpha cleavage. The structures of the alkyl chain in the cation could be readily established.

A wide number of compounds are used as surfactants, including amines, amine salts, quaternary amines, amine oxides, polyethoxylated quaternary amines, and amphoteric amines. Most commercial surfactant mixtures reflect the mixed nature of the fatty acids from which they are synthesized. Since the desired activity does not depend upon the purity of the sample, the mixtures are often complex. In addition, the charged nature of the sample has in the past precluded many standard methods of analysis. Lyon et al.[612] used FAB, itself well suited for the analysis of ionic organic compounds, and daughter ion MS/MS experiments (triple-sector EBE mass spectrometer) to characterize the individual components of the complex mixture introduced into the source of the mass spectrometer. Homolog distribution and molecular weight information were obtained directly from the positive ion FAB mass spectra. Daughter ion MS/MS spectra of the molecular cations provided the detailed structural information necessary for deduction of surfactant type, length and number of alkyl chains, the presence of isomers, and the degree of ethoxylation.

Anionic surfactants have been studied by daughter ion MS/MS with a triple-sector instrument.[613] Molecular weight distributions of commercial surfactants could be obtained from either the positive or the negative ion FAB mass spectrum, with daughter ion MS/MS spectra used to confirm the structural identification. Investigations included alkyl and alkylaryl sulfates and sulfonates, alcohol ether sulfates, alpha olefin sulfonates, fatty acid salts, sulfosuccinate diesters, and N-acylated amino acids. In general, each surfactant class produces daughter ions characteristic of that functional grouping. The utility of the parent ion MS/MS scan for all of the components in a surfactant mixture that are of a particular type is apparent.

MS/MS has been used to examine fluoroalkanesulfonates, mixtures of which are in general commercial use as anionic surfactants.[614] The ionic nature of these compounds has in the past complicated the analytical methods necessary to characterize these commercial mixtures. FAB was used to produce high mass positive and negative cluster ions from these compounds. The presence of homologous impurities in the nominally pure salts was established with the observation of the appropriate cluster ion series. Daughter ion MS/MS spectra contained a number of equally spaced fragment ions separated by 50 daltons, corresponding to sequential losses of CF_2 along the alkyl chain. The length of the chain is readily determined by counting the number of fragment ions apparent in the daughter ion MS/MS spectrum.

Ionic surfactants present in surface waters (3 L of water extracted for surfactants, with the residue concentrated to 50 μL, sufficient for 2–3 mass spectrometric analyses) have been analyzed by MS/MS.[615] Field desorption was used as the ionization method since it is well suited for the analysis of the ionic surfactants. At progressively higher heating currents applied to the FD emitter, nonionic surfactants (sodium-cationized molecular ions) are desorbed first, followed by the cationic surfactants (direct cation desorption), and finally the anionic surfactants (as a cluster ion composed of two cations with a single anion). For each compound group, daughter ion MS/MS spectra provide confirmation of structure based on comparison with the MS/MS spectra of standard compounds.

6.3.3. Polymers

The analysis of polymers has always been a difficult problem. A descriptive parameter of great importance is the molecular weight distribution of the components in a polymer mixture, since the distribution establishes the physical properties observed. Mass spectrometry has been used in the determination of the molecular weights of the individual components; the breadth of application depends upon the success of the chosen ionization method in producing characteristic ions from the polymer sample. It is of further interest to determine the structures of the compounds involved, or the variants of the structures included in the mixture, since the polymerization reaction is often not synthetically precise. It is in these experiments that MS/MS has begun to play a role.

An early application was the use of daughter ion MS/MS spectra in conjunction with direct pyrolysis mass spectrometry[616] to characterize poly(carboxypiperazine) in a synthetic polymer that begins to degrade at 270°C. The structures of pyrolyzed fragments were individually characterized by high-energy collision-induced dissociations obtained on a BE instrument. Structures were assigned by comparison of the

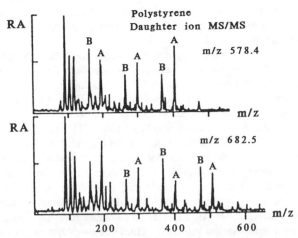

Figure 6-8 ■ Daughter ion MS/MS spectra (BE instrument) of molecular ions of polystyrene at m/z 578.4 (top) and m/z 682.5 (bottom). The parent ions were formed by field desorption. Ions in the A series are represented by the formula $m/z = 104n + 91$, where n is the number of repeat units of polystyrene and 104 is the mass of $C_7H_6CH_2$. Ions in the B series are represented by the formula $m/z = 104n + 57$.

spectra thus obtained with the spectra of synthetic standards of the expected thermal oligomers.

Craig and Derrick[617–619] have made extensive use of MS/MS with field desorption mass spectrometry. The focus of these studies has been the demonstration of the method, but also a study of the amount of energy deposited in an ion during the process of field desorption. Daughter ion MS/MS spectra were measured (BE instrument) for the molecular ions of polystyrene M^+ ranging in mass from 474 to above m/z 4000. Figure 6-8 compares the daughter ion MS/MS spectra measured for two polystyrene molecular ions. In general, the dissociations observed on CID differed from those observed in metastable dissociations, with a greater degree of dissociation to lower mass fragment ions observed in CID. The distinctive feature of the MS/MS spectra for the polystyrene molecular ions is the predominance of low mass fragment ions, a trend that becomes more pronounced as the mass of the molecular ion increases. This result contrasts with recent results on the CID of large peptide ions, which show abundant high mass fragment ions in the daughter ion MS/MS spectra, but relatively few low mass daughter ions. In the case of polystyrene, the ions observed are in fact closed-shell ions postulated to arise as the result of a number of direct cleavage reactions, initiated randomly along the length of the polystyrene molecule. Such a situation contrasts with the daughter ion MS/MS spectra of protonated molecules formed from peptides, in which the proton may still be distributed along the sequence, but interacts most strongly with the basic sites of the amino acids. The contrast is between a parent ion in which the charge is initially localized at one site in the polystyrene molecular ion and can act to initiate cleavage reactions from that point, and a protonated molecule of a high-mass peptide, in which the proton is delocalized over a number of different sites and for which charge-site initiated cleavage might not be expected. If the model holds true, additional emphasis is placed on the choice of ionization method to provide a parent ion either susceptible or resistant to dissociation, as the situation warrants.

Kiplinger and Bursey[620] have used the daughter ion MS/MS spectra of organic polymers to study the mechanism of the collision-induced dissociation itself. A four-sector (BEEB) instrument was used to study the daughter ion MS/MS spectra of the $(M + H)^+$ and $(M - H)^-$ ions of polyethylene glycols ionized by fast atom bombardment. Both the metastable ion dissociations and the CID processes produce a series of daughter ions 44 daltons apart, corresponding to the (CH_2CH_2O) repeating unit in the polymer. For parent ions at both the low and the high end of the polyethylene glycol mass range, loss of 44 daltons from the parent ion is a low abundance process and loss of 88 daltons produces the base peak in the spectrum. The abundances of all other daughter ions are low. The relative endothermicities of the dissociation reactions for the parent ions of the polyethylene glycols could be estimated. Bursey argues that a decrease in the relative amount of dissociation of higher mass ions holds true only with less endothermic fragmentation processes. Daughter ions that require more excitation of the parent ion for their formation and those that may represent sequential cleavage or drastic rearrangement reactions are not so affected and may eventually dominate the daughter ion MS/MS spectra of high mass ions. Further experiments are necessary to test this hypothesis, including careful experiments in which the initial structure of the dissociating ion can be reasonably controlled.

Shushan has described a commercial triple-quadrupole mass spectrometer interfaced to a thermogravimetric analysis (TGA) system. The first report[621] of this combination in 1983 emphasized that conventional atmospheric pressure ionization mass spectra could be recorded during the course of the thermogravimetric analysis. Individual ionic components evolved during the heating could be selected for daughter ion MS/MS analysis in real time. In such an experiment involving an NBR polymer with 20% bound acrylonitrile, the evolution of cyanobutane during the decomposition of the polymer was unambiguously determined.

Complete computer control of the system was described in 1984.[622,623] A styrene–isoprene–styrene polymer was analyzed by the system. The mass spectral data were interpreted to indicate that weight losses from the polymer were due to the decomposition of polyisoprene and polystyrene; the MS/MS data showed that the structures of the compounds formed on decomposition of the polyisoprene included monomeric isoprene and oligomers such as monoterpenes, sequiterpenes, and diterpenes. Another analytical method applied to the characterization of polymers and coatings is dynamic mechanical analysis (DMA), essentially a method that measures the deformation of materials subjected to vibration. The curing process, for instance, can be followed by DMA. Since the process of curing often includes the evaporation of volatile compounds from the compound mixture, such a process can also be followed by mass spectrometry. A complete system was described by Shushan[624] and was used to show that the kinetics of a polyurethane curing reaction could be monitored by following the evolution of methylethylketoxime (the structure determined unambiguously by MS/MS) during a thermosetting procedure.

Considering the widespread use of mass spectrometry in polymer analysis,[625] it is somewhat surprising that MS/MS has not been more widely applied. One reason may be the preponderance of multiquadrupole instruments in the commercial MS/MS market. Until only recently, the mass range of these instruments was limited to 1800 daltons, while the requisite mass range for polymer analysis is much higher. Although daughter ion, parent ion, and neutral loss MS/MS experiments can be completed on sector instruments, the spectra may be limited in resolution or subject to artifacts. With the development of multisector instruments, and new, higher mass range quadrupoles, this limitation may disappear in the next few years. A second reason for the slow infiltration of MS/MS into polymer analysis may be based on the limitations of the usual ionization methods of electron and chemical ionization. Many of the higher mass polymers cannot be ionized by such methods. Desorption ionization methods such as FAB and SIMS are now available and can be used with MS/MS, but many of the measurements in polymer analysis must be referenced against standard methods and much development work must be completed in order to make an accurate transition from one method to another. Biochemical analysis is not encumbered by a long tradition of mass spectrometric studies against which newer techniques must be compared, and MS/MS applications in this area have seemed to progress much more rapidly (see Section 6.6).

An interesting aspect of polymer analysis is the study of polymers that form in the gas phase, such as in the source of a mass spectrometer, as contrasted with the study of polymer samples. Gleria et al.[626,627] have used high-collision-energy daughter ion MS/MS spectra to study the gas-phase polymerization reaction of polyphosphazenes. Mass spectrometric polymerization results from the introduction of a fairly large sample into the ion source of the mass spectrometer. The empirical formula of the sample, octachlorocyclotetraphospatetraene, is $N_4P_4Cl_8$ (mass of 460 daltons). Gas-phase polymerization leads to ions of m/z as high as 1345 daltons, corresponding to ions of formula $N_nP_nCl_{n-1}$, where n varies between 5 and 12. Daughter ion MS/MS spectra were used to show that the structure of the oligomeric ions were consistent with a polymerization of $N_4P_4Cl_8$ units rather than with the alternative mechanism that involved the polymerization of $N_3P_3Cl_6$. It was further demonstrated that two stable and distinct structures of an ion at m/z 460 of empirical formula $N_4P_4Cl_8$ exist; one is the result of a direct electron ionization, and the other a dissociation product of an oligomerization reaction.

Doyle and Campana[628,629] have studied the structures of the cyclic oligomers of methylnitramine and formaldehyde units. Daughter ion MS/MS spectra (BE instrument) of $[(CH_2NNO_2)_nH]^+$ ($n = 3$–12) cluster ions formed from hexahydro-1,3,5-trinitro-1,3,5-triazine gas-phase oligomerization were measured.[630] The patterns of abundances in the daughter ion MS/MS spectra were interpreted as a structure of the oligomer consisting of "core" units bound through hydrogen to remaining monomers. This interpretive approach was refined in a recent study.[631] The daughter ion MS/MS spectra of oligomeric ions from paraformaldehyde (linear forms) and of the same ions from trioxane and tetroxocane (cyclic forms of formaldehyde) were compared. The daughter ion MS/MS spectrum of the linear form contains fragment ions corresponding to cleavage at each bond in the backbone, in contrast to the daughter ion MS/MS spectra of the cyclic forms, which contains a series of fragment ions corresponding to loss of the monomeric cyclic units. Figure 6-9 illustrates the distinction in the daughter ion MS/MS spectra for the parent ion at m/z 271, representing $[(CH_2O)_9H]^+$. The data suggest that the oligomerization in the source produce cluster ions rather than polymeric ions.

In addition to the characterization of polymers directly, MS/MS has been used to investigate the products of thermal degradation of polymers. The work of Shushan

Daughter ion MS/MS

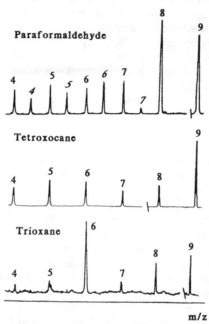

Figure 6-9 ■ Daughter ion MS/MS spectra (BE instrument) of m/z 271 ions corresponding to $(CH_2O)_9H^+$ ions formed in the electron ionization mass spectrum of paraformaldehyde (top), tetroxocane (middle), and trioxane (bottom). All ion abundances to the left of the slash are multipled by a factor of 10. Ions identified by block numerals correspond to ions in the series $[(CH_2O)_nH^+]$ for n ranging from 4 to 9. Ions identified by italic numerals correspond to $[(CH_2O)_nCH_2H^+]$ for n ranging from 4 to 7.

described previously falls into this category. Pyrolysis mass spectrometry has also been used with MS/MS to characterize the degradation products of several polymers. Linked scans (B/E) have been used to study the thermal degradation products of aliphatic and aromatic polyamides,[632] mostly to confirm the structures of low mass ions found in the mass spectrum. Ballistreri et al.[633] have described the experiment in detail. The analysis of low molecular weight polymers formed during the synthesis of high molecular weight oligomers has been approached by Ballistreri et al.[634] using daughter ion and parent ion MS/MS scans.

6.3.4. Rubber and Rubber Additives

Commercial rubber materials are complex mixtures of base materials and additives that provide desired characteristics such as stability to oxidation. Identification and quantification of these additives is a problem to which mass spectrometry has often been applied.[635,636] In recent work, Lattimer[637] has used MS/MS to study organic additives in rubber compounds. In an application of fractional distillation from a direct insertion probe, daughter ion MS/MS spectra were used to show that the ion at m/z 256 (electron ionization) evaporated from the sample with no sample heating was due to S_8^+. The ion at the same integer mass that was evaporated at a probe temperature of 150°C originates in palmitic acid, a known component in the recipe of the rubber compound.

Other MS/MS experiments were useful in the analysis of the rubber sample. A characteristic daughter ion at m/z 353 is formed from components of a t-octylphenol/formaldehyde tackifier additive. A parent ion MS/MS spectrum for this selected ion indicates several ionic precursors present in the mixture. Incidentally, the rubber sample was not extracted for these experiments, but a small sample was placed directly in the crucible of the heated direct insertion probe. In a neutral loss MS/MS experiment, a distinctive loss of 71 daltons (C_5H_{11}) from aromatic compounds containing a t-octyl group was used to identify members of that specific functional classification in the rubber mixture. Indicated components include dioctyldiphenylamine, trioctyldiphenylamine, and several other resin components.

6.3.5. Agricultural Products

Most analyses of pesticide and herbicide formulations are based on GC/MS. Some preliminary applications of MS/MS to the analysis of such products have recently appeared. Cardaciotto et al.[638] describe the use of daughter ion MS/MS in a study of the manufacturing process of ASSERT, a postemergence herbicide. Registration of an herbicide for use requires that the structure of the herbicide and the by-products produced during its manufacture be known. The production of this herbicide involves an esterification process that can result in the formation of structural isomers differing in the site of methylation. The by-products in the commercial product are present at a level of about 1%. For this work, the compounds were separated and concentrated by liquid chromatography, and MS/MS spectra were used in conjunction with NMR data for structural differentiation.

Roach and Carson[639] have measured the daughter ion MS/MS spectra of organophosphorus pesticides, prior to the analysis of pesticide residues in foods such as lettuce and strawberries. The daughter ion MS/MS spectra were easily interpreted based on extensive mass spectrometric analyses of these compounds. Organic solvent extracts of lettuce and strawberries were fortified with pesticides at the expected levels of residues, and the analysis compared for direct analysis of the crude extracts and analysis of the extracts that had been subjected to a simple charcoal column chromatographic cleanup. Chemical noise originating from the matrix in the former set of samples effectively prevented the identification of pesticides at the 0.2-ppm level. Even in cases in which there was no actual isobaric interference between the ion $(M + H)^+$ from the pesticide and an intense signal from the matrix, interference was still observed because of a limit in the abundance sensitivity of the MS/MS instrument. Saturated signals of one mass to charge were apparent in adjacent mass channels, providing an instrument noise consequence of the chemical noise in the system. With cleanup by the charcoal column, pesticides could be identified at the desired 0.2-ppm level.

6.4. Foods and Flavors Applications

The use of gas chromatography/mass spectrometry in food and flavor analyses is now well established.[640-642] Of course, the volatile aromas from food are of interest, contributing to the perceived taste as well as aroma. Gas chromatographic separation is well suited to the separation of these volatile components and GC/MS necessary in their identification. The complexity of the problem is underscored by the listing of the 500 components identified in the aromas of fresh coffee, or the 114 volatile compounds emitted by mangoes, or the quantitative differences in the identities and distributions of compounds emitted by mangoes of different species. Many food volatiles are members of homologous series of compounds, and their mass spectra can thus be remarkably similar. In such cases, the separation of the gas chromatography is indispensable. Having identified the more concentrated members of a homologous series, the food chemist is often alerted to the presence of the more trace components, which can be identified upon careful inspection of the data and comparison with standards, but would not be identified without such direction.

The bulk of the mass of food is composed of more polar materials such as sugars, amino acids, phenolics, lipids, fats, and other compounds. These nonvolatile compounds cannot be separated by gas chromatography, and derivatization to increase volatility is seldom practiced because of matrix effects and the increasing contamination of the sample with increased sample handling. Applications of LC/MS to this field are now expanding, especially with the introduction in the past few years of microbore columns of high resolving power. Again, LC/MS is the method of choice, and indeed the only analytical method with the demonstrated ability to identify the hundreds of compounds separated by the chromatography. The comparative attributes of the direct liquid introduction interface, the thermospray interface, and the electrospray interface are currently under investigation; each performs well within certain compound polarity ranges. Games has described recent LC/MS work in the area of food and flavor analysis.[643]

By contrast, the use of MS/MS in this area is less widespread. In part this has been due to the longer availability of commercial GC/MS instruments as opposed to MS/MS instruments, but also in no small part due to the enormous success of the GC/MS method itself, and more recently of the LC/MS combination. Both of these methods have evolved around the approach that a complex mixture is first separated and then the individual components identified in a sequential fashion. The contribution of MS/MS in this area is likely to focus on the abilities of the method to pinpoint compounds in homologous series and those in specific compound classes. The experimental approach is therefore somewhat at variance with the usual protocol. Just as in many other fields (particularly pharmaceutical analysis), success will catalyze further application. With a judicious amount of sample cleanup and preparation, MS/MS analysis for classes of compounds in food and flavor matrices may become commonplace in the next few years.

One advantage of MS/MS in the analysis of food-related compounds is the reduction in the amount of sample preparation necessary. In GC/MS, sample treatment is often extensive. In preparation for GC/MS analysis of nutmeg, Harvey[644] ground 100 mg of nutmeg to a fine powder and extracted for 1 h with ethyl acetate. The extract was filtered, frozen to precipitate triglycerides, filtered again, and then derivatized overnight (trimethylsilylation) to increase the volatility of the compounds of interest. By contrast, in the MS/MS analysis of nutmeg by Davis and Cooks,[645] 10 to 50 mg of ground nutmeg are loaded into the direct insertion probe of the mass spectrometer and vaporized by a short heating program.

A constant concern in the analysis of flavor components is alteration and contamination of the sample; losses of volatile components are a major problem. In MS/MS, sample handling is reduced and chances for contamination minimized. Sample carry-over, a problem during extraction procedures for GC/MS, evolves into a problem of source contamination in MS/MS. This problem was severe in early MS/MS work, but removable ion source volumes are commercially available and greatly reduce the extent of the problem. Problems peculiar to MS/MS in the analysis of food and flavor components include matrix effects that occur because a relatively large amount of sample is introduced into the source. This problem is less severe for the volatile compounds emitted by the sample, as these can be evaporated into the ion source without heating the direct insertion probe. Those compounds which require sample heating are seldom introduced into the source at a steady rate, and the chemical matrix in the source of the mass spectrometer is constantly changing. A second problem with MS/MS analysis is that the homogenization of a sample that occurs in GC/MS pretreatment may not occur. Sample inhomogeneities thus become of much greater concern. Sample-to-sample variation is already fairly high in samples of natural origin, and the extensive application of MS/MS to these compounds may require more careful sampling procedures. A simple form of sample pretreatment is often employed in MS/MS to concentrate the sample and to preserve the cleanliness of the ionization source.

6.4.1. Food Components

Labows and Shushan[646] have reviewed the direct analysis of food aromas by a commercial MS/MS (QQQ) system using an atmospheric pressure ionization source. With atmospheric pressure ionization, the inlet is simplified to an all-glass gas flow device that channels volatile components emitted by food materials into the mass spectrometer. Losses due to sample preparation are minimized, as are absorption or decomposition problems associated with chromatographic fractionation. Profiles of aroma compounds obtained by this method are claimed to be more accurate than those obtained using other analytical methods. Because of the high discrimination against chemical noise in the MS/MS system, detection limits can be very low, reported as 0.5 ppb for ethyl butyrate, 0.8 ppb for linalool, and 45 ppb for limonene. These limits were established with daughter ion MS/MS spectra.

The system has been used for many applications in the area of food analysis. In one demonstration, a sample collection device (a glass funnel) is placed over the sample and air drawn over the food, carrying the volatiles into the source of the mass spectrometer. Various fruits have been sniffed, and the volatiles emitted from bacon sizzling in a fry pan have been analyzed by the API MS/MS instrument. Figure 6-10 shows the correlation between the daughter ion MS/MS spectrum of authentic nootkatone (with the ion at m/z 219 selected as the parent ion) and the ion at the same mass from one of the volatile compounds emitted directly from a grapefruit. The match between the two spectra confirms the presence of this targeted compound in the emitted volatiles. By replacing the grapefruit with an orange, confirmation of the emission of this compound from oranges can also be obtained. The experiment is complete within a minute. Note that the unit mass resolution of the triple-quadrupole instrument allows an accurate assignment of abundances for daughter ions of adjacent mass in these MS/MS spectra. A close examination of the spectra show that the match between the authentic and the target compound is not perfect. Either the instrumental parameters were not constant or there is an additional component at m/z 219 in the volatiles emitted by the grapefruit. It is at this stage that a simple prefractionation experiment

Daughter ion MS/MS m/z 219

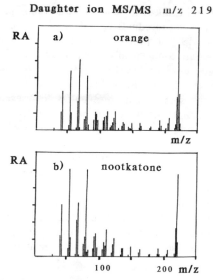

Figure 6-10 ■ (*a*) Daughter ion MS/MS spectrum of the parent ion at *m/z* 219 formed in the mass spectrum of the volatiles from an orange compared to (*b*) the daughter ion MS/MS spectrum of the parent ion at *m/z* 219 of authentic nootkatone ($C_{15}H_{22}O$). Nootkatone is also a flavor component of grapefruit.

or an alternative ionization method becomes necessary to establish the number of components present at this mass.

Labows and Shushan[646] described other MS/MS experiments that provided functional group information. Under the conditions of their experiment, dissociation to form *m/z* 18 (NH_4^+) is indicative of amines, and *m/z* 19 (H_3O^+) is a typical fragment ion from alcohols. Thus, a parent ion scan for these characteristic daughter ions pinpoints compounds with these functional groups in the emitted volatiles. Acetate esters produce daughter ions at *m/z* 43 and *m/z* 61. A parent ion scan for the latter produces the MS/MS spectrum shown in Figure 6-11, which is the sum of all the parent ions of all of the acetate esters in the volatiles emitted from a banana. The base peak at *m/z* 131 represents the parent ion of isoamylacetate, widely known for its characteristic banana odor.

A neutral loss (44 daltons) MS/MS experiment is used to follow the loss of carbon dioxide from (M − H)⁻ ions of carboxylic acids. The spectrum so obtained, with a Teewurst sausage as the sample, contains ions at *m/z* 87, 89, and 121, identified as originating from butanoic or pyruvic acid, lactic acid, and benzoic acid, respectively. The identities of these ions are confirmed by examining the daughter ion spectra of the authentic compounds.

Direct analyses of volatiles has been suggested as a means of screening food products that might otherwise pass agricultural borders. The sensitivity seems to be sufficiently high for this purpose, and the MS/MS analysis possesses the requisite speed and specificity for real-time analysis. A comprehensive study of the interferences that might be expected in such use would be needed to evaluate the suitability of this technique.

Parent ion MS/MS   m/z   61

Figure 6-11 ■ Parent ion MS/MS spectrum of the characteristic daughter ion at m/z 61 formed in the dissociation of acetate esters. The sample consists of the mixture of volatile compounds emitted from a banana. The most intense peak in the parent ion MS/MS spectrum represents isoamylacetate, the volatile compound that provides bananas with their characteristic odor.

Davis and Cooks[645] have studied the composition of nutmeg using MS/MS. Nutmeg has been extensively studied because of the large number of psychoactive species alleged to be present. This particular MS/MS study is noteworthy because of the use of both high-energy and low-energy collisions (on two different instruments) to acquire daughter ion MS/MS spectra. Programmed thermal desorption from the direct insertion probe was used for a crude distillation of the sample into the source of the mass spectrometer. Isobutane was used as the reagent gas to create an abundant protonated molecule for each component with a minimum of fragmentation. Charge exchange ionization was also used to form parent ions $M^{+\cdot}$ for a complementary daughter ion MS/MS analysis.

Comparison of the daughter ion MS/MS spectra of authentic 4-allyl-2,6-dimethoxyphenol at m/z 195^+ and the same mass ion from the nutmeg sample is given in Figure 6-12. The spectra are sufficiently similar that the presence of this compound in nutmeg was confirmed. The sharp peak at 97.5 on the mass scale results from a charge-stripping reaction (see Section 3.5.2.2) and is often characteristic of nitrogen- and oxygen-containing compounds; in this case, this ion is the product of the oxidation of singly charged 195^+ to doubly charged 195^{2+} as a result of the high-energy collision. The match between the daughter ion MS/MS spectra is close but not exact. In general, this situation indicates that not all of the ion current at the selected mass, in this case m/z 195, is due to this compound alone.

It is known that nutmeg contains diphenylpropanoids of both cyclic and acyclic forms. The acyclic form fragments to characteristic daughter ions at m/z 193 and the cyclic form to daughter ions at m/z 203. Davis and Cooks used these daughter ions in a parent ion MS/MS experiment to examine the entire nutmeg mixture for parent ions of these classes of diphenylpropanoids. Figure 6-13 shows the result of this experiment for the two characteristic daughter ions. Parent ions at 357^+, 371^+, 387^+, and 401^+ are indicated to be acyclic diphenylpropanoids. Cyclic forms are indicated in the parent ion scan at 327^+, 341^+, 355^+, 357^+, 371^+, and 375^+. The common parent ions 355^+, 357^+, and 371^+ are clearly indicated as consisting of both forms of diphenylpropanoid structures. The parent ion at 355^+ is thought to be a dehydrodiphenylpropanoid derivative of myristicin, identified for the first time in nutmeg.

In many cases, particular targeted components in food products are of interest. Walther et al.[647] have described an assay of caffeine in beverages such as coffee, tea,

Daughter ion MS/MS

a) **4-allyl-2,5-dimethoxyphenol**

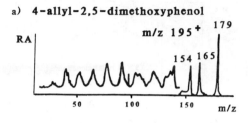

b) **nutmeg m/z 195$^+$**

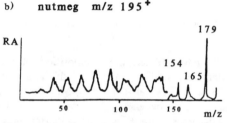

Figure 6-12 ■ (*a*) Daughter ion MS/MS spectrum (BE instrument) of the parent ion at *m/z* 195 generated from authentic 4-allyl-2,5-dimethoxyphenol by chemical ionization compared to (*b*) the daughter ion MS/MS spectrum of the *m/z* 195 ion created directly from a sample of nutmeg.

Parent ion MS/MS

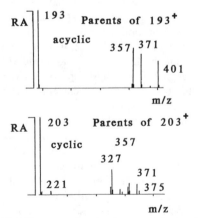

Figure 6-13 ■ Parent ion MS/MS spectra for the characteristic daughter ions of the acyclic and cyclic forms of the diphenylpropanoids. The overlap of parent ion masses in these two parent ion MS/MS spectra show that the ions at those masses are a mixture of the cyclic and acyclic forms of the compound.

and cola drinks with parent ion MS/MS spectra. A deuterium-labeled internal standard was used to provide a reproducibility better than 5%. Regular coffee was found to contain 900 mg/L of caffeine; decaffeinated coffee still contained 44 mg/L. Cola drinks ranged in value from 73 to 158 mg/L of caffeine. One milliliter of the beverage was analyzed, with a one-step extraction of the caffeine into an organic solvent.

Fraisse et al.[648] have described the use of daughter ion MS/MS for the analysis of vanillin in beans of *Vanilla plantifolia* and *V. tahitensis*. Comparisons of analytical performance were made with a high-performance liquid chromatographic assay and a GC/MS assay. The MS/MS assay was based on a multiple-reaction monitoring experiment, and since it required no sample preparation, was the fastest to use. The coefficient of variance for the MS/MS work was reported to be 10%, while the variance for the HPLC experiment (UV detection) was 2%. No variance for the GC/MS analysis was reported, but a linear range of calibration over the range of 4 ng to 1 mg was established.

Better analytical techniques applied to the analysis of food products result in the discovery of new endogenous components. Such is the case with arsenic-containing compounds in fish, lobster, and shrimp. Lau et al.[649] have applied various MS/MS experiments to the confirmation of arsenobetaine in sole, haddock, lobster, and shrimp, and to the identification of arsenocholine in shrimp. Cleaned-up extracts of the sample were ionized by fast atom bombardment, which produces abundant ion signals for these precharged organic compounds. Daughter ion MS/MS spectra confirmed the structure of the compounds, and parent ion MS/MS spectra indicated that the ions observed were not fragments of higher mass arsenic-containing compounds.

6.4.2. Food Additives

Many food products are processed and the identity and amounts of modifiers added are of interest. Many agricultural products retain traces of pesticides or herbicides that may have been applied in the field, and the levels of these residues are also typically of interest. MS/MS has been applied to both of these problems. Sphon et al.[650] have briefly described several applications of MS/MS to the general problem of contaminant analysis in foods, including the determination of aflatoxin B_1 in catfish food down to the 10-ppb level and the confirmation of the presence of aldicarb sulfoxide and sulfone (insecticides) in watermelon at the 50-ppb level.

Oxatetracycline is routinely administered to cows to treat the common respiratory infections to which they are prone. Residual amounts of this drug accumulate in the milk and the meat and must be monitored in these matrices at very low levels. Traldi et al.[651] have described the daughter ion MS/MS spectra obtained on a BE instrument, selecting as the parent ion the $M^{+\cdot}$ ion formed in direct exposure probe electron ionization. The limit of detection for oxytetracycline as determined from standard samples is 5 pg; for meat or milk extracts, the limit of detection is 1 ng. The total time required for sample manipulation and analysis was 10 min. The authors propose the daughter ion MS/MS experiment as the method of choice for both the qualitative and quantitative analysis of oxytetracycline in meat and milk residues. A similar method for the analysis of flukicidal compounds by daughter ion MS/MS has been proposed by the same authors.[652]

Brumley et al.[653] have described the use of daughter ion MS/MS (BE instrument) for the identification of sulphonamide drugs in swine liver. The protonated molecules are formed by chemical ionization; daughter ion MS/MS spectra of 18 sulphonamide drugs have been measured and are used to confirm the identity of the drugs found in liver extracts. Liver was homogenized, extracted with an organic solvent, back-extracted into an acidified aqueous solution, washed with hexane, neutralized and extracted into methylene chloride, and finally dissolved in methanol for analysis. Recoveries through this step were on the order of 90%. Even with this level of sample cleanup, a large number of coextractives are carried along that interfere with the normal assay methods, and the presence of these compounds prompted the use of MS/MS. The authors conclude that with this level of cleanup, sulphonamide drugs in liver can be reliably determined at the 0.1-ppm level, using the direct insertion probe for sample introduction. The authors caution that full daughter ion MS/MS spectra should be recorded, as opposed to a selected or multiple-reaction monitoring experiment, so that the level of interferences in the quantitative determination can be more accurately gauged.

Trehy et al.[654] have recently described the use of short open tubular columns in conjunction with daughter ion MS/MS for the quantitative determination of the pesticide aldicarb and its toxic residues. The higher sample flux that elutes from a short chromatographic column as compared to a longer chromatographic column of greater separatory power provides a lower limit of detection in the MS/MS experiment. The loss in selectivity between components that might otherwise be chromatographically separated is compensated by the use of MS/MS instead of mass spectrometric detection. Methodology for the determination of aldicarb in citrus fruit samples was described, with a total instrumental (QQQ) analysis time per sample of 120 s and a limit of detection of 20 ppb. Andrzejewski et al.[655] have described the use of daughter ion MS/MS spectra for the determination of N-nitrosamines in food products such as cured meats, bacons, and cheese, and the experiment has also been used for the analysis of these compounds in rubber baby-bottle nipples.

Farrow et al.[656] have described the analysis of carotenoids in food extracts with liquid chromatography/mass spectrometry (LC/MS) and LC/MS/MS. In the combination of chromatography with MS/MS, these experiments offer the same analytical advantages as do the GC/MS/MS experiments previously described. Further work of this type can be expected as commercial or custom LC/MS interfaces are added to MS/MS instrumentation. In this particular example, the food extracts contain such high levels of fatty acids that the determination of carotenoids was difficult even with selected ion monitoring on the LC/MS system. The protocol ultimately developed included multiple-reaction monitoring daughter ion MS/MS to determine unambiguously the identity and level of carotenoids in the mixture. Differentiation of alpha and beta carotene was possible through an examination of the full daughter ion MS/MS spectrum of each. Neutral loss MS/MS experiments (loss of 92 as the toluene structure) were also used for a class-specific identification of all of the carotenoids present in the original mixture.

6.5. Forensic Chemistry Applications

The routine identification of drugs in physiological fluids by forensic laboratories usually involves a two-step process that includes an initial screening step, followed by confirmation with GC/MS. The most onerous requirements of this two-step process are the time consumed and the need to maintain legal quality procedures and documentation throughout. Fetterolf[657,658] has described the development of MS/MS procedures that can be used to advantage in forensic laboratories. Of particular advantage is the minimization of sample preparation involved in the MS/MS analysis. For instance, the analysis of cocaine often involves the simultaneous search for metabolites of cocaine. The presence of cocaine in a urine sample without the presence of the metabolites of cocaine is evidence that the cocaine was added to the sample after it was obtained. The major metabolite of cocaine is benzoylecognine, which is very hydrophilic and difficult to extract from the aqueous sample. In addition, while cocaine can be analyzed directly by GC/MS, benzoylecognine must be derivatized to increase its volatility. Using a direct daughter ion MS/MS analysis, Fetterolf[657] has described a method for the rapid determination of cocaine and its metabolites in urine at the 10-ng/mL level, with an analysis time of 1 min/sample and minimal sample preparation. Figure 6-14 compares the daughter ion MS/MS spectrum of cocaine and benzoylecognine as standards and as present in urine. In addition to the selectivity afforded by MS/MS, each of these compounds has a distinctive evaporation profile from the direct insertion probe. For other analyses, such as the screening and identification of barbiturates and amphetamines, or of cannabanoids, MS/MS methods

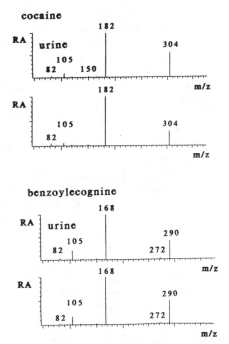

Figure 6-14 ■ Daughter ion MS/MS spectra (multiquadrupole instrument) for the parent ion at m/z 304 $(M + H)^+$ of cocaine as a standard and detected in a urine sample (top), and daughter ion MS/MS spectra of a cocaine metabolite benzoylecognine with the parent ion at m/z 290 $(M + H)^+$ as a standard and in a urine sample (bottom).

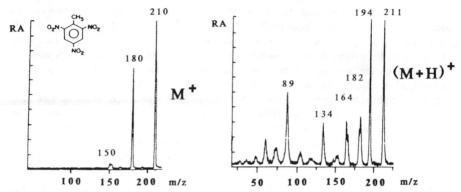

Figure 6-15 ▪ Daughter ion MS/MS spectra of the (top) molecular ion of trinitrotoluene formed by electron ionization and the (bottom) protonated molecule of trinitrotoluene formed by isobutane chemical ionization.

avoid the need to fractionate the sample into acid- and base-soluble fractions and to carry out separate analyses.

McLuckey et al.[659] have described the analysis of explosives by MS/MS. Electron ionization, isobutane chemical ionization, and negative ion chemical ionization were used to form ions from samples of nitrobenzene, heterocyclic nitramine, and nitrate ester explosives. Assays for these materials include determinations in postexplosion debris, in biological matrices (has an individual been in contact with these materials?), or in the atmosphere, as part of security screening at airports or secure installations. The problems associated with these experiments are familiar—sensitivity must be extraordinarily high, specificity must be high, and the speed of analysis must also be high. For instance, security screening at an airport must operate under a very tight time constraint to avoid unacceptable delays.

Figure 6-15 compares the daughter ion MS/MS spectra obtained from the molecular ion of trinitrotoluene (TNT) formed by electron ionization and the protonated molecule of TNT formed by isobutane chemical ionization. Data were measured with a BE instrument; in this case, the limited mass resolution of the daughter ions is not of significance. The dissociation pathways to all of the daughter ions seem straightforward and are in accord with previous MS/MS work on these and similar compounds.[119,660,661] Similarities in the daughter ion MS/MS spectra obtained for each compound class and for each ionization method make feasible MS/MS experiments designed to screen for each compound class. For example, all protonated nitrotoluenes dissociate by loss of $OH^\cdot$ and $NO_2^\cdot$, so that neutral loss MS/MS experiments for losses of 17 and 46 daltons, respectively, could be used to identify possible nitroaromatic components in a sample mixture. Substituted nitroaromatic molecular anions formed in negative ion chemical ionization characteristically lose $OH^\cdot$ and $NO^\cdot$, and a complementary neutral loss MS/MS experiment is feasible under these conditions as well.

Daughter ion MS/MS spectra were obtained for most of the ions found in the electron ionization mass spectrum of "China White," an illicit narcotic whose structural elucidation became a classical problem in analytical spectroscopy.[518] Spectra were obtained on a triple-sector EBE instrument; the daughter ion MS/MS spectra were compared to reference spectra in a library file (see Section 5.3.3). The reference spectra were daughter ion MS/MS spectra obtained for small fragment ions generated from other ionic precursors. Identity of spectra provided substructures that were assembled to produce the overall structure of the drug. In this case, the approach was not specifically to create an abundant molecular ion to be selected for dissociation and to interpret its spectrum from first principles, but rather to build up the molecular structure from detailed analyses of the structures of the smaller fragment ions. Such an approach will succeed as long as there are no significant ion structural rearrangements, and the fragment ions examined are the complementary products of simple bond cleavages.

Fetterolf[662] has published a report in which daughter ion MS/MS analysis was used to identify some of the organic compounds formed in the decomposition of human hair in a pyrolysis electron ionization source. Analysis of human hair for traces of drug use, or identification of the source of hair, is often part of a forensic chemical investigation. Additional applications of this type can be expected in the next few years. In an early application, daughter ion MS/MS has been used for the rapid determination of traces of morphine in the hair of heroin addicts.[663] Morphine is traceable in biological fluids such as blood or urine for only a few days after the use of heroin, placing severe time constraints on the sampling procedure. On the other hand, morphine in the bloodstream is apparently deposited into the growing hair shaft, and the record of its deposition remains. Prior to the development of the MS/MS method, Pelli et al.[663] noted that only radioimmunoassay and liquid chromatography with fluorometric detection had exhibited the requisite specificity and sensitivity for the analysis of morphine in hair. A hair sample is washed, digested, and extracted in order to concentrate the materials from the hair. A total of 1 to 5 g of hair was used in the analysis to produce 1 mL of sample; 5 μL of this extract was placed in the source of the mass spectrometer (BE) operated in the electron ionization mode. Detection limits were 1 to 10 fg for the pure compound, increasing by a factor of 50 in the hair matrix to 50 to 500 fg for the determination of morphine in the hair extract.

6.6. Petroleum and Petroleum Products Applications

The petroleum industry was the initial site of application of mass spectrometry as it was developed in the late 1940s and early 1950s, and its members have been among the first to invest in MS/MS instrumentation. Due to the competitive nature of the business, a large number of applications of the method remain unpublished. As Kondrat[664] has stated, the task facing an analytical chemist in the petroleum industry is often the determination of a single component or group of components in a complex matrix consisting of hydrocarbons mixed with sulfur-, nitrogen-, and oxygen-containing compounds, seasoned with a variety of trace metal-containing compounds. Over time, a number of mass spectrometric methods have been developed, perfected, and standardized.[665] These methods provide a tremendous amount of information about a petroleum-based sample. On a day-to-day basis, this level of detailed information is not required. Instead, methods have been developed for the functional group screening of petroleums, for the identification of trace targeted components such as biomarkers in petroleum, or for the identification of catalyst poisons, particularly those compounds that contain sulfur in its various forms. The capabilities of MS/MS in each of these areas fuel its applications in the petroleum industry.

6.6.1. Geochemical Applications

Biomarkers, typically consisting of compound classes such as terpanes or steranes, are compounds present in petroleums that can be used to correlate a particular sample of petroleum and its source. Oil exploration is greatly aided by chemical evaluation of the petroleum sample, particularly with respect to the nature and distribution of biomarker compounds found within the crude oil mixture. The chemical separation of these compounds from the other components in the mixture is an arduous task, and the investigation of the distribution of these compounds in petroleum samples constituted one of the first applications of MS/MS to problems in the petroleum field.

Gallegos[666] described the application of metastable ion MS/MS methods to the characterization of these compounds in an extract from Green River Shale. Molecular ions from terpanes formed in an electron ionization source typically dissociate in a metastable process to a common skeletal ion at m/z 191. Using an accelerating-voltage scan experiment for the parent ions of this characteristic daughter ion, Gallegos identified parent ions corresponding to C_{31}, C_{30}, C_{29} steranes and C_{20} and C_{21} terpanes. A similar experiment for a daughter ion at m/z 217 provided the distribution of parent ions for steranes. Metastable daughter ion MS/MS experiments have been used to characterize the complex mixture of steroids in crude oils and sedimentary rock extracts.[667] Again, electron ionization was used to create the $M^{+\cdot}$ ions from steranes separated by gas chromatography, and the metastable ion transitions of these ions were monitored. The most common markers are C_{27}, C_{28}, and C_{29} steranes, all of which fragment to a common daughter ion at m/z 217. Successive transitions ($372^+ \rightarrow$, $386^+ \rightarrow$, and $400^+ \rightarrow 217^+$) were monitored during a gas chromatographic separation to emphasize the distribution of these steranes in crude oil fractions.

Philip[668] has provided a recent overview of the use of MS/MS in geochemistry related to oil exploration, specifically with respect to the biomarker hydrocarbons found in crude oil. While much of the geochemical information generated by the analytical laboratories of the oil companies remains proprietary, the widespread use of GC/MS for the analysis of biomarkers is now complemented by use of sophisticated MS/MS experiments. The analytical characteristic of MS/MS which seem to be most valuable in this work is that of speed of analysis. An aliquot of the crude oil is introduced directly into the ion source of the mass spectrometer, and a profile of the biomarker ion distribution is obtained using parent ion MS/MS spectra of daughter ions characteristic of selected compound classes. Description of a particular example follows. However, a general and extremely valuable analytical result provided by the use of MS/MS is the extension of biomarker series to include compounds with molecular weights far higher than those previously observed by GC/MS. These higher mass biomarker molecules ($m/z > 600$ daltons) are volatile (they *do* evaporate from the direct insertion probe), but probably elute so slowly through a GC column that their peaks are broadened to the level of baseline. Alternately, insufficient time may be allowed for their elution. Once analyzed by MS/MS, crude oil samples that appear to be similar are further analyzed by GC/MS or GC/MS/MS methods. The MS/MS experiment pinpoints those samples that can most benefit from the more extensive analysis.

Hopanes and tricyclic terpanes are commonly encountered biomarkers present in crude oil as degradation products of C_{35} compounds present in the source material. Examination of the daughter ion MS/MS spectra of a number of hopanes and tricyclic terpanes established the fact that a characteristic daughter ion at m/z 191 is representative of molecular structures in these series (hopanes at C_nH_{2n-6} and tricyclic terpanes at C_nH_{2n-4}). A parent ion MS/MS spectrum of m/z 191 generates the molecular ion distribution for all parent ions that undergo reaction to that particular daughter ion. Crude oils are designated as similar or different based on the relative abundances of the parent ions corresponding to the various members of the hopane and tricyclic terpane series. For an accurate description of the molecular ion profiles, the data must be summed over the complete evaporation of the crude oil sample from the direct insertion probe. Otherwise, fractional distillation may lead to misleading relative abundance data.

Other compounds that have been investigated as biomarkers in petroleum samples include the porphyrins and the metalloporphyrins.[669] In work aimed at the analysis of metalloporphyrins in biological and geological samples, hydrogen chemical ionization was used with daughter ion MS/MS experiments to study the structures of the porphyrins and their hydrogenation products. Rather than synthesize the hydrogenated forms of the porphyrins outside of the mass spectrometer, hydrogen chemical ionization was used to create not only M^+, but also $(M + nH)^+$, where $n = 2, 4,$ or 6. These adduct ions formed in the source of the mass spectrometer were selected as parent ions in the daughter ion MS/MS experiment (triple-quadrupole instrument). The structures of the hydrogenated forms were deduced from the MS/MS spectra, and the effect of the central metal in directing the fragmentation of the parent ions was established.

Photodissociation rather than collision-induced dissociation has been used by Fukuda and Campana[670] in a study of the daughter ion MS/MS spectra of a series of synthetic porphyrins. The instrument used was a BE mass spectrometer that was modified to allow coaxial ion/photon beam overlap in the second reaction region. Daughter ion MS/MS spectra of porphyrins ionized by FAB provided structural information; meso-substituted porphyrins also provided interpretable photodissociation daughter ion MS/MS spectra. Significantly, neither adduct ions of the porphyrin molecules formed as the result of solvent irradiation nor interfering glycerol cluster ions produce photodissociation products with the wavelengths of laser light used, illustrating a measure of selectivity of this experiment over collisional activation techniques.

Johnson et al.[671] described the application of MS/MS to the study of high carbon number porphyrins isolated from oil. A crude oil sample was separated by column chromatography to isolate a fraction containing the metalloporphyrins. After demetallation, thin-layer chromatography was used to fractionate further the sample. Electron ionization mass spectrometry was used to establish the carbon number ranges and general classes of porphyrins found in each fraction, followed by electron ionization MS/MS analysis based on neutral loss scans to provide a qualitative fingerprint of the lengths of the carbon chains attached to the porphyrin skeleton. Ratios of alkylated isomers of the high carbon number porphyrins may be characteristic of the source of the oil.

6.6.2. Fuel Characterization

As mentioned in the Introduction, the evaluation of petroleum fuels and fuel-derived products for targeted compounds is another large applications area for MS/MS in the petroleum industry. Of all the analytical techniques available to the fuels chemist, MS/MS is unique in the ability to complete a functional group analysis of these materials at low levels of detection commensurate with the normal levels of mixture components. The first application of MS/MS to petroleum-derived samples was that of Zakett.[672] Isobutane and ammonia chemical ionization were used to generate molecular weight profiles from a coal liquid, and daughter ion MS/MS spectra (BE instrument) were used to characterize the compounds present. Comparison of the measured spectra with those of the authentic samples provided a means of confirmation of the presence of tetrahydroquinoline, tetrahydroisoquinoline, 5,6-benzoquinoline, and acridine, among other compounds. The presence of a charge-stripping peak (M^{2+}) was noted to be especially characteristic for nitrogen-containing compounds in CID with the high-energy collision conditions used. These compounds could also be identified by their preferential ionization in ammonia chemical ionization as compared to the more general isobutane chemical ionization. Daughter ion MS/MS experiments were next applied to the analysis of polycyclic aromatic hydrocarbons in solvent-refined coal.[320] In these examples, the use of charge inversion reactions was particularly advantageous. Many of the polycyclic aromatic hydrocarbons can be analyzed by negative ion chemical ionization mass spectrometry. However, the daughter ion MS/MS spectrum of the parent negative ion is uninformative. Since high-energy collisions on a BE instrument were used, the parent negative ion could be oxidized to the parent positive ion in a charge inversion reaction (see Section 3.5.2.1), and the daughter ion MS/MS spectrum of the positive analog of the parent ion recorded. Once compounds were identified, a standard addition experiment was used to provide quantitative information.[673] These early uses of MS/MS in the analysis of petroleum and fuel samples have been reviewed.[451]

Hunt and Shabanowitz[674] described a combination of daughter ion, parent ion, and neutral loss MS/MS experiments implemented on a triple-quadrupole mass spectrometer to determine organosulfur compound classes directly in crude petroleum distillates. Protonated molecules were formed by chemical ionization from thiophenols, dithienyls, trithienyls, and dibenzothiophene derivatives. All of these ions dissociate by loss of HS. A neutral loss (33 daltons) MS/MS spectrum provides the parent ion distribution for these compound classes. Chemical ionization using nitric acid as the reagent gas produces $(M + NO)^+$ ions of particularly high abundance for aliphatic sulfones, which dissociate in low-energy collisions to NO^+. A parent ion MS/MS spectrum of NO^+ thus provides the distribution of molecular weights for the aliphatic sulfones present in the complex mixture. Sample size for these experiments was 1 mg, with the levels of individually identified compounds ranging from 50 to 400 ng. Total analysis time was about 20 min.

Aza- and amino-polynuclear aromatic hydrocarbons in coal-derived liquids were studied by daughter ion MS/MS using a triple-quadrupole and a BE instrument.[251] Chemical ionization was used to create $(M + H)^+$ ions from a solvent-refined coal fraction introduced on the direct insertion probe. In general, different fragment ions were observed in the high-collision energy (sector) and the low-collision energy (quadrupole) daughter ion MS/MS spectra. Low mass daughter ions that result from highly endothermic fragmentation processes are not observed in the low-energy daughter ion MS/MS spectra. Differentiation of isomers was possible with both types of MS/MS spectra, although with a different strategy. Reproducibility of the high-energy daughter ion MS/MS spectra was better than 5%. The low-energy spectra exhibited a greater dependence on instrument parameters such as collision gas pressure and collision energy. Even when these values were held as constant as possible, the reproducibility worsened to 10%.

A nitrogen-base extract of a coal liquid has been examined for partially hydrogenated nitrogen-containing polynuclear aromatic hydrocarbons using MS/MS on a triple-quadrupole instrument.[675] Such compounds are implicated in the mutagenicity of these coal liquids. Hydrogenation of these compounds is generally required before the coal liquid can serve as a fuel source. In the experiments described, neutral loss MS/MS spectra coupled with trifluoroacetyl derivatization of the sample allowed the differentiation of the partially hydrogenated components from other nitrogen-containing components in the mixture. Specifically, primary and secondary amines react with the derivatization reagent to form the trifluoroacetyl derivatives while tertiary amines do not. For the first few members of each series, partially hydrogenated nitrogen heterocycles can be differentiated from primary amino polynuclear aromatic hydrocarbons, both derivatized, by the much higher relative abundance for the product of the loss of 28 daltons (ethylene) observed in the daughter ion MS/MS spectra of the former. Furthermore, primary amines of two types can be distinguished by a parent ion MS/MS spectrum. Ring–NH_2 compounds do not dissociate to the daughter ion at m/z 30, $CH_2NH_2^+$, characteristic of ring–CH_2NH_2 compounds. The use of daughter ion, parent ion, and neutral loss MS/MS experiments provide significant information concerning the presence of partially hydrogenated nitrogen-containing compounds in coal liquids introduced directly into the source of the mass spectrometer.

Wood et al.[676] have described a chemical reduction scheme applied to the characterization of a distillate of coal-derived liquid with special emphasis on the sulfur-containing components. These compounds have a magnified environmental effect due to their transformation during combustion into species that may ultimately contribute to the problem of acid rain. The effect of these compounds on the longevity of refining catalysts is also of some concern. Daughter and parent ion MS/MS spectra were obtained with a triple-quadrupole instrument. Reductive cleavage of aromatic heterocycles containing the thiophene moiety was accomplished with calcium to yield thiophenols. The negative ion daughter ion MS/MS spectra of thiophenols, with either M⁻ · or (M − H)⁻ selected as the parent ion, contain characteristic ions at m/z 122, corresponding to the thiatropylium ion $C_7H_6S^-$. Parent ion MS/MS spectra for this selected ion provide the distribution of alkylbenzothiophenes in the distillate mixture directly. Similarly, sulfones can be reduced to sulfinic acids, which produce SO_2^- upon collision-induced dissociation; a parent ion MS/MS spectrum thus establishes the distribution of this functional group in the mixture as well.[677–679]

This work is especially significant in that it includes a chemical derivatization step designed to take advantage of the functional group specificity of the parent ion MS/MS experiment. No quantitative data were described in this study, which emphasized the qualitative distribution of functional groups directly within fairly complex mixtures. The actual time of analysis was a few minutes, compared with a sample preparation time of several days. Parallel preparation of samples coupled with the short instrumental analysis time suggests a widespread use of the derivatization MS/MS experiment in future functional group screening work.

Wong[680] has described the use of MS/MS for the determination of sulfur compounds present in oil produced from shale. The limitations (federal and state) on the permissible sulfur content vary widely, but adequate analytical methodology for such an assay has not been developed. Detection limits for the targeted volatile sulfur compounds were in the 1-to-10-ppm range, including such compounds as methanethiol, carbonyl sulfide, thiophene, and carbon disulfide. A significant effort was invested to construct a completely computerized triple-quadrupole mass spectrometer so that on-line analysis could be completed. For a rapidly changing gas stream, analytical results for 10 volatile sulfur compounds could be acquired and processed in about 30 s. For kinetic studies, selected reaction monitoring could provide data on the relative concentrations of the same 10 compounds in as short a time as 540 ms. Sulfur compounds in petroleum products have been of interest for some time. McLafferty and Bockhoff[681] demonstrated that thiophene could be detected in gasoline samples at the 25-ppm level. Morgan et al.[682] used electron ionization to form molecular ions of sulfur-containing compounds, which could then be distinguished from hydrocarbons on the basis of their dissociations by loss of HS and H_2S. Sulfur compounds in crude petroleum fractions typically dissociate in low-energy collisions by loss of HS; Hunt and Shabanowitz[674] used neutral loss MS/MS spectra to determine the distribution of the sulfur-containing compounds in crude petroleum fractions. Interestingly, this experiment also involved the use of an isotope-resolved MS/MS experiment in that the loss of $H^{32}S$ from the molecular ion at mass M must always be accompanied by loss of $H^{34}S$ from the molecular ion at mass $M + 2$, and the ratio between the values associated with these losses must be within a fairly narrow abundance ratio.

MS/MS has also been applied to the analysis of diesel particulates and soots emitted from gasoline and diesel engines. The studies described here are those that sample directly from the combustion products rather than from environmental sinks in which these compounds ultimately reside. Wood et al.[683] described the characterization of diesel particulates by MS/MS. Positive ion chemical ionization mass spectrometry was used to form the protonated molecules of each of the components present in a collected particulate sample. Daughter ion MS/MS experiments were used to provide the identification of the individual compounds via comparison with MS/MS spectra of standards. Such experiments provide confirmation of the presence of compounds such as methylphenanthrene and 2-methylnaphthalene in the diesel particulates. Under the conditions used in the MS/MS analysis (low-energy collisions on a triple-quadrupole instrument), it was found that most of the hydroxy compounds in the diesel particulate fraction dissociated by loss of water. A neutral loss MS/MS experiment (18 daltons) could therefore be used to identify hydroxy compounds as a class in the sample. Although the identification of all the individual components was not attempted, the distribution and abundances of hydroxy compounds could be followed with changes in diesel engine speed (rpm) or with changes in the fuel-to-air ratio. Figure 6-16 compares the MS/MS spectra (neutral loss of 18 daltons) measured at a constant engine speed of 1500 rpm as the fuel-to-air ratio was changed from 0.3 to 0.5. With a path from the operating engine to the MS/MS instrument, such experiments could be completed in real time.

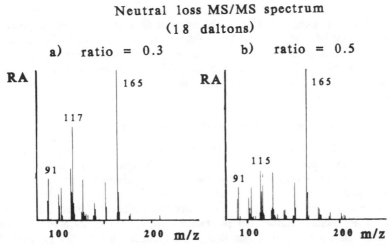

Figure 6-16 ■ Neutral loss MS/MS spectra (loss of 18 daltons) in a search for hydroxy compounds produced in diesel engine combustion. Spectrum (*a*) is obtained when the engine's fuel/air ratio is 0.3, and spectrum (*b*) is obtained when the fuel/air ratio is 0.5.

Neutral loss MS/MS and selected reaction monitoring daughter ion MS/MS experiments have been used by Henderson et al.[684,685] to determine nitro-polynuclear aromatic hydrocarbons in extracts of soot emitted from diesel engines. Isobutane chemical ionization was used to form the protonated molecules; the characteristic dissociation for this class of compounds was the neutral loss of OH (17 daltons). The analysis was performed on 50 to 100 μg of material, with a reported reproducibility of 25% for duplicate samples. Polynuclear aromatic hydrocarbons were added to the diesel fuel and seemed to promote the formation of additional polynuclear aromatic hydrocarbons during the combustion process.

Doretti et al.[686] have also studied diesel particulates by daughter ion MS/MS experiments, in particular the differentiation of anthracene and phenanthrene. Despite the difficulties discussed previously in the use of charge-stripping peaks for quantitative analysis, under a given set of experimental and instrumental conditions, these authors found that the ratios of the various doubly charged ions formed in the high-collision-energy CID daughter ion MS/MS spectra of these two isomers could be used to differentiate and quantitate these samples. The rest of the daughter ion MS/MS spectrum did not contain discriminatory information. Sensitivities of 5 to 10 ng of anthracene or phenanthrene per milligram of diesel particulate were achieved. The precision of the determination of the molar ratio of phenanthrene to anthracene was less than 1%. A total of 5 min of instrument time was required for the analysis.

Missler et al.[687] have used low-energy ion/molecule reactions in both a hybrid mass spectrometer (EBQQ) and an ion trap mass spectrometer to differentiate isomeric olefins found in gasoline. Butadiene was used as the collision gas to form addition products with ionic reactants that contain the double bond; the reaction is formally a Diels–Alder condensation. Isomeric compounds that did not contain the double bond also reacted with the butadiene, but the pathways as shown in the daughter ion MS/MS spectrum were different.

Niwa et al.[688] investigated the distribution of straight-chain fatty acids and fatty acid esters in coal extracts with a combination of field desorption mass spectrometry and daughter ion MS/MS analysis. The extracts of the lignites contain the straight-chain fatty acids with C_{20} to C_{34} and the long-chain fatty acid esters with C_{48} to C_{64}. Daughter ion MS/MS spectra showed that the smaller members of the series are composed of one predominant isomeric form, while the larger compounds contain mixtures of isomers.

6.6.3. Dating Techniques with Accelerator Mass Spectrometry

In petroleum-related research, it is often of interest to date the samples under study. Accelerator mass spectrometry is a specialized technique used to measure the abundances of isotopes of several elements for which the abundance ratios are used as a means to establish the age of the material from which the sample is drawn.[689-692] The most commonly measured atom is ^{14}C, since the ratios of the various isotopes of carbon in an organic sample can be unambiguously tied to its age. The principal isobaric interferent in the determination is the presence of ^{14}N. Several methods are used for the differentiation of the signal from ^{14}C from that of ^{14}N, including the acceleration of the atomic ions to very high energies followed by charge stripping. A high yield of multiply charged ions formed from singly charged ions is obtained when the sample beam is directed through a charge-stripping foil. Energy-loss-rate measurements in a special detection system also serve to differentiate these ions, as does the use of negative rather than positive ions as the original source of sample ions, since $^{14}C^-$ is stable while $^{14}N^-$ is not.

The primary advantage of accelerator mass spectrometry in dating measurements is the fact that the method measures all of the ions in the sample. The dating of samples by accelerator mass spectrometry is a general method. Although many of the applications have been to ^{14}C-based dating, other isotopes such as ^3H, ^{10}Be, ^{41}Ca, and ^{53}Mn have been used in specialized applications.[693,694] A discussion of accelerator mass spectrometry may seem to be out of place in a book of mass spectrometry/mass spectrometry, but the experiment is based on changes in the charge of an ion, just as in the common daughter ion or parent ion MS/MS experiments.) Specifically, for the measurement of ^{14}C, accelerator mass spectrometry is based on the analysis of all of the ^{14}C ions that can be produced from the sample. In contrast, the method of radioactive counting of the decay of ^{14}C samples only those atoms that choose to decay during the integration time of the experiment; for reasonable experimental duration, the number of decaying ions is but a small fraction of the total number of carbon atoms available. As a consequence, the sample sizes required for dating by accelerator mass spectrometry can be several orders of magnitude smaller (milligrams rather than grams) than those required in the radioactivity-based dating method. The significance of this reduction in requisite sample size cannot be understated considering the rarity of the archaeological artifacts that are the samples for these investigations.

6.7. Bioorganic Applications

Chapter 4 deals with the application of MS/MS to the study of small organic ion structures and the interconversion and differentiation of isomeric forms by MS/MS experiments. This chapter describes applications of MS/MS to organic and biological problems with an emphasis on the elucidation of the behavior of classes of compounds or the analysis of complex mixtures of these compounds. The evaluation of MS/MS data from smaller organic structures can include a comparison to thermochemical data or the correlation of MS/MS data with data derived from theoretical calculations. By contrast, application of MS/MS to larger organic and bioorganic structures is a more empirical pursuit, involving the comparison of the MS/MS spectrum of the sample to that of the standard when available or rationalizations of the dissociation chemistry observed from similar compounds.

Problems of biochemical interest have always presented a challenge to mass spectrometry. For many years, the approaches were few and only partially successful, due mainly to the limitations of the ionization methods in use. With the advent of the desorption ionization and nebulization ionization methods, and fast atom bombardment in particular, the creation of ions from large, thermally fragile biomolecules is possible, even if the mechanisms of these ionization methods are not completely understood. A second challenge, and one that remains, is the mass range of the instruments used in the analyses and the interpretation of the information contained in the mass spectrum of a high-mass biomolecule. The progression from analysis of biomolecules with a mass of perhaps 1500 daltons to those up to 5000 daltons in mass, and now extending past 10,000 daltons, is a result of the synergism between ionization developments, on the one hand, and advances in mass analysis technology, particularly magnetic sector and time-of-flight mass spectrometers, on the other. The nature of the information that can be obtained in a mass spectrometric experiment changes dramatically as the mass of the ions is increased. A present challenge is to extend the uses of mass spectrometry in these ranges beyond the determination of molecular weight to include the structural analysis of high mass biomolecules. Such extension presages the use of sophisticated data-processing schemes and interpretation of mass spectra based on pattern recognition algorithms, such as in the computer programs developed for peptide sequence analysis by MS/MS. In many cases, the ionization method that must be used for the analysis of these biomolecules is a relatively soft method; stable molecular ions are formed, but the degree of fragmentation in the source is limited, and the amount of structural information that can be deduced is thereby limited. In many cases, the burden of structural analysis of these high mass ions has shifted to MS/MS.

6.7.1. Biological Compound Classes

Summarized in this section are studies that have used MS/MS to establish the dissociation patterns exhibited for samples of similar structure. The goal of these studies is to understand the mechanisms through which the energy of the activation process is converted into structurally related dissociations of the parent ion, and to use that understanding to rationalize the behavior of an otherwise unknown sample and establish its structure. A result of this approach has been the criticism often leveled at MS/MS that it provides the spectrum and the structure of known rather than unknown compounds. The appearance of such criticism is expected for a technique just entering its second decade of application. Much the same criticism was leveled at the work of organic mass spectrometry as that method was first developed in terms of spectral/structure correlations.

6.7.1.1. Carbohydrates and Saccharides. Structural elucidation of complex carbohydrate structure has been a particularly vexing biochemical problem to which mass spectrometry has begun to make important contributions. To date, most of the work has been in the structural confirmation of saccharides resolved by gas chromatography. The importance of the structural identification problem is such that a more extensive and diverse contribution by mass spectrometry should be expected. For instance, the properties of biological surfaces are to a large extent controlled by the surface topography, in turn determined by the presence and mobility of glycoproteins and glycolipids. The sequence, linkage, and conformations of these carbohydrates are all of intense interest, and usually the sample can be isolated only in vanishingly small amounts. The strategy for sequence determination of polysaccharides has typically involved dual steps of permethylation and controlled degradation to a family of oligomers. Components of the mixture were then separated by gas chromatography, and the individual eluants ionized by electron or chemical ionization mass spectrometry.

The use of MS/MS experiments in the structural analysis of carbohydrates and saccharides has been a natural adjunct to the increased deployment of SIMS, FAB, and laser desorption techniques for the ionization of these compounds. As mentioned in the introduction to this section, these "soft" ionization methods produce mass spectra in which the even-electron molecular ion is the predominant ion, and the relative amount of fragmentation is reduced. The determination of molecular weight is thus made more simple, but detailed structural analysis becomes concomitantly more difficult. Daughter ion MS/MS experiments using the protonated molecule, for example, as the parent ion, regain the ability to establish the structure of the saccharide. The use of mass spectrometric techniques in the analysis of complex carbohydrate structure has been reviewed by Reinhold and Carr.[695]

Some of the earliest work on the collision-induced dissociations of saccharides was reported by Rollgen. The daughter ion MS/MS spectrum of the $(M + Na)^+$ ion formed from glucose by field desorption was first reported in 1975.[696] Rollgen et al.[496] later compared the daughter ion MS/MS spectra of $(M + Li)^+$, $(M + Na)^+$, and $(M + K)^+$ ions of monosaccharides formed by field desorption. This comparison is presented in Figure 6-17. Rollgen suggests that the nature of the dissociations induced by the high-energy collisional activation is essentially independent of the identity of the alkali ion, and that the alkali ion is retained in most of the ions in the daughter ion MS/MS spectrum. This early result suggested that the alkali metal could be chosen as convenient to create a parent ion $(M + Li)^+$, $(M + Na)^+$, or $(M + K)^+$ without significant changes in the daughter ion MS/MS spectrum. This result stands in contrast to results obtained for daughter ion MS/MS analysis of peptides cationized by various alkali metals, in which the choice of the alkali has a significant effect (see Section 6.7.1.6).

de Jong et al.[697] reported that the daughter ion MS/MS spectra of protonated molecules of permethylated disaccharides formed by methane chemical ionization could not be used to differentiate between the isomeric disaccharides. A second report[698] measured the daughter ion MS/MS spectra of the same ions now formed by ammonia and trimethylamine chemical ionization, showing that the anomeric disaccharides could be distinguished on the basis of the daughter ion MS/MS spectra. It was necessary to use both reagent gases to differentiate the isomers of the 22 disaccharides studied. Furthermore, the introduction of water into the trimethylamine reagent gas plasma accentuated the differences observed in the daughter ion MS/MS spectra. Although the authors attribute this effect to a change in the internal energy of the

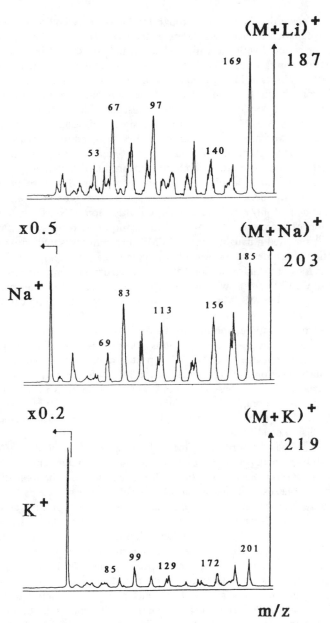

Figure 6-17 ■ Comparison of the daughter ion MS/MS spectra of $(M + Li)^+$, $(M + Na)^+$, and $(M + K)^+$ ions of D-glucose created by field desorption. The position on the mass scale of the parent ion is indicated by the arrow.

protonated molecule that perseveres through the high-energy collisional activation, an alternative explanation invokes two forms of the $(M + H)^+$ ion, the first formed in a proton transfer reaction with the trimethylamine and the second in a distinct proton transfer reaction with water. The authors describe a flowchart that uses the ion abundance ratios observed in the daughter ion MS/MS spectra, in conjunction with information from the electron ionization and methane chemical ionization mass spectra, which can be used to identify an unknown disaccharide isomer.

The daughter ion MS/MS spectrum of the simple disaccharide sucrose has been reported by several workers, often as part of studies designed to showcase instrumental or ionization method advances. Aubagnac et al.[699] reported that the $(M + Na)^+$ ion from sucrose formed by fast atom bombardment produced daughter ions at m/z 185 and m/z 203, identified as a split between the two sugar units with the sodium ion associated with either species, although loss of water from the ion at m/z 203 would also yield the ion at m/z 185. A similar daughter ion MS/MS spectrum of the $(M + Na)^+$ ion of sucrose created by field desorption was reported by Puzo et al.[700] Cerny et al.[701] reported a daughter ion MS/MS spectrum measured with more refined instrumentation. Field desorption was used to create the $(M + Na)^+$ ion from sucrose, and the major ions in the daughter ion MS/MS spectrum were interpreted with respect to the structure. In agreement with the results of Aubagnac et al., the most abundant ions in the daughter ion MS/MS spectrum correspond to the sodium-cationized sugar units.

Laser desorption with a continuous wave infrared laser was used to provide $(M + Li)^+$, $(M + Na)^+$, and $(M + K)^+$ ions for mono- and disaccharides.[702] The low-collision-energy daughter ion MS/MS spectra were then measured on a triple-quadrupole mass spectrometer. The daughter ion MS/MS spectrum of the $(M + Na)^+$ ion sucrose measured under these conditions contained the ions at m/z 185 and 203 described previously, along with a sodium ion of low abundance. In general, the lithiated adduct ions provide more complex daughter ion MS/MS spectra under low-energy collision conditions.

Puzo et al.[703] have described a method using daughter ion MS/MS spectra (obtained with high-energy collisions on a BE instrument) to differentiate the stereoisomers of various hexoses. Fast atom bombardment was used as the ionization method, and a cluster ion (hexose − cation − matrix molecule)$^+$ was chosen as the parent ion, with the cation selected from lithium, sodium, potassium, or rubidium, and the matrix molecule from either glycerol or diethanolamine. The cation affinity sequence of the various hexoses was established and a relationship determined between the structures and the configurations of the different aldohexoses and their ability to retain the alkali cation in a collision-induced dissociation process. Fournie and Puzo[704] have also used daughter ion MS/MS spectra to differentiate the stereoisomeric forms of N-acetyl-hexosamines. Again, the dissociations of the cluster ion formed between the sample, a cation, and a molecule of solvent (the cation is lithium, sodium, potassium, or rubidium) were investigated. The relative distribution of the central cation between the solvent molecule (glycerol or diethanolamine) and the sugar could be interpreted to differentiate stereoisomeric compounds such as 2-acetamido-2-deoxy-D-sugar, in which the sugar was glucose, mannose, or galactose. The stereoisomerism is reflected in the cation affinity of the sugar as deduced from the unimolecular dissociations of the cluster ion. These authors note that the reproducibility of the fragment ion ratios was improved when the results of unimolecular rather than collision-induced dissociations were used. These experiments are an application of the fundamental experiment described in Section 4.4.3.1 on the dissociation of ion-bound dimers.

In recent work, Puzo et al.[705] have used daughter ion MS/MS spectra to differentiate anomeric methyl glycosides, selecting as the parent ion the $(M + \text{cation} + \text{glycerol}))^+$ ion formed by fast atom bombardment of a sodium or lithium iodide-doped glycerol solution of the glycoside. Competition between reformation of the cationized glycoside molecule and formation of the cationized solvent molecule after collisional activation leads to characteristic daughter ion ratios that were used to distinguish between glucopyranosides and galactopyranosides.

Tondeur et al.[706] have studied the daughter ion MS/MS spectra of hexose and pentose monosaccharides ionized from glycerol or diethanolamine solutions by fast atom bombardment mass spectrometry. In the mass spectra, protonated molecules were usually absent, replaced by abundant ions $(M + S + H)^+$ in which S represents a molecule of the solvent. For all of the sugars investigated, high-energy collision-induced dissociation of the $(M + S + H)^+$ parent ion, with glycerol as the solvent, provided relatively abundant daughter ions that could be interpreted with respect to the structure of the sugar. With diethanolamine as the solvent, the daughter ion MS/MS spectrum contained only one abundant ion corresponding to protonated diethanolamine. The results of Tondeur are in agreement with the earlier work of Puzo and Prome[707] that showed that the $(M + S + H)^+$ ions of trehalose with triethanolamine solvent dissociated to the protonated molecule of the solvent, while the dissociation of the glycerol-solvated ion provided useful structural information. Puzo and Prome[708] also showed that the addition of alkali iodide salts to the solution produced abundant ions $(M + S + \text{alkali})^+$ in fast atom bombardment mass spectrometry. Collision-induced dissociations of these cluster ions provided an important pathway for formation of the cationized sugar molecules in the positive ion fast atom bombardment mass spectrum. For the nonreducing sugar trehalose, these dissociations were investigated in some detail for the cations lithium, sodium, potassium, rubidium, and cesium, and for the solvents glycerol, diethanolamine, and triethanolamine. The ratio of the ion abundances between the cationized trehalose and the cationized solvent molecule is independent of the identity of the alkali ion and independent of the solvent for glycerol and diethanolamine. For triethanolamine, formation of the cationized solvent is enhanced. The authors suggest that the higher basicity of the triethanolamine matrix leads to this effect; the important parameter is actually the higher cation affinity of the triethanolamine molecule as compared to that of diethanolamine or glycerol.

Carr et al.[709] have reported daughter ion MS/MS spectra of isomeric hexasaccharide alditols and several N-linked oligosaccharides, with measurements completed on a four-sector instrument of BEEB geometry. High-energy collision-induced dissociations occurred between the electric sectors. A linked scan (see Chapter 2) is necessary to analyze the daughter ions formed in this region, since the experiment is equivalent to analyzing the daughter ions formed in the first reaction region of an EB mass spectrometer. In addition, although the parent ions could be selected with high resolution through the first two sectors of the double-focusing instrument, this is not the usual mode of operation, since the ion transmission through the instrument is greatly attenuated with increased resolution. Selection of the monoisotopic parent ion with an exclusive composition of ^{12}C, ^{1}H, and ^{16}O requires only unit mass resolution; the lower resolution of the linked-scan daughter ion MS/MS spectrum becomes of significance only when the masses of the daughter ions are within a few daltons of each other. In order to interpret the daughter ion MS/MS spectra for these compounds, and for higher mass peptides as well, it is usually necessary to select the monoisotopic parent ion. As the mass of the parent ion increases, the relative abundance contribution of this ion to the molecular ion envelope decreases. The practical limit for the daughter ion MS/MS analysis of biomolecules due to this reason is about 4000 daltons. The limitations in present-day instrumentation make experiments with parent ions of masses higher than this difficult.

Carr et al.[709] used fast atom bombardment to create the protonated molecules of the saccharides, then selected as the parent ions in a high-energy collision-induced dissociation experiment. Typically, the positive ion FAB mass spectrum contained an abundant protonated molecule, but no distinguishable structure-related fragments other than the loss of a terminal galactose fragment. The ions in the daughter ion MS/MS spectra are much more abundant and readily interpreted in terms of the structure of the saccharide. Glycosidic bonds are cleaved and dehydration also occurs. Some isomeric differentiation was possible. For example, the relative abundances of the daughter ions formed from the parent ion of a linear hexasaccharide alditol were distinct from those formed from the isomeric branched hexasaccharide. The authors note that derivatization reactions (used to enhance the detectability of the saccharide in a liquid chromatography/spectroscopic detector separation) created a product in which the derivative directs the fragmentation of the protonated molecule of the hexasaccharide, as only a limited amount of dissociation originating in the nonreducing, derivatized end of the linear molecule is observed. Figure 6-18 is the daughter ion MS/MS spectrum of a hexasaccharide derivative, ionized with fast atom bombardment to produce the protonated molecule. The rationalization of the dissociation pattern is also provided in the figure. Note that none of the daughter ions exhibit an isotopic distribution since the monoisotopic parent in was selected for the collision-induced dissociation.

Chen et al.[710] have recently described the use of daughter ion MS/MS to determine the structures of steroid oligosaccharides that contained from two to four sugar units. The most significant ion in the positive ion FAB mass spectrum was the protonated molecule $(M + H)^+$, and in some cases, the signals from the FAB solvent interfered with the assignment of the fragment ions in the FAB mass spectrum. Daughter ion MS/MS of the $(M + H)^+$ ion provided an unambiguous identification of the fragment ions, allowing the identification and sequence of the sugar units in the molecule to be determined. Furthermore, the daughter ion MS/MS spectra of related compounds exhibited common daughter ions that promise future use in parent ion MS/MS experiments to characterize mixtures of these compounds, which occur naturally in Chinese medicinal herbs.

Daughter ion MS/MS spectra measured on a QQQ instrument were used to study a series of linkage-isomeric fucosyl-lactosamine compounds by Pamidi et al.[711] Both protonated molecules and abundant fragment ions in the positive ion chemical ionization mass spectrum were selected as parent ions to determine the extent to which the linkage information was retained. Differences in the relative abundances of the daughter ions could be correlated with the linkage structures of both the protonated molecules and the fragment ions. In general, the differences in relative abundance were factors of 2 or 3. A similar experiment was carried out by Guevremeont and Wright[712]; in this case, an MS/MS/MS experiment was completed on a hybrid instrument of BEQQ geometry. Hexose and 2-amino-2-deoxy-hexose stereoisomers were studied. It was found that the daughter ion MS/MS experiments allowed the differentiation of the

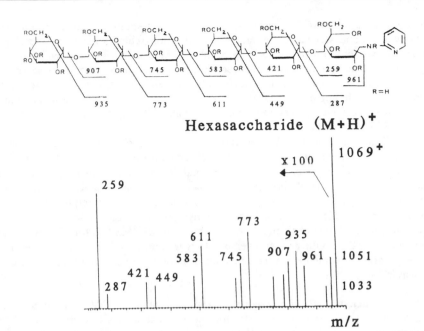

Figure 6-18 ■ Daughter ion MS/MS spectrum of the protonated molecule at m/z 1069 formed from a hexasaccharide derivative by fast atom bombardment. The structure of the hexasaccharide is included in the figure, along with the rationalizations for the dissociations to produce the daughter ions.

peracetylated saccharides more easily than for the underivatized sugars. Daughter ion MS/MS allowed the differentiation of derivatized glucose, galactose, and mannose, but allose was not easily distinguished from galactose. Glucosamine, galactosamine, and mannosamine can also be distinguished.

Granddaughter ions generated in an MS/MS/MS experiment show intriguing differences. Some of the daughter ions retain isomeric structural information while others do not. For instance, acetate is lost from the peracetylated sugars to form a fragment ion at m/z 331, and this fragments in a first MS/MS experiment to form daughter ions at m/z 271, 211, and 169. The granddaughter ions of the m/z 271 ion from acetylated glucose and galactose retain isomeric information, but the dissociations of the ion at m/z 211 from glucose and galactose yield identical granddaughter MS/MS spectra.

The preceding examples illustrate a few specific applications of MS/MS in the structural analysis of carbohydrates and saccharides, but it is also accurate to note that applications of MS/MS in peptide sequencing analysis have been much more rapidly developed. To this point, there appear to be two major reasons for the slower development of MS/MS in carbohydrate and saccharide studies. The first is related to the ionization methods available. Fast atom bombardment has been extensively used in the creation of $(M + H)^+$ or $(M + Na)^+$ ions from neutral molecules of these compounds. In some instances, as in the ionization of hexasaccharide alditols,[709] sensitivity was sufficient to record MS/MS spectra from 5 μg of material. In many other situations, it seems as if FAB does not provide a good sputtered ion signal for large carbohydrates and saccharides, although derivatization schemes have been developed by several workers to help address this problem. These schemes often involve perme- thylation and peracetylation of oligosaccharides, for instance, and increase sputter yields as well as increasing the degree of fragmentation in the FAB spectrum. Additional developments in new derivatization reactions and new FAB and liquid SIMS matrices can be expected in the near future and will help to accelerate the growth of MS/MS in this area.

Carr[709] has suggested that a greater degree of dissociation is observed for an oligosaccharide parent ion compared to that observed for a peptide parent ion of equal mass, since the cleavage of the glycosidic bond is more facile than cleavage of the secondary amide bonds of the peptide backbone. Additional studies of model com- pounds are required to determine if this is true. However, interpretation of the daughter ion MS/MS spectra of large carbohydrates and saccharides is likely to be more difficult than the interpretation of daughter ion MS/MS spectra of peptides (see Section 6.7.1.6) due to the greater diversity of sugar structures that can be present in an oligosaccharide as compared to the constrained set of 20 or so amino acids that can be present in the peptide.

6.7.1.2. Nucleosides and Nucleotides. The study of nucleosides and nucleotides is one of the earliest uses of MS/MS for biochemical analysis. During the 1970s, predominant theories of cancer etiology focused on the role of damage to DNA and RNA molecules in the cell, and in particular, on the methylation of the nucleotides and nucleosides by environmental and toxic agents that could be linked to cancer. Much effort was expended on the development of methods for the separation of methylated nucleosides and nucleotides from their naturally occurring counterparts. The attraction of MS/MS in these studies resided in the decreased sample preparation, and in the decreased chances for sample contamination, in a direct analysis of biological mixture for the presence of these particular compounds. There have been notable successes, but also some misleading results, as can be expected in the development of any analytical methodology.

Levsen and Schulten[713] described the analysis of pyrolysis products of deoxyribonu- cleic acid by daughter ion MS/MS in 1976. The early date places this as one of the first applications of MS/MS to mixture analysis of any type of sample. Herring DNA was placed in a glass tube connected directly to the source of a BE mass spectrometer, and the sample was heated to 600°C within 1 min. The volatile products of the pyrolysis were swept into an electron ionization source, and daughter ion MS/MS spectra were recorded of parent ions with masses of less than 100 daltons. The six most intense ions in the pyrolysis mass spectrum were studied and their structures assigned by a comparison of the daughter ion MS/MS spectra to those of standards. Since the peaks observed in the pyrolysis mass spectrum were those of the stable molecular ions of small organic molecules, the identification was straightforward, and such compounds as furan, methylfuran, and furfuryl alcohol were identified as the products of the pyroly- sis.

Schoen at al.[714] described the pyrolysis of salmon sperm DNA on the direct insertion probe and the measurement of daughter ion MS/MS spectra on a BE instrument. The data were interpreted to show the presence of methylated bases in the original salmon sperm sample. Identification of 1-methyladenine was of particular importance. It was later concluded[715] that pyrolysis of a synthetic sample of poly(dAdT)–poly(dAdT) also produces detectable quantities of 1-methyladenine. Therefore, despite the match in the daughter ion MS/MS spectra, the presence of 1-methyladenine on the original experiments could be an artifact of the sample preparation procedure. This is an instance of a matrix effect in which thermally induced reactions of the sample in the ion source produce the ionic surrogate of a new molecule not present in the original sample.

Straub and Burlingame[716] used a combination of field desorption mass spectrometry and daughter ion MS/MS to study modified polynucleotides. Conventional derivatization procedures and high-resolution electron ionization mass spectrometry had been used by these workers to study the interaction of the carcinogen benzo-[a]-pyrene with DNA. Successful derivatization and analyses required micrograms of the sample. Furthermore, the derivatization necessary to increase the volatility of the sample increased the mass of the sample beyond the mass range of the instrument then in use and limited the study to that of the monomers of methylated DNA. Present-day instruments are less limited in terms of the accessible mass range, but trimethylsilylation at multiple sites within the sample molecule does produce a significant increase in the mass range that must be scanned during the course of the mass spectrometric analysis. With daughter ion MS/MS and field desorption mass spectrometry, monomers and dimers of the modified DNA could be analyzed, and the amount of sample necessary was reduced to the point at which the minor products of the chemical reaction could be identified.

Thomson et al.[717] described the technique of liquid ion evaporation as an ionization method coupled with a triple-quadrupole MS/MS instrument. One of the samples analyzed was adenosine triphosphate. The base peak in the negative ion mass spectrum was that corresponding to the $(M - 2)^{2-}$ molecular ion of the disodium salt. The mass of this ion is 505 daltons, but the signal is observed at a position on the mass-to-charge scale of 252.5 because of the presence of two charges. Since most of the daughter ions will be singly charged, the daughter ion MS/MS spectrum must be recorded for mass-to-charge values *higher* than that of the parent ion. The monomeric metaphosphate ion at m/z 79 is observed in the daughter ion MS/MS spectrum, as is a daughter ion at m/z 426 that corresponds to $(M - PO_3)^-$.

Sindona et al. have published a series of papers describing the applications of daughter ion MS/MS to oligonucleotides. The first studies[718,719] used fast atom bombardment mass spectrometry to produce ions from isomeric 2'-deoxyoligonucleotide salts. Specifically, $3' \rightarrow 5'$- and $5' \rightarrow 5'$-dithymidylic anions were analyzed as their triethylammonium salts. Both the negative ion FAB mass spectrum and the daughter ion MS/MS spectrum of the anion of the salt provided the ability to distinguish between the $3' \rightarrow 5'$- and the $5' \rightarrow 5''$-isomers. Fast atom bombardment and daughter ion MS/MS have also been applied to the study of deoxyoligonucleotide triethylammonium salts such as d-TACC and d-GGTA.[720] The negative ion FAB mass spectrum of each of these compounds contained an intense signal for the anion of the salt; this ion was selected as the parent ion in the MS/MS experiment. Interpretation of the mass spectrum itself did not allow a clear identification of the first nucleotide of the sequence that bears a 3'-phosphate linkage. Furthermore, the possibility of structural rearrangements in many ionization methods for the analysis of the oligonucleotides was specifically mentioned. The daughter ion MS/MS spectra provided a clear structural determination of the sequence of the oligonucleotide, and the use of fast atom bombardment ionization was necessary to avoid the structural rearrangements that might otherwise take place.

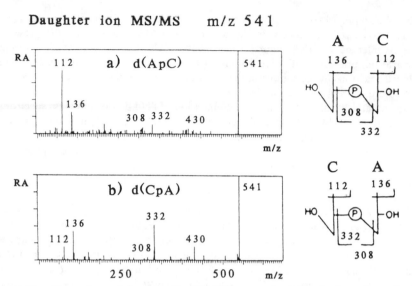

Figure 6-19 ■ Comparison of daughter ion MS/MS spectra for isomeric deoxyribonucleoside phosphates d(ApC) [spectrum (*a*)] and d(CpA) [spectrum (*b*)).

Linscheid and Burlingame[721] used linked scanning on an EB instrument to study the daughter ion MS/MS spectra of parent ions generated from six protonated dinucleoside monophosphates formed by field desorption. The data provided a distinction between isomeric sequences in both the ribo- and the deoxyribo-nucleotide series. Figure 6-19 compares the daughter ion MS/MS spectra obtained for the isomeric deoxyribonucleoside phosphates d(ApC) and d(CpA). The intensities of the fragment ions were used to discriminate between the structures of these isomers.

The work by Linscheid and Burlingame showed that patterns of intensities in the daughter ion MS/MS spectra could be used to sequence oligonucleotides. Extensions to a pattern of systematic recognition required a thorough study of a large number of nucleosides and nucleotides. In 1984, Crow et al.[722] described the positive and negative ion fast atom bombardment spectra and the daughter ion MS/MS spectra measured for the protonated and deprotonated parent molecular ions for 30 nucleosides and 2 isomeric dinucleotides. The systematic spectral interpretation of the positive and negative ion FAB mass spectra was reported. Although the $(M + H)^+$ and BH_2^+ (B is the base) ions were always abundant in the positive ion spectrum, fragment ions corresponding to the sugar portion of the molecule were difficult to discern above the background level of signals from the glycerol matrix. Similarly, the $(M - H)^-$ and B^- ions were predominant in the negative ion FAB mass spectrum, but the sugar ions were low in abundance and intermittent in appearance.

The daughter ion MS/MS spectra of both the positive $(M + H)^+$ and negative $(M - H)^-$ ions of the nucleosides convey much the same information as do the respective mass spectra. The major advantage of the MS/MS technique is that any ambiguity concerning the source of fragment ions is removed, and the ions of low relative abundance are seen more clearly in the daughter ion MS/MS spectra because of the reduction in chemical noise inherent in the mass selection of the parent ion. The

Daughter ion MS/MS

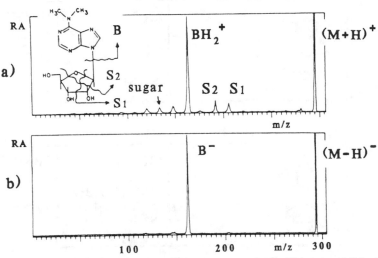

Figure 6-20 ■ Comparison of daughter ion MS/MS spectra of $(M + H)^+$ (*a*) and $(M - H)^-$ (*b*) parent ions from N^6, N^6-dimethyladenosine.

base peak in the daughter ion MS/MS spectra of each of the protonated molecules $(M + H)^+$ is almost always the BH_2^+ ion; daughter ions corresponding to the sugar portion of the molecule are usually present. The daughter ion MS/MS spectra of the $(M - H)^-$ ions tend to be of lower intensity and are dominated to a greater degree by the BH^- ion, and the other daughter ions, including those that originate in the sugar moiety, are of low relative abundance. Figure 6-20 compares the daughter ion MS/MS spectra of the $(M + H)^+$ and $(M - H)^-$ ions of N^6, N^6-dimethyladenosine and illustrates these effects.

Several substituted nucleosides were studied by Crow et al., and the daughter ion MS/MS spectra could be useful in differentiating the position of the substitution. Figure 6-21 is an illustration of this capability for 5- and 6-bromotubercidin. The main features of the daughter ion MS/MS spectrum are similar, but the 5-isomer exhibits a small but reproducible loss of HBr. The rationalized mechanism places the added proton in the amine function, with loss of HBr possible in the 5-isomer, but not in the 6-isomer.

In such a broad study of 30 nucleosides, mass overlaps between the sample ions and the abundant ions in the positive or negative mass spectrum of glycerol matrix are inevitable. Such overlaps occur for BH_2^+ in which the base is cytidine or uridine, and an interference from $(3G-2(H_2O))^+$ can be expected, where G is the glycerol molecule. For a substituted nucleoside, 2-phenylethyl tubercidin, the mass of the protonated molecule overlaps with the abundant signal for $(4G + H)^+$ at m/z 369. Figure 6-22 is an illustration of the high-resolution parent ion selection capability of the EBE instrument used by Crow et al. in an experiment that separates the parent ions and measures the daughter ion MS/MS spectra of each component individually. The top MS/MS spectrum (*A*) is that measured when the parent ion is selected with a

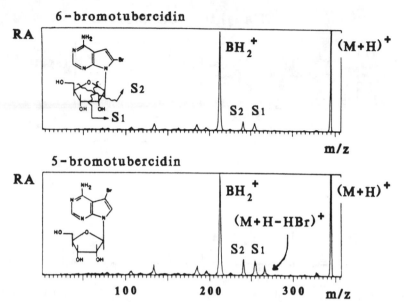

Figure 6-21 ■ Daughter ion MS/MS spectra of the $(M + H)^+$ parent ions from 6- (top) and 5-bromotubercidin (bottom), demonstrating the differentiation of these isomers.

resolution of 2000. At this resolution, the signals from both the sample and the glycerol matrix overlap, and the measured daughter ion MS/MS spectrum is a combination of the two spectra. At a parent ion resolution of 10,000, the contributions from the sample (spectrum B) and the glycerol tetramer (spectrum C) can be separately measured. The ion at m/z 277 in the daughter ion MS/MS spectrum of the glycerol tetramer interferes with and almost masks the signal from one of the sugar fragment ions at m/z 279. Higher resolution in the parent ion selection resolves this interference; alternatively, unit resolution in the measurement of the daughter ion MS/MS spectrum, as is available with triple-quadrupole and hybrid instruments, would also provide a solution in this particular case.

Cerny et al.[723] later expanded the fast atom bombardment MS/MS technique to the study of mono- and dinucleotides. Negative ion FAB mass spectra contain the $(M - H)^-$ ions that were selected as the parent ions in this study. The $(M - H)^-$ ions of all the dinucleotides eliminate the base (B) as BH, and this loss is facilitated from the 3′- rather than the 5′-terminus. Isomeric dinucleotides can be distinguished on the basis of this dissociation. Comparison of the relative abundances of the daughter ions in the MS/MS spectra generated by loss of BH generates a scale of the inherent gas phase basicities of the nucleoside base anions, inferred to be $C^- > A^-$, T^-, $> G^-$ (Cytidine, Adenosine, Thymidine, and Guanosine). A logical approach to the determination of the structure of an unknown dinucleotide is to measure the negative ion fast atom bombardment mass spectrum to establish the molecular mass, followed by a daughter ion MS/MS spectrum of the $(M - H)^-$ ion. Identification of the bases can be made from the dominant loss of BH and the formation of B^- and other sequence ions.

Daughter ion MS/MS

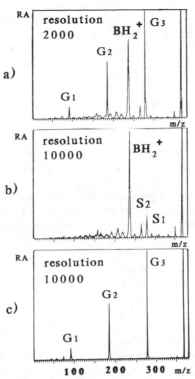

Figure 6-22 ■ High-resolution parent ion selection for daughter ion MS/MS in an EBE instrument. (*a*) Parent ion resolution of 2000, giving a daughter ion MS/MS spectrum with contributions from dissociations of both the sample and glycerol-derived ions at the same nominal parent ion mass. (*b*) Same daughter ion MS/MS spectrum at a resolution of 10,000, derived from sample ions. (*c*) Same daughter ion MS/MS spectrum at a resolution of 10,000, derived from glycerol ions. The parent ion is the ion at m/z 369 from 2-phenylethyltubercidin. S_1, S_2, and BH_2^+ refer to fragments as shown in Figure 6-21. B represents the base moiety.

The base located at the 3′-terminus is apparent because of the greater facility of the BH loss for this substituent.

Kingston et al.[724] studied the FAB mass spectra and daughter ion MS/MS spectra of cyclic nucleotides adenosine 3′,5′-cyclic monophosphate (AMP), cytidine 3′,5′-cyclic monophosphate (CMP), and guanosine 3′,5′-cyclic monophosphate (GMP), thought to play a central role in mammalian cellular regulation. The positive ion FAB mass spectra provide structural information, but cannot be used to distinguish the 3′,5′-isomers from the 2′,3′-cyclic isomers. The daughter ion MS/MS spectra measured on a BE instrument, on the other hand, can distinguish these isomers for both the cyclic CMP and the cyclic GMP. Spectra could be obtained and a qualitative identification made with 1 to 5 μg of the sample present on the FAB probe. Quantitative evaluation was attempted with internal standards, but was unsuccessful due to poor solubility of the standards in the glycerol solution. As in the paper of Crow et al.,[722] Kingston et al. note that the matrix background signals from glycerol can introduce interferences into the daughter ion MS/MS spectra. In addition to isobaric interferences in which there is an explicit overlap at the integral mass of a signal from the sample with that of the matrix, Kingston et al. discuss a more subtle interference that arises through the dissociation of higher mass glycerol-containing ions in the first reaction region of the BE instrument.[26,27] These interferences are more prevalent in the higher kinetic energy portion of the daughter ion MS/MS spectrum (measured with a scan of the electric sector voltage) and are also more prevalent in the daughter ion MS/MS spectra of lower mass parent ions.

In a related study, Kingston et al.[725] have used high-collision-energy daughter ion MS/MS to discriminate between the positional isomers of nucleoside monophosphates. Three positional isomers of uridine, cytidine, adenosine, and guanosine monophosphates were used as model compounds in this study. Fast atom bombardment ionization was used to create the protonated molecule $(M + H)^+$ selected as the parent ion. The ratios of the characteristic daughter ions corresponding to cleavages between the base, the sugar, and the phosphate were reproducibly different for the various positional isomers.

Kralj et al.[726] measured the positive ion FAB mass spectra of 13 synthetic nucleosides. Daughter ion MS/MS spectra were measured with high-energy collisions in the third reaction region of an EBE instrument. Many of the dissociations of the $(M + H)^+$ ions were the same as those observed in the FAB mass spectrum itself. However, as in the previous work of Crow,[722] the daughter ion MS/MS spectra contained additional ions resulting from dissociations of the sugar moiety. Also of interest in this study was the ability of the daughter ion MS/MS spectrum to differentiate isomers that provide identical positive ion FAB mass spectra. Differences in peak intensities in the daughter ion MS/MS spectrum were used for this purpose, with the largest differences in peak intensities corresponding to those daughter ions that are thought to originate in a multistep process.

The methylation of calf thymus DNA produces a distribution of methylated nucleosides. Ashworth et al.[727] described a complete liquid chromatographic separation for 15 methylated and naturally occurring nucleosides and several daughter ion MS/MS spectra (low collision energy, triple-quadrupole mass spectrometer) for the methylated nucleosides. The nucleosides could be qualitatively identified at a level of 10 ng (36 pmol) by daughter ion MS/MS; at this level, the chemical noise in the mass spectrum was sufficiently high (even in a cleaned-up sample) to prevent quantitation. Deuterated samples were used as the internal standards, and a comparison of ion intensities in the two daughter ion MS/MS spectra provided the quantitative value. In this case, chemical ionization was used to produce the parent ions $(M + H)^+$. In a later study from the same laboratory, Isern-Flecha et al.[728] characterized the structure of an alkylated dinucleotide ionized by desorption chemical ionization. Desorption chemical ionization is an ionization method used for the analysis of nonvolatile and thermally fragile molecules (see Chapter 5) and typically produces a large $(M + H)^+$ or $(M - H)^-$ ion in the mass spectrum. Advantages relative to fast atom bombardment, which is typically used for the ionization of this class of biocompounds, are the simplicity of the instrumental hardware, the ability to select several different gases as the chemical ionization reagent gas, and the freedom of the mass spectrum from the chemical noise derived from irradiation of the liquid matrix used in FAB. These advantages can afford better sensitivity in the analysis of compounds that can be analyzed by the direct desorption method.

Both positive and negative ion desorption chemical ionization mass spectra were reported for thymidyl ($3' \rightarrow 5'$)-thymidine methyl phosphotriester. A balance between the heating rate of the probe and the scanning rate of the mass spectrometer was necessary to obtain reproducible mass spectra, with a reported detection limit of 10 ng (positive ions, full scan) and 1 ng (negative ions, full scan). Daughter ion MS/MS experiments did not provide a decrease in the limit of detection, as the limiting factor is not the chemical noise in the spectrum but rather the irreproducibility of the total ion flux emanating from the direct exposure probe. The daughter ion MS/MS spectra were used to establish the dissociation pathways. Dissociations of the $(M + H)^+$ ion produce the protonated molecule of thymidine, loss of water from this ion, the protonated base, and the protonated mononucleotide. Methylation in the phosphate group can be established by the presence of a characteristic ion at m/z 113 that is absent in the daughter ion MS/MS spectra of other nucleosides.

The daughter ion MS/MS spectra of standard nucleosides and nucleotides are now well understood, and many studies are now focused on the analysis of modified compounds. Tomer et al.[729] have used daughter ion MS/MS with fast atom bombardment mass spectrometry to study adducts of guanosine, adenosine, uridine, or thymidine bound to various pyrrolizidine alkaloid metabolites. About 5 μg of these materials were required for the MS/MS analysis, which had as its goal the comparison of the daughter ion MS/MS spectra obtained from parent ions $(M - H)^-$, $(M + H)^+$, and $(M + K)^+$, all formed in the FAB source. In general, the daughter ion MS/MS spectra of $(M - H)^-$ parent ions of these adducts proved to be the least informative, containing the base anion as the predominant peak in the mass spectrum. The collision-induced dissociations of the positive ions provided much more structural information. Differences in the partitioning of ion current between the daughter ions from the base, the adduct species, and the sugar provided information about the preferred sites of association for the proton versus those for the potassium cation. Interpretation of the daughter ion MS/MS spectra provided the structural information necessary to draw conclusions about the nature and sites of modification of the base, sugar, and alkaloid.

Mallis et al.[730] have studied the daughter ion MS/MS spectra of a number of nucleotides, and agree with Tomer et al.[729] that the MS/MS spectrum of the $(M - H)^-$ ion of the parent nucleotide is generally uninformative. However, only a limited amount of dissociation could be observed from the $(M + H)^+$ ion. Fortunately, the positive ion fast atom bombardment mass spectra of phosphorylated nucleotides contain abundant ions corresponding to $(M + Na)^+$ ions, and the daughter ion MS/MS spectra of these ions contain increased abundances of structurally significant ions. Mallis et al.[730] used daughter ion MS/MS spectra to monitor the sites of incorporation of ^{18}O labeled atoms in the phosphorylated nucleotides. Shifts in the masses of the daughter ions formed by characteristic losses of moieties associated with the phosphate group could be interpreted to provide the location of the label. A triple-sector EBE instrument was used, and the mass resolution of the daughter ions was relatively low, and in several cases just sufficient for the determination. Analysis with a triple-quadrupole or hybrid instrument (providing that low-energy collisions activate the appropriate dissociation pathways) should provide confirmation of the results obtained.

The biological activities of nucleotides and nucleosides are diverse. Clifford et al.[731] have used daughter ion MS/MS to establish the occurrence of a specific reaction between retinyl phosphate and guanosine 5'-diphospho-D-mannose to form retinyl phosphate mannose. This reaction occurs in vitro within liver membranes. The product is fairly unstable (being an intermediate itself in glycoprotein biosynthesis), and MS/MS experiments confirmed its synthesis by analysis of a mixture after simple methanolic extraction.

Use of mass spectrometry in the area of nucleotide and nucleoside analysis dates back to the early 1960s. Nucleosides were found to sublime from a direct insertion probe, and good mass spectra could be obtained. With nucleotides, derivatization by trimethylsilylation was necessary to increase the volatility of the compounds to the point at which they could be analyzed by electron and chemical ionization methods. Polynucleotide studies were aided by the development of field desorption mass spectrometry, which provided the means to create ions from truly nonvolatile sample molecules. The extent of the studies appears to be limited by the versatility of the ionization methods available. Fast atom bombardment and liquid secondary ion mass spectrometry are very effective ionization methods for nucleosides and nucleotides. If there is any drawback, it would seem that the amount of fragmentation observed in the mass spectrum itself is below the level at which interpretation of the spectrum could provide detailed structural information. MS/MS provides this information, and the next few years will document a greatly expanding use of MS/MS experiments in this field, coupled with desorption ionization methods.

6.7.1.3. Fatty Acids and Lipids. Jensen et al.[732] have recently described an experimental methodology that pinpoints the location of double bonds in polyunsaturated fatty acids. Earlier work[733,734] had established that the daughter ion MS/MS spectra of $(M - H)^-$ ions from unsaturated fatty acids formed by FAB ionization contain a distinctive and reproducible set of daughter ions corresponding to sequential losses of C_2H_{2n+2} units from the alkyl terminus of the fatty acid anion. Dissociation occurs readily along the alkyl chain, yielding the daughter ions in a smoothly varying progression of relative abundances. The position of a double bond is apparent from an interruption in this pattern. Figure 6-23 compares the daughter ion MS/MS spectra of stearic acid and oleic acid (high-energy collisions, EBE instrument). The difference in the pattern of intensities is clear, and the masses of the daughter ions for which the relative abundances are diminished provide the location of the double bond within the alkyl structure.

The daughter ion MS/MS spectra of multiply unsaturated acids quickly becomes too complex for the precise location of the double bonds within the molecular ion structure. Reduction of the double bonds in such compounds with a deuterium-labeled reagent, deuterodiimide N_2D_2, reestablishes the simplicity of the daughter ion MS/MS spectrum in terms of the smooth progression of abundances. With this simplification, the location of any and all double bonds in the fatty acid can be established by the appropriate shift in the daughter ion masses. Using this methodology, branched fatty acids can be distinguished from isomeric straight-chain fatty acids[735] and can be used for the determination of the structures of more complex phosphatidylserines and phosphatidylcholines.[736]

Cervilla and Puzo[737] have also developed methods for the determination of the location of the double bond in monounsaturated fatty acids, with the ultimate applica-

Daughter ion MS/MS

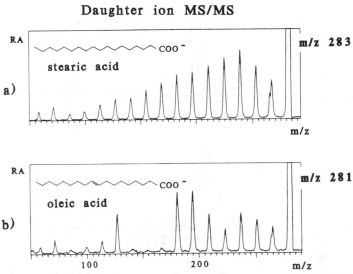

Figure 6-23 ■ Differences in the daughter ion MS/MS spectra of (*a*) stearic and (*b*) oleic acids (high-energy collisions, EBE instrument).

tion to the analysis of these compounds from bacterial sources. They have described a method in which the alkene is converted into an amino alcohol. The protonated molecule of this derivative is formed by chemical ionization and selected in a daughter ion MS/MS experiment (high-energy collisions, BE instrument). Localization of the proton on the heteroatoms facilitates dissociation at the carbon on either side of the original double bond. The two largest ions in the daughter ion MS/MS spectra correspond to the two complementary portions of the original molecule, termed A and B. To some extent, mixtures of fatty acid isomers could be characterized, as long as the A^+/B^+ ion pairs dominate the daughter ion MS/MS spectrum. The method was used to characterize the monounsaturated fatty acids found in *Mycobacterium phlei* following an extraction, a chromatographic cleanup, and derivatization to the amino alcohols.

Mycobacteria taxonomy has involved the determination of mycolic acid alkyl chain lengths, with the mycolic acid (an alpha-alkyl-beta-hydroxylated fatty acid) found in the cell wall of the bacteria. In an experimental protocol developed around chemical ionization MS/MS, the structures of the fatty acids (derivatized to the amino alcohols) were determined directly from a complex mixture.[738] The appropriate bacterial fractions were purified by column chromatography, but the use of the daughter ion MS/MS experiment was crucial in determining the structures of the fatty acid homologs found in unfractionated mixtures. About 50 μg of sample were required for the analysis. Fast atom bombardment produced a stable ion current of the protonated molecules of the sample; a single scan of the daughter ions formed as the result of collision-induced dissociation provided sufficient information for the direct determination of the alkyl chain lengths.

Peake and Gross[739] have described a method for locating double and triple bonds in alkenes, alkynes, dienes, fatty acids, fatty acid esters, and alkenyl acetates. An earlier mechanistic study[740] had shown that atomic Fe^+ adds oxidatively to double bonds in a

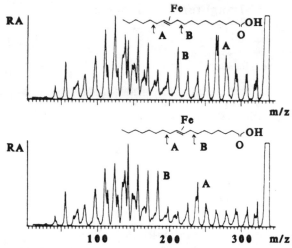

Figure 6-24 ■ Daughter ion MS/MS spectra of the Fe^+-adducts of two isomers of octadecenoic acid. The empirical formula of the parent ion in each case is $FeC_{18}H_{34}O_2^+$. Daughter ions corresponding to cleavages at points A and B are indicated in each of the spectra.

gas-phase chemical ionization reaction and that the daughter ion MS/MS spectrum of the Fe^+-complex dissociates by loss of a small olefin, the mass of which reveals the location of the double bond in the original sample molecule. In addition to the general use for the location of double bonds, the use of Fe^+ chemical ionization has been applied to the study of fatty acids and fatty acid esters that contain a double bond. Interpretation of the data is the same as that described previously for the daughter ion MS/MS spectra of the $(M - H)^-$ ions of the fatty acids created by fast atom bombardment. The location of the double bond is reflected in the disruption of the smoothly varying ion abundances in the daughter ion MS/MS spectrum. Figure 6-24 illustrates the daughter ion MS/MS spectra for two isomers of octadecenoic acid. Daughter ions corresponding to cleavages at points A and B are indicated in each of the spectra.

Finally, it should be noted that the strategies developed for the location of double bonds in simple fatty acids can often be used without modification for fatty acid esters. The daughter ion MS/MS spectra of the parent ion $RCOO^-$, or alternatively $(M - CH_3)^-$ for methyl esters, is identical with that obtained from the $(M - H)^-$ of the corresponding fatty acid.[741] Bambagiotti et al.[742] have noted, however, that the pattern of daughter ion abundances created in high-energy collisions on a BE instrument are not reproduced in low-energy collisions on a hybrid instrument and that daughter ion MS/MS analysis under low-collision-energy conditions could not be used to determine the location of the double bond in fatty acids or fatty acid esters.

Various methods of MS/MS analysis have been applied to the characterization of lipids from biological sources. Metastable ion mass spectra were used by Batrakov et al.[743-746] to study triglycerides present in microbial fractions. No separation of the triglyceride fraction was required, as the contributions of the individual components in the mixture to the measured spectrum could be established. Isomers with different fatty acid compositions could also be distinguished through analysis of the metastable ion mass spectra. Compounds for which both metastable ion and CID daughter ion MS/MS experiments proved useful include phosphatidylglycerols, lecithins, phosphatidylethanolamines, and N-methyl- and N,N-dimethyl-phosphatidylethanolamines. For each analysis, 100 to 200 μg of the lipid fraction was required.

As MS/MS methodology has become more widely available, basic studies of the chemical systems have evolved into direct applications to problems of predominantly biochemical interest. Typical of these developments is a study by Tomer et al.[747] that used daughter ion MS/MS experiments to analyze a mixture of two thermally labile, zwitterionic ornithine-containing lipids isolated from *Thiobacillus thiooxidans*. Data were measured on an EBE instrument, and fast atom bombardment was used to create both $(M + H)^+$ and $(M - H)^-$ ions for study by MS/MS. Twenty micrograms of sample was used in experiments that provided the exact mass measurement of the molecular ions, and thus a suggested empirical formula, and recorded the daughter ion MS/MS spectra of both the $(M + H)^+$ and $(M - H)^-$ ions. The MS/MS data provided structural information that identified the minor component of the two as a homolog of the major component.

Sherman et al.[748] have used MS/MS data measured with a BEB instrument to determine fatty acid ester compositions for phosphoinositides and related compounds. Daughter ion MS/MS data acquired through collisions in the second reaction region of the instrument, with analysis by only the electric sector, provided insufficient daughter ion mass resolution for the fatty acid compositions to be determined unambiguously. With collisional activation in the third reaction region of the instrument and mass analysis of the daughter ions by the final magnetic sector, improved mass resolution of the daughter ions could be attained, and the presence of several isobaric components in the parent ion beam could be resolved by the unique daughter ions formed upon collisional activation. Second reaction region CID with linked scanning of the EB portion of the instrument would have provided even better resolution.

6.7.1.4. Steroids. The analysis of steroids by mass spectrometry has been a topic of long-standing interest[749] and almost routine success. As problems in steroid analysis became more complicated, new mass spectrometric methods were brought to bear. The results of electron ionization mass spectrometry were supplemented by high-resolution mass spectrometry and gas chromatography/mass spectrometry. Chemical ionization mass spectrometry was joined by the methods of fast atom bombardment and secondary ion mass spectrometry for the analysis of more thermally fragile steroid molecules. MS/MS was appropriated for use in steroid analysis almost as soon as it was first reported as a general analytical technique and continues its proliferation. Much early work used the dissociations of metastable ions to provide information about the detailed structures of the steroid molecules or about the distributions of various forms of steroids in biological mixtures. Collisional activation was adopted in some cases, but the MS/MS analysis of steroids has most often relied solely on metastable ion dissociations. The discussion of MS/MS in this section first presents an overview of the MS/MS methods based on metastable analysis and then moves to CID-based MS/MS methods of steroid analysis.

Smith et al.[750] studied the dissociations of metastable ions of steroids occurring in the second reaction region of a BE mass spectrometer, and they applied the technique to the analysis of simple mixtures of steroids prepared by mixtures of each of the pure components. Samples were introduced via the direct insertion probe of the instrument, and electron ionization was used to form $M^{+\cdot}$ ions selected for daughter ion MS/MS analysis. Under constant conditions, the short-term reproducibility for the most abundant daughter ions (as measured by the measured abundances relative to the abundance of the parent ion) was 1% to 2%. Long-term reproducibility (day-to-day) was 10%, with the increase due to drifts in operating conditions of the mass spectrometer that were not monitored or controlled, and to variations in the cleanliness of the ion source. After these figures of merit were established, daughter ion MS/MS spectra were recorded for estrone, estradiol, and estriol standards, and then for each of these compounds present in a ternary synthetic mixture. Smith found that the abundances of the daughter ions in the MS/MS spectrum were reproduced in the abundances of the same ions of the MS/MS spectra recorded from the mixture. Assuming linear addition of ion abundances, these MS/MS spectra could be used as a quantitative tool in the estimation of concentrations of mixture components. This work, although limited in scope, first established the possibilities for MS/MS analysis of mixtures of these compounds.

Gaskell and Millington[751] used a selected reaction monitoring experiment based on the metastable ion dissociation of 5-alpha-dihydrotestosterone as its tert-butyl-dimethylsilylether derivative. Derivatization was necessary to increase the volatility of the compound for a gas chromatographic separation. The mass spectrometer used as the detector was set to monitor the metastable dissociation of the parent ion of this compound at m/z 347 $(M - C_4H_9)$ to the daughter ion at m/z 271 [M − C_4H_9—$H(CH_3)_2SiOH$]. Since the intensity of the daughter ion was 10% that of the original parent ion beam, excellent sensitivity in the analysis for this particular compound could be achieved. Detection limits of 20 pg were given with a reproducibility of 10%. Specific quantitation of this one compound in the presence of other closely related and isomeric compounds is achieved through a combination of the gas chromatographic retention time window and the selected reaction monitoring experiment.

Gaskell et al.[752] continued this investigation into a general study of stereoisomeric androstanediols. Linked scanning with an EB instrument was used to measure daughter ion MS/MS spectra for the tert-butyldimethylsilyl ether derivatives of these compounds. These derivatives are often chosen for the analysis of steroids because of the dominance of the ion fragmentation in an electron ionization source to yield $(M - C_4H_9)^+$ ions. A large ion is thus available in the spectrum with the steroid nucleus intact; these ions were chosen for daughter ion MS/MS experiments. Table 6-2 summarizes the daughter ion MS/MS spectra obtained for eight isomeric androstanediols. In each case, the parent ion is at m/z 463, representing the $(M - C_4H_9)^+$ ion of the derivatized molecule. The dissociation pathways that lead to each of the daughter ions are shown in Scheme 6-1, with separate parent ion MS/MS experiments used to confirm the several precursors of the daughter ion at m/z 255. Examination of the pattern of ion abundances in Table 6-2 indicates that the stereochemistry of the steroid has an important effect on the relative contributions of the competing dissociations and that these isomers at least can be distinguished from one another in an MS/MS experiment. Such capability is an advantage when differences based on stereochemistry are not apparent in the mass spectrum because of the presence of interfering ions that skew the pattern of fragment ion abundances.

Fast atom bombardment ionization became available in 1981 and was adopted as the ionization method of choice with daughter ion MS/MS for the analysis of steroids.[753] In particular, steroid sulphates produce strong signals in the negative ion

Table 6-2 ■ Daughter Ion MS/MS Spectra for Parent Ions of *m/z* 463 Derived from *t*-Butyldimethylsilyl Ethers of Isomeric Androstanediols

Steroid name	*m/z*	387	373	345	331	255
5-alpha-A-3-alpha,17-alpha-diol TBDMS		35	13	4	100	20
5-alpha-A-3-alpha,17-beta-diol TBDMS		100	6	1	45	24
5-alpha-A-3-beta,17-alpha-diol TBDMS		100	3	1	41	70
5-alpha-A-3-beta,17-beta-diol TBDMS		100	5	1	11	16
5-beta-A-3-alpha,17-alpha-diol TBDMS		24	97	24	100	28
5-beta-A-3-alpha,17-beta-diol TBDMS		100	9	3	30	40
5-beta-A-3-beta,17-alpha-diol TBDMS		21	99	14	100	28
5-beta-A-3-beta,17-beta-diol TBDMS		100	13	2	84	21

(Header spanning columns 387–255: "Daughter ion relative abundances")

Origins of the daughter ions are explained in Scheme 6-1.
Data from S. J. Gaskell, A. W. Pike and D. S. Millington, *Biomed. Mass Spectrom.*, 1979, **6**, 78.

FAB mass spectrum that represent the steroid sulfate anion. Full spectra can be measured with less than 15 ng of the steroid. Metastable ion dissociations are recorded in the daughter ion MS/MS spectra and allow the differentiation, for example, of dehydroepiandrosterone sulfate and testosterone sulfate. Analysis of extracts from biological samples produce matrix effects that can partially or even completely suppress the signal from the steroid sulfate known to be present. A simple extraction from the complex matrix based on immunoadsorption is sufficient to remove the material that gives rise to the matrix effect. A second problem in the assay was the overlap in mass between the sample ions and those derived from the glycerol background. This

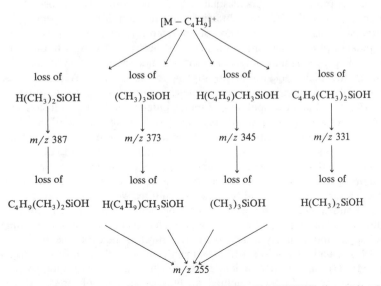

Scheme 1 ■ Four competitive paths of dissociation for androstanediol TBDMS ether. (After S. J. Gaskell, A. W Pike and D. S. Millington, *Biomed. Mass Spectrom.*, 1979, **6**, 78.)

interference was avoided with the use of the pentafluorobenzyloxime derivative of the steroid. No increase in the sensitivity of the analysis was noted, as the formation of the $(M - H)^-$ ion is not an electron capture process for whichpentafluoro derivatives have often been used, but the mass of the ion to be investigated was increased to a point at which signal from the matrix background no longer caused a significant problem.

Maquestiau et al.[754] used daughter ion MS/MS spectra based on the metastable dissociations of ions created in an electron ionization source to study the steroid contents of extracts from marine animals. Marine invertebrates contain extremely complex mixtures of sterols, the identification of which is necessary for an understanding of their physiological function. For instance, 47 different sterols were identified in a GC/MS study of a Caribbean coral. This is the sort of complex mixture analysis for which MS/MS is expected to be particularly useful. The general nature of the compounds under investigation is known, many of the compounds are of closely related structures, and the mixture is of such complexity that extensive sample preparation becomes a hindrance in terms of the time involved and also in the increased chances for sample contamination or transformation. Maquestiau introduced the sterol fractions from six marine vertebrates via the direct insertion probe into the electron ionization source of a BE mass spectrometer. The molecular ions of many different sterols could be identified in the mass spectrum, but the overlap in the fragment ions made the unambiguous identification of any of them impossible. Daughter ion MS/MS spectra were recorded first with the metastable dissociations of these ions as they occur in the second reaction region of the mass spectrometer. In each case, a sufficient number of abundant daughter ions could be observed so that a fingerprint spectrum for each sterol could be established by comparison with the daughter ion MS/MS spectra of standard sterols. The sample size introduced into the source was 1 to 2 mg; some of the molecular ions of the steroids were present with a relative abundance that implied an individual sterol sample size of about 1 μg.

In this initial study, Maquestiau et al.[754] also compared the use of metastable ion dissociations with CID for obtaining daughter ion MS/MS spectra. No significant difference was noted for many of the sterol isomers in terms of the nature or abundance of daughter ions in the MS/MS spectrum. A danger in these comparisons is that CID processes may in fact be occurring under nominally metastable ion conditions due to the high background pressure in the instrument. A rigorous assessment of metastable ion abundances requires the completion of a pressure-resolved experiment (Section 5.3.1) and extrapolation of ion abundances to "zero" collision gas pressure. In a few cases such as analysis of gorgosterol and fucosterol, the use of the collision gas introduced into the reaction region of the mass spectrometer increased the number of daughter ions observed so that the identification of the sterol could be made with a greater certainty. MS/MS experiments were later used by these same workers to determine the contents of sterol fractions from organisms in tissue culture.[755, 756] Temperature and light conditions for the cultures were changed to vary the degrees of sexual and asexual morphogenesis, or exogenous chemicals were introduced into the culture system to induce the same changes.Here again, the minimized sample handling and speed and accuracy of analysis were central to the experiments.

Most daughter ion MS/MS experiments involve selection of a molecular ion $M^{+\cdot}$ or $(M + H)^+$ as the parent ion. Bozorgzadeh et al.[757] have adopted a different and simple approach in the structural analysis of steroids of similar structure. Several steroids of similar tetracyclic carbon skeletons were investigated by daughter ion

MS/MS; the differences in the structures of these steroids resided in the side chain attached to the carbon skeleton. Many of the ions in the daughter ion MS/MS spectra of the molecular ions could thus be expected to be similar and of little use in determination of the structure of the side chain. However, each of the electron ionization mass spectra of these compounds contained ions that are derived directly from the side chain. Despite the fact that these ions are sometimes of low relative abundance (less than 10%) in the mass spectrum, they can be selected as parent ions in a daughter ion MS/MS experiment. The daughter ion MS/MS spectra so obtained could be compared to those of standards from steroids of known side-chain structure or even to standards of much smaller and simpler organic compounds that correspond only to the putative structure of the side chain. Five steroid structures with different side chains were studied, and the daughter ion MS/MS spectra of each of the side-chain derived ions was distinctive. These experiments suggest a novel approach to the structural elucidation of closely related compounds such as the steroids.

As mentioned in Chapter 2, the use of a BE instrument for MS/MS experiments involves a decreased daughter ion resolution when CID occurs in the second reaction region of the instrument. The kinetic energy release associated with the dissociation broadens the width of the daughter ion signal that is measured as kinetic energy by the electric sector. In general, the values of translational energy release are not used in studies of mixture analysis, but are useful in detailed studies of ion structure. Larka et al.[758] have used the translational energy release of metastable ions as a means of distinction between steroid epimers. The loss of a methyl group from the molecular ion was chosen for the measurement. The data show that the T values for the epimeric pairs are reproducibly different, as shown in Table 6-3. The values measured are sufficiently different that measurement of the kinetic energy release for either one of the pure compounds should be adequate for its identification.

Table 6-3 ■ Measured Translational Energy Releases for the Loss of Methyl from the Molecular Ion $M^{+\cdot}$ for Some Epimeric Steroids

Steroid	T (meV)
5-alpha,14-alpha androstane	35 ± 1
5-beta,14-alpha androstane	28 ± 1
5-alpha,14-beta androstane	31 ± 1
5-beta,14-beta androstane	25 ± 1
5-alpha pregnane	51 ± 2
5-beta pregnane	34 ± 1
5-alpha-androstane-3-one	33 ± 1
5-beta-androstane-3-one	39 ± 1
5-alpha-androstane-17-one	75 ± 2
5-beta-androstane-17-one	37 ± 1
5-alpha-androstane-3,17-dione	124 ± 3
5-beta-androstane-3,17-dione	101 ± 3
5-alpha-androstane-3,11,17-trione	159 ± 4
5-beta-androstane-3,11,17-trione	121 ± 3

Data from E. A. Larka, I. Howe and J. H. Beynon, *Org. Mass Spectrom.* 1981, **16**, 465.

Larka et al.[759] have used daughter ion MS/MS spectra to distinguish ring junction stereochemistry in a number of steroids. Daughter ion MS/MS spectra were recorded for nine pairs of steroid isomers that differed only in the stereochemistry at two of the ring junctions. Reproducible differences in the MS/MS spectra are observed that allow the distinction between the isomers. Furthermore, these differences are more pronounced than those observed in the electron ionization mass spectra of these same compounds. This difference arises because metastable ion dissociations are sensitive to changes in the critical energy of the reaction, which is itself a function of the stereochemistry of the structure of the molecule. In general, the 5-beta isomers are less stable than the 5-alpha isomers, and the loss of the 19-methyl group relieves a greater amount of strain energy at the A/B junction, leading to a lower critical energy for the dissociation reaction. At the C/D ring junction, the 14-beta isomer is more stable. Changes in the stereochemistry at each of the ring junctions will change the critical energies for all of the metastable dissociations that involve ring cleavage, but the effect of the stereochemical change appears to be greatest on the dissociation reaction of lowest critical energy that corresponds to the loss of methyl from the molecular ion.

Proctor et al.[760] showed that two configurational steroid isomers, 14-alpha-equilenin and 14-beta-equilenin, could be distinguished on the basis of differences in the ion abundances recorded in the metastable ion daughter ion MS/MS spectra. The molecular ions of these compounds dissociate in the source through a sequence of consecutive reactions that involve losses of methyl and carbon monoxide. In the case of the 14-alpha-compound, these losses could take place in either order, while in the 14-beta-compound, loss of methyl occurred first followed by the loss of carbon monoxide. Consecutive metastable ion dissociations were studied to clarify the reaction sequence. Since the data were recorded on a BE instrument, the translational energy releases for each of the dissociations were measured and were shown to be useful in determining the number of distinct ion structures formed in each step of the reaction sequence.

The use of translational energy release in the determination of the stereochemistry of steroids was studied further by Zaretskii et al.[761] in a study of the epimeric 3-keto- and 3-hydroxysteroids. Again, unimolecular dissociations of metastable ions involve ions that contain a narrow range of internal energies just in excess of the critical energy for the reaction; the behavior of metastable ions is therefore more sensitive to stereochemical differences than are the dissociations that result from CID. The results obtained were complex, and no simple rules for predicting the relative values of the kinetic energy release were forthcoming from this study. Of the compounds studied, in 3-hydroxysteroids, the 5-beta-isomers release smaller amounts of translational energy in the dissociation by loss of methyl. In most of the 3-ketosteroids studied, larger translational energy release values are associated with methyl loss from the 5-beta-isomers. Exceptions occur when additional keto groups are introduced onto the steroid skeleton at the 11- or 17-positions.

Recent work with MS/MS in the analysis of steroids has concentrated on a more rigorous treatment of the mechanisms of the metastable dissociations, on the one hand, and more sophisticated applications, on the other. In a detailed elucidation of the mechanism several years ago, Brown and Djerassi[762] synthesized a number of deuterium-labeled ketosteroids and studied the mechanisms of fragmentation in the electron ionization mass spectra and in the metastable dissociations studied by daughter ion MS/MS. These studies are necessary because of the increasing number of steroids of novel structure isolated from marine organisms and the ability of analytical methods, and mass spectrometry in particular, to determine their presence and their structure even at trace levels of occurrence.

As the mass spectrometric methodology becomes more refined, applications to more complicated systems become possible. A good example of this work is the study of steroids produced by microbial fermentation of sterols using daughter ion MS/MS experiments.[763] In this work, Prome et al. characterized the microbial transformation of simple sterols such as cholesterol into more complicated steroids such as pregnane derivatives. Although the yield of the microbial synthesis was low, the direct bioconversion was of interest since it offered a potential alternative route to several chemical steps in the widely used laboratory synthesis. Metastable dissociations of parent molecular ions generated by electron ionization were recorded by scanning the electric sector voltage in a BE instrument. For higher sensitivity, single- and multiple-reaction monitoring was used. Results reproducible to 10% were obtained with sample sizes of 1 to 1000 ng. Ten minutes was required for the analysis of each fermentation broth sample. The major products of microbial conversion of cholesterol, such as androsta-1,4-diene-3,17-dione and androsta-4-ene-3,17-dione, were unambiguously identified in the sample. As the concentration of the transformation products decreased to below 5%, identification became more tenuous due to the lower signal levels and the increased contribution of interfering ions in the abundances of the diagnostic daughter ions. In addition, since more time was required to integrate the signal, changes in the character of the ions formed in the source became apparent as the sample evaporated from the direct insertion probe. Matrix effects were also apparent in the quantification of even the major components. A standard addition method used to create a calibration curve provided satisfactory results. In summary, Prome et al.[763] found that daughter ion MS/MS could be used for the direct analysis of the steroid contents of the fermentation broths. GC/MS analysis using selected ion monitoring could be up to 100 times more sensitive than the MS/MS method used, but was sufficiently time-intensive that it was used only when the quantitation of the minor components in the synthetic mixture became of interest. The determination of major components by MS/MS provided information on a time scale more appropriate to the monitoring of the biotransformation process.

Daughter ion MS/MS analyses of steroids in conjunction with CID experiments were described by Kruger et al.[764] using a BE instrument. Standard samples investigated were testosterone, corticosterone, norgestrol, estradiol, dehydroepiandrosterone, cholesterol, hydrocortisone, and ethynylestradiol. Furthermore, direct analysis of a urine sample (2 μL on the direct insertion probe) was used to identify dehydroepiandrosterone present at about the 10-ng level. Higher levels of this particular compound are diagnostic for ovarian tumors. Such demonstrations, now a decade old, were the impetus for the development of MS/MS for complex mixture analysis.

More recently, Cheng et al.[765] used daughter ion MS/MS experiments to derive structural and stereochemical information from a number of steroids. The objects of the study were oxygenated steroids containing 3- and 17-substituents. Daughter ion MS/MS spectra were recorded on many of the lower mass ions in the mass spectra that correspond to the side chain of the structure and also of the intact molecular ions formed by electron ionization. Assignment of the stereochemistry of the hydroxyl group and of the ring stereochemistry could be based on the daughter ion MS/MS spectra of the parent molecular ions. Comparison of the daughter ion MS/MS spectra of the side-chain-derived ions were compared to spectra generated from standards of known structure and to the MS/MS spectra of the same ions derived from smaller organic precursors.

Unger[766] has also used daughter ion MS/MS experiments in the differentiation of structural isomers of steroids. In this study, two isomeric ethylthio-steroids were investigated. Differences in the abundances of the daughter ions that correspond to the loss of this particular substituent provided a means for the differentiation of the isomers. Fast atom bombardment mass spectrometry was used to create the protonated molecule selected for the daughter ion MS/MS analysis. Ten micrograms of sample was enough for the determination, with a reproducibility of ion abundances of 5%. Liehr et al.[767] studied the daughter ion MS/MS spectra of isomeric bile acid salts, based on steroid-type structures. Fast atom bombardment mass spectrometry was used to create $(M + Na)^+$ and $(M + H)^+$ ions of the salts, where M represents the sodium salt of the sulfonate for a series of bile acid salts that include sodium taurourdexycholate (1), sodium taurochenodeoxycholate (2), sodium taurodeoxycholate (3), sodium taurohyodeoxycholate (4), sodium glycoodeoxycholate (5), and sodium glycochenodeoxycholate (6). Of these compounds, the first four are isomers of M with a molecular weight of 521 and two were isomers with a molecular weight of 471. The daughter ion MS/MS spectra could be used to distinguish salt 3 from 1, 2, and 4. Salt 4 could be distinguished from salts 1 and 2 with some difficulty, but isomeric compounds 1 and 2 could not be distinguished. Differentiation of the latter two isomers could, however, be made based on the daughter ion MS/MS spectra of the $(M + Na)^+$ ion.

With the advent of hybrid geometry instruments for MS/MS, relatively high parent ion selection could be combined with unit mass resolution of the daughter ions (see Section 2.5), providing the specificity needed for difficult steroid analyses. Daughter ion MS/MS analysis was used in combination with gas chromatographic separation by Gaskell et al.[768] to detect a steroid, oestradiol-17-beta (E_2) in an extract of blood plasma. This steroid is involved in the biochemistry of ovarian tumors, and a number of clinical assays call for its specific detection in the presence of all other normal steroid components of blood. This assay had previously been completed both by high-resolution mass spectrometry using selected ion monitoring and by daughter ion MS/MS using metastable ion dissociations. A comparison of the analytical techniques was therefore possible. The target steroid was derivatized to a bis-tertbutyldimethylsilyl ether to make it volatile enough for the gas chromatographic separation. Single-reaction monitoring was used to measure the abundance of the transition from the parent ion at m/z 500 (formed by electron ionization) to the abundant daughter ion at m/z 443. This ion represents loss of the butyl group, which, as explained earlier, is a facile loss that regenerates the intact steroid nucleus and is a common loss in both metastable and CID daughter ion MS/MS studies. Low-pg detection limits for this steroid (1–5 pg) were possible with this analytical combination using a parent ion selection resolution of 1000. An analytical interferent was apparent at this resolution that was distinct in terms of its retention time. A parent ion selection resolution of 5000 was sufficient to eliminate the detection of this interferent, but with a loss in sensitivity that was not specified. The sensitivity of the experiment is approximately the same as that found in the daughter ion MS/MS experiment using metastable ion dissociations.

Daughter ion MS/MS spectra based on both unimolecular dissociations and collisional activation have been recorded by Guenat and Gaskell[769] for pairs of isomeric steroids using a four-sector instrument (BEEB). Parent ions were selected by the first two sectors at low resolution and then passed into the reaction region located between the electric sectors. Linked scanning of the final two sectors (B/E) provides the spectrum of daughter ions resulting from dissociations in this particular reaction region of the instrument. Two isomeric pairs of isomeric glucoronides were studied. The first pair consisted of two isomeric pregnan-triol-one-3-glucoronide isomers. Daughter ion MS/MS spectra of the (M − H)⁻ or the (M + Na)⁺ ions formed by fast atom bombardment did not differentiate between the structural isomers, but the daughter ion MS/MS spectra of the (M + H)⁺ ions in both unimolecular dissociations and CID provided distinct spectra. Differentiation between the second pair of isomers is shown in Figure 6-25, which contrasts the daughter ion MS/MS spectra of 5-alpha- and 5-beta-androsterone-3-glucoronide obtained at collision energies of 8 keV and 200 eV (obtained by floating the collision cell in the intersector reaction region). The lower-collision-energy daughter ion MS/MS spectrum does not provide a clear distinction between these isomers; the difference in the high-collision-energy daughter ion MS/MS spectrum can be used for an unambiguous identification.

Further work by Cole et al.[770] with the four-sector BEEB instrument explored the effect of experimental conditions such as collision gas pressure, collision energy, and analyte concentration on the daughter ion MS/MS spectra measured for a pair of model stereoisomeric steroid conjugates, androsterone glucoronide and etiocholanone glucoronide. Again, daughter ions were formed in the third reaction region of the BEEB instrument and analyzed by a linked scan of the final two sectors. Experimental parameters that were varied included the attenuation of the parent ion beam and the

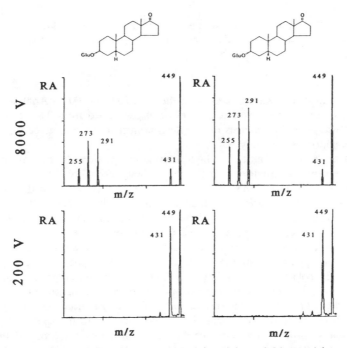

Figure 6-25 ■ Daughter ion MS/MS spectra of 5-alpha- (left) and 5-beta- (right) androsterone glucoronide obtained at collision energies of 8000 V (top) and 200 V (bottom).

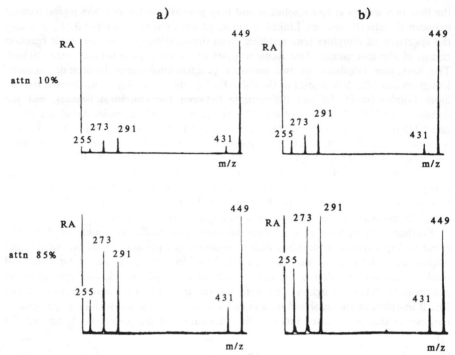

Figure 6-26 ■ Comparison of the daughter ion MS/MS spectra of $(M+H)^+$ ions of two isomeric steroids (*a*) androsterone glucoronide and (*b*) etiocholanone glucoronide at parent ion beam attenuations of 10% (top) and 85% (bottom).

collision energy, which was varied over a range of 200 to 8000 V. Analyte concentrations were those of sample solutions in glycerol and varied from 0.68 to 68 m*M*. No correction was made for changes in sample concentration due to glycerol evaporation during the course of the analysis.

In this work, it was found that high internal energy depositions was necessary to differentiate isomeric structures. This point is shown explicitly in Figure 6-26, which compares the daughter ion MS/MS spectra of the $(M+H)^+$ ions of the two isomeric steroids obtained with a parent ion attenuation by the collision gas of 10% (*a*) and 85% (*b*). Reproducible differences in the daughter ion MS/MS spectra of the two isomers become more distinct at the higher collision gas pressures.

Changes in the collision energy across the range of 200 to 8000 V had the expected results on the appearance of the daughter ion MS/MS spectra of these two compounds. As the collision cell was floated to higher and higher potentials, thus decreasing the energy of the collision between the parent ion and the neutral gas molecule, the relative contribution to the spectrum of metastable dissociations occurring outside of the collision cell is decreased. Therefore, the contribution to the spectrum of low-energy dissociative processes) was reduced. As the collision energy dropped still further, the amount of internal energy deposited into the parent ion was decreased and these same ions increased in relative abundance.

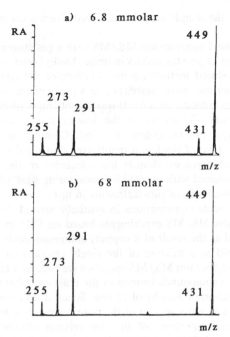

Figure 6-27 ■ Comparison of daughter ion MS/MS spectra for androsterone glucoronide at concentrations in a glycerol solution of (*a*) 6.8 m*M* and (*b*) 68 m*M*.

Perhaps the most interesting of the experiments were those that dealt with the effect of analyte concentration on the daughter ion MS/MS spectrum. Sample concentration can be controlled in these analytical experiments, but in the analysis of unknowns or complex mixtures, this variable fluctuates across a wide range of values. Figure 6-27 compares the daughter ion MS/MS of androsterone glucoronide as a solution in glycerol at 6.8 and 68 m*M* concentrations. The daughter ion MS/MS spectrum for etiocholanone glucoronide shows no change with this shift in analyte concentration and is also identical for lower concentrations. The daughter ion MS/MS spectrum for androsterone glucoronide, as shown in Figure 6-27, changes significantly as the analyte concentration is increased. In particular, the ratio of the daughter ions at m/z 273 and m/z 291, a criterion that could be used to distinguish between the isomers at all other concentrations, has reversed, with the ion at m/z 291 now more abundant in the daughter ion MS/MS spectrum. The authors attribute this effect to two possible causes. The first is the formation of a surface monolayer for this steroid sample in which the protonated molecules $(M + H)^+$ are different in structure from the $(M + H)^+$ ions formed when the sample interacts with the glycerol matrix. The internal energy of the $(M + H)^+$ ions may also be dependent upon the formation process itself. Sample ions desorbed from a glycerol-sample mix may be solvated by glycerol molecules. Energy from the $(M + H)^+$ ion is lost through a process of desolvation. In any case, the effects of sample concentration on ion abundances in the daughter ion MS/MS spectrum can be significant even in the case of a pure sample and will certainly be so in the instance in which the sample is analyzed directly from a matrix that may vary in its own composition.

Schweer et al.[771] used daughter ion MS/MS with a gas chromatographic separation to determine the levels of prostaglandins in urine. Methyl ester/trimethylsilyl ethers of the steroids were produced to increase the volatility of the compounds under study. Characteristic daughter ions were established in a study of the model compounds, and then single reaction monitoring on a triple-quadrupole mass spectrometer was used for the GC/MS/MS assay. Sensitivity in the low-pg range was established for the detection of the target prostaglandins in urine. Comparison of the daughter ion MS/MS results with those of single-ion monitoring illustrated the susceptibility of the latter to coeluting interferences, despite the selectivity of the gas chromatographic separation. Results obtained with selected ion monitoring were a factor of 5 too high in the measurement of the levels of prostaglandins in urine.

MS/MS has also made contributions in synthetic steroid chemistry. Pelli et al.[772] have used daughter ion MS/MS experiments based on CID to differentiate epimeric androsterones formed as the result of a copper/aluminum oxide catalyzed hydrogenation of a diene steroid to a mixture of the singly unsaturated compounds. Relative abundances in the daughter ion MS/MS spectra were used as a clean assessment of the molar ratio of the two compounds formed in the reaction, 5-beta-androstan-3-beta-ol-17-one and 5-beta-androstan-3-alpha-ol-17-one. Spectral differences are also observed in the electron ionization mass spectra of the pure compounds, but were lost when the electron ionization mass spectrum of the raw mixture obtained after the catalytic hydrogenation was recorded. Molar ratios of up to 5 to 1 in either component could be quantitatively assessed.

The daughter ion MS/MS spectra of the steroid glycosides were measured in a systematic study carried out by Crow et al.[773] Collisional activation of the $(M + H)^+$ or $(M - H)^-$ ions formed by fast atom bombardment produces daughter ions that can be related to sequential losses of the glycosidic moieties in a reproducible pattern that can be interpreted in terms of the sequence of the glycosidic chain. Homologous glycosidic chains can in favorable cases also be established. Compounds studied include cymarin, digoxin, and digitoxin; the latter two are of intense clinical interest due to their widespread use as cardioactive drugs.

In summary, MS/MS analyses of steroids have complemented a full range of ionization methods used to form ions from the neutral steroid molecules. Interpretation of daughter ion MS/MS spectra has been very carefully considered in work that identified stereochemical isomers. With proper control of instrumental and experimental conditions, such detailed interpretation is possible. Successes with steroid analysis might serve as an excellent example to encourage more careful work in other areas of application.

6.7.1.5. Bioamines. The determination of biologically active amines has long been an application of mass spectrometry.[774,775] In this section, the application of MS/MS to the analysis of bioamines is described. A distinction is made between naturally occurring amines, described in this section, and pharmaceutical compounds and their metabolites that may also be amines; the analysis of these compounds is discussed in Section 6.8. Biogenic amines is a general term that has come to refer to those amines that occur naturally as a result of metabolic processes in animals, plants, and microorganisms. They are generally compounds of low molecular weight and may be aliphatic, alicyclic, or simple heterocyclic compounds. As a term, biogenic amines has almost become synonymous with neurotransmitters such as catecholamine or dopamine, but in truth also includes such compounds as the natural histamines and other amines that may not have an explicitly defined neurochemical function.

Despite the interest in these compounds and the compelling need for a rapid, selective, and sensitive analytical method, MS/MS experiments have not been widely used in the analysis of biogenic amines. This stands in contrast to the tremendous amount of work describing MS/MS analysis of saccharides and carbohydrates, or steroids, or peptides. This paucity of application can be laid to the relative expense of MS/MS instrumentation, to the general tradition for such work to be done in clinical laboratories that do not have access to such instrumentation, and above all, to the success of the GC/MS method itself, which routinely requires 10 to 100 ng of material for the identification of an entirely unknown bioamine, but which also provides pg sensitivity in experiments that assay the levels of particular targeted bioamines.

An early application of MS/MS by Addeo et al.[776] was based on unimolecular dissociations of metastable ions formed in a low-energy electron ionization source; the samples were dansyl derivatives of 20 biogenic amines chosen as model compounds. Each of the parent ions of the dansyl derivatives dissociated to form a characteristic daughter ion at m/z 171 corresponding to the dimethylaminonaphthalene ion. A parent ion MS/MS spectrum obtained by a defocusing technique provided the molecular weights of all of the dansyl derivatives of the bioamines in a mixture, without a separation of these compounds prior to the mixture being introduced into the source. The data are shown in Figure 6-28. The method was applied to the analysis of amines in wines and beers and gave analytical results similar to those obtained with the usual single stage of mass spectrometric analysis.

Twelve years later, analysis of biogenic amines in wine was again based on MS/MS experiments. Walther et al.[777] prepared the dansyl derivatives of the amines found in 0.5 mL of wine and noted the same abundant dissociation of the parent ion formed by electron ionization to the dimethylaminonaphthalene ion at m/z 171. With the addition of a collision gas to the second reaction region of a BE instrument, the same

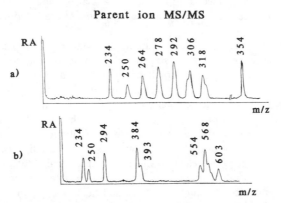

Figure 6-28 ■ Parent ion MS/MS spectrum for m/z 171, characteristic of dansyl derivatives of biogenic amines, measured at a probe temperature of (a) 100°C for "volatile" amines and (b) 180°C for "nonvolatile" amines. The spectrum was measured using a linked scan.

dissociations are observed. The absolute intensity of the ion at m/z 171 is increased, and still more abundant daughter ions at m/z 169 and m/z 170 are recorded. A linked scan for the parent ions of the characteristic daughter ion at m/z 169 provided the parent molecular ions of the dansylated amines in the mixture introduced via the direct insertion probe. Amines that are not usually found in wines were used as internal standards to show that a red wine sample contained 3 to 26 mg/L of various biogenic amines such as putrescine, histamine, and tyramine. The standard deviation of the result was 12% to 17%, and the time required for 10 replicate analyses was 30 min. Most of the variation in the intensity of the signal was thought to be due to variations in the rate of sample evaporation from the direct insertion probe. With deuterated internal standards of the amines themselves, standard deviations at the same level of sensitivity decreased to 3% to 5%.

Tryptolines in crude brain extracts have been analyzed directly by daughter ion MS/MS on a triple-quadrupole mass spectrometer.[778] These compounds had long been suspected to be present in brain tissue, as they are the products of a condensation reaction between indoleamines and aldehydes. A variety of analytical methods has been used for their assay, but they have lacked the required level of specificity or were subject to the formation of artifacts. In this case, targeted compounds could be identified ahead of time, and the MS/MS spectra obtained under standard instrumental conditions could be obtained. Speed of analysis was not an attribute in this particular analysis. Rather it was the ability of MS/MS to focus upon the target compound, and its ability to select and measure a chemical reaction characteristic of it, that made MS/MS the method of choice in this analysis.

Three forms of tryptoline were derivatized with heptafluorobutyrylimidazole to increase the efficiency of the formation of negative ions. Negative chemical ionization (methane moderator gas) was used to create $M^{-\cdot}$ and $(M - HF)^{-}$ parent ions for MS/MS experiments. Each sample extract was passed through a short-packed GC column to eliminate matrix interferences. For maximum sensitivity in the analysis, selected reaction monitoring was used to trace the occurrence of a selected parent ion to daughter ion transition. Figure 6-29 illustrates the magnitudes of the signals derived from such measurements, here with 1.4 pg of derivatized tryptoline. The comparison with an experiment using selected ion monitoring is also made in the figure. Limits of detection for the selected reaction monitoring experiment are listed as 0.1 to 0.2 pg.

6.7.1.6. Peptides. Biemann[779] and Biemann and Martin[780] have summarized the use of mass spectrometric methods in the sequencing of proteins. More accurately, MS/MS is used in the sequence analysis of small peptides derived from proteins. Increases in the accessible mass ranges of modern instruments and advances in ionization methods have made it possible to observe ions from peptides with masses as high as 25,000 daltons. The highest mass parent ion for which daughter ion MS/MS spectra have been obtained to elucidate the sequence is currently about 2600 daltons; this upper limit will certainly increase over the next few years. Determination of the sequence by daughter ion MS/MS requires selection of the parent ion and activation, most often by a collision with a neutral target gas molecule, but in a few cases by absorption of photons. The cross sections for these reactions and the transmission or retention of ions through these additional stages of analysis is low.

The key to the sequencing of a much larger peptide is a careful experiment in which enzymes are used to digest the larger protein. Cleavage of the protein into smaller

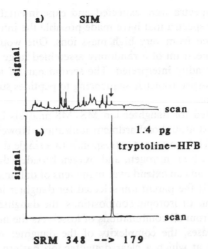

Figure 6-29 ■ (a) Selected ion monitoring (SIM) for m/z 348 (M − HF)⁻ (GC/MS) and (b) selected reaction monitoring (SRM) of m/z 348 → m/z 179 (GC/MS/MS) for analysis of 1.4 pg of tryptoline present as the heptafluorobutyrylimidazole derivative in an extract of rat brain. (Courtesy of R. A. Yost and J. V. Johnson.)

peptides occurs at specific sites, depending on the specificity of the digestion enzyme. A second digestion enzyme of different specificity is used to generate a series of overlapping small peptides. Often, high-performance liquid chromatography is used to separate the complex mixture of small peptides into fractions that contain only a few peptide fragments. Each of the fractions is analyzed by fast atom bombardment mass spectrometry, the parent molecular ions [usually (M + H)⁺] identified, and sequenced by daughter ion MS/MS. The complete data array is examined for overlays in the sequence determinations, and in this manner, with a high degree of certainty, the sequence of the entire protein can be generated. In favorable instances, only the mass spectrometric analysis is needed for the entire analysis. For instance, Johnson and Biemann[781] have recently published the complete primary structure of thioredoxin from *Chromatium vinosum* determined solely by mass spectrometry. The molecular mass of this protein is 11,750.2 daltons, and the primary structure contains 107 amino acid residues. Only the isomeric identity of eight of 16 leucine/isoleucine residues were left undetermined. The entire analysis required only 100 nmol (about 1 mg) of the thioredoxin. This material must be pure to avoid complications in the analysis of the digests and the sequences of the peptides so derived.

The presentation in this section first reviews many of the fundamental early studies that showed that MS/MS could be used for the analysis of peptides, and for sequencing in particular. Such work involved the application of ionization methods such as chemical ionization for smaller peptides, and progressed to larger peptides with the increased use of field desorption ionization, and accelerated rapidly with the use of fast atom bombardment, which seems to be ideally suited to these sorts of analyses. Much early work also explored the systematics of the gas-phase unimolecular and collision-induced dissociations of the molecular ions. Since peptides are for the most part composed of the 20 or so common amino acids, systematic patterns of losses in the daughter ion MS/MS spectra were expected and experimentally confirmed. It is the systematic nature of the spectra that have made possible the interpretation of daughter ion MS/MS spectra even from very high mass ions. One would not expect that the daughter ion MS/MS spectrum of a randomly assembled large organic ion of equivalent mass would be as readily interpreted. The second part of this section describes a few applications of mass-spectrometric sequencing to peptides such as insect hormones, neuropeptides, and toxins.

The parent ion selected for daughter ion MS/MS analysis is usually $(M + H)^+$ or $(M - H)^-$ formed by fast atom bombardment ionization. However, as the mass of the peptide increases, $(M + H)^+$ no longer corresponds to a single dominant ion. Contributions from isotopes of carbon, nitrogen, and oxygen broaden the molecular ion signal into an isotopic cluster than can extend over many tens of daltons on the mass-to-charge scale of the instrument. If the parent ion selected for daughter ion MS/MS analysis is of several different atomic or isotopic compositions, the daughter ion MS/MS spectra contain ions that result from all combinations of losses. As the number of compositions of the parent ion increases, the complexity of the daughter ion MS/MS spectrum increases past the point at which a clear daughter ion pattern can be interpreted in terms of the sequence of the peptide. For this reason, the monoisotopic parent ion $(M + H)^+$ of the peptide, consisting exclusively of ^{12}C, ^1H, ^{16}O and the like, is selected as the parent ion, despite the fact that other ions in the isotopic cluster may be of higher absolute intensity. The dilution of ion current in an isotopic envelope with increasing mass places an upper limit on the mass of the parent ion that can be selected in terms of the requisite intensity for the measurement of the MS/MS spectrum. This limit can be expected to be about 3000 daltons in the present generation of instruments.

Discussions of the mass spectra of peptides must begin with a definition of the nomenclature for the fragment ions observed in the spectra. The nomenclature proposed by Roepstorff and Fohlman[782] is almost consistently used and will be used in the present discussion with only one minor revision suggested by Biemann and Martin.[780] The essentials of this systematic nomenclature are given in Figure 6-30. The three possible cleavage points of the peptide backbone are called A, B, and C when the cleavage is such that the charge is retained on the N-terminal fragment of the peptide. Analogously, the cleavage points are called X, Y, and Z when the charge is retained on

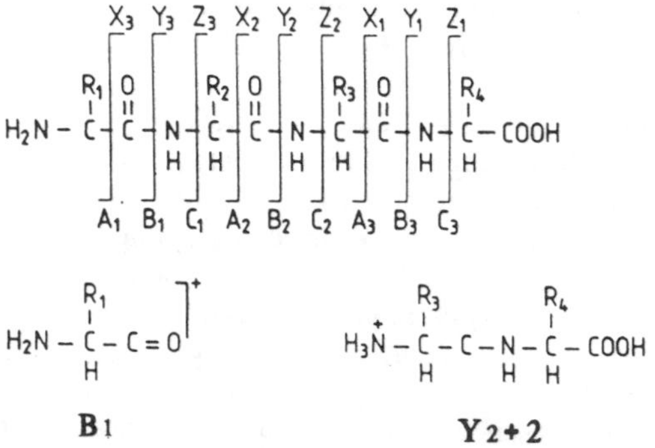

Figure 6-30 ■ Systematic nomenclature for fragment ions formed from peptides.

the C-terminal fragment of the peptide. A subscript number indicates which peptide bond is cleaved, counting from the N-terminus or C-terminus, respectively, and thus also provides a count for the number of amino acid residues in the designated ion. One advantage of the nomenclature is that the hydrogen transfers that take place both to and from a particular fragment ion can be noted without ambiguity. As suggested by Biemann and Martin, such hydrogen transfers are indicated by the notation -2, -1, $+1$, $+2$, etc. Thus, the "Y" ion in Figure 6-30 is in this case formed by the addition of two hydrogens and is thus designated $Y_2 + 2$. The naming of in-chain fragment ions is based on this letter code nomenclature, with an indication of the cleavages involved in the formation of the fragment ion. Much of the early work does not follow this precise nomenclature, and readers must exercise care in the interpretation of the spectra, especially with a proliferation of daughter ions identified by codes rather than masses.

Fundamental Studies. Many of the first uses of MS/MS in the sequence analysis of peptides used metastable transitions to establish sequence ion to sequence ion reactions, and the relative positions of amino acid residues in small peptide ions. In 1974, Levsen et al.[783] used both metastable and CID daughter ion MS/MS to provide an unambiguous differentiation of leucine and isoleucine in small peptides. This problem surfaced very early in the study of peptides by mass spectrometry because the isobaric doublet of leucine and isoleucine is unique in the common amino acids, the remainder of which (other than the structurally distinct lysine and glutamine) can be assigned on the basis of their unique masses. Levsen et al. found that the metastable ion daughter ion MS/MS spectrum of the immonium ion $(YN^+H{=}CHC_4H_9)$ found in the electron ionization mass spectra of all the small peptides studied, provided an unambiguous differentiation between a single leucine or isoleucine in the peptide.

The differentiation of leucine and isoleucine by MS/MS has been reinvestigated with regularity as new ionization methods for the analysis of peptides have been introduced. Characteristic side-chain reactions permitted the differentiation of leucine and isoleucine in small peptides (di- and tripeptides) for which the $(M - H)^-$ ions are formed by negative ion chemical ionization and selected as the parent ions in a high-collision-energy CID experiments.[784] Since most peptide analysis now uses fast atom bombardment as the ionization method, Aubagnac et al.[785] determined that the CID daughter ion MS/MS spectra of the immonium ion at m/z 86 from leucine and isoleucine were reproducibly different. Kinoshita et al.[786] have also demonstrated differences in FAB/MS/MS spectra due to the presence of leucine or isoleucine. This work was extended by Heerma and Bathelt,[787] who showed that the immonium ion in question at m/z 86 was relatively abundant in all positive ion FAB spectra, regardless of the position of the leucine or isoleucine in the peptide. The work was also extended to the study of the daughter ion MS/MS spectrum of norleucine, a synthetic amino acid often found in manipulated peptides. The daughter ion MS/MS spectra were sufficiently different to differentiate all three amino acids. The methodology is rigorously limited to peptides in which only one leucine, isoleucine, or norleucine is present. In peptides in which more than one of these residues is present, the daughter ion MS/MS spectrum is a combination of the individual spectra, although the contribution of the isomer nearest the N-terminus of the peptide seems to be favored.

As mentioned previously, much of the early MS/MS work involved the analysis of the dissociations of metastable ions. For example, the ratio of abundances for a reaction occurring as a metastable reaction versus the same reaction occurring in the source of the mass spectrometer was used to establish the identities of several small dipeptides.[788] For these studies, the peptides were derivatized to increase their volatility to a level sufficient for electron ionization mass spectrometry.

Derivatized peptides ionized by electron ionization were studied by Schlunegger et al.[788, 789] using both metastable ion dissociations and CID in two different reaction regions of a BE mass spectrometer. The daughter ion MS/MS spectra of isomeric N-acetyl-LEU-ALA-VAL-methyl ester and N-acetyl-VAL-ALA-LEU-methyl ester were shown to be differentiable in the earlier work.[788] The later work[789] showed that artifact peaks could be significant in the daughter ion MS/MS spectra obtained by linked scanning even for molecules as low in mass as the tetrapeptides, but also showed that the use of CA to increase the abundances of the characteristic daughter ions was a promising adjunct to the usual metastable ion analysis. This observation was reinforced by Steinauer and Schlunegger[790] in a general comparison of the daughter ion MS/MS spectra of tripeptides obtained with, and without, collisional activation. In general, CID spectra displayed greater abundances and diversity of structurally informative ions.

The potential of metastable mapping for the sequencing of peptides was discussed by Farncombe at al.[791] Metastable mapping is a technique[792] in which the measured ion intensity distributions as a function of a range of possible B and E sector values are stored as an array in the data system of the mass spectrometer. The array then contains the entire MS/MS data field. Linked scans of the B and E parameters are then simulated by the data system in an off-line experiment. The daughter ion, parent ion, and neutral loss MS/MS spectra are generated by processing of data already stored with the matrix in the data system. These spectra contain, of course, the same information as would be derived from discrete experiments, with the difference only in the manner in which the data are measured and passed into the data system. Metastable mapping was applied by Farncombe to the sequencing of methyl esters of three N-aceytl peptides: TRP-ALA-LEU, GLU(OMe)-ALA-LEU, and VAL-ILE-GLY-LEU. Two milligrams of the derivatized sample introduced into the electron ionization source of the mass spectrometer provided a stable signal for approximately 50 min. The extended time was required to scan the range of values of the B and E fields. The results of the analysis for N-acetyl-GLU(OMe)-ALA-LEU-methyl ester is shown in Figure 6-31. Several of the parent ion to daughter ion transitions are indicated on this two-dimensional chart of ion abundances. The advantage of such a display is that all parent ion/daughter ion pairs and the relative abundances of the ions are simultaneously displayed.

In 1980, Weber and Levsen[793] used field desorption (FD) to analyze underivatized peptides directly, with collisional activation used to increase the relative abundances of ions which could be interpreted for structural information. Di- and tripeptides were examined, and differentiation of the isomeric forms GLY-ALA and ALA-GLY peptides via their daughter ion MS/MS spectra was demonstrated. Shortly thereafter, Matsuo et al.[794] published a method in which daughter ion MS/MS spectra were used to sequence peptides via the $(M + H)^+$ and $(M + Na)^+$ ions formed by field desorption; several different penta- and hexapeptides were studied. A data system was used to collect data from multiple linked scans (approximately 50–100 scans). In general, daughter ion MS/MS spectra of the $(M + H)^+$ ions proved most useful, providing an abundant series of daughter ions that could be related to the sequence of residues. Daughter ion MS/MS spectra of the $(M + Na)^+$ ions typically contained a predominant daughter ion corresponding to the loss of the sodium atom. The FD mass spectra

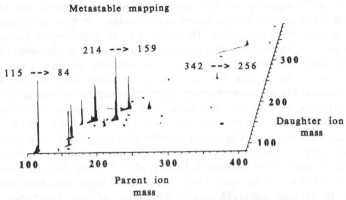

Figure 6-31 ■ Metastable mapping experiment for *N*-acetyl-GLU(OMe)-ALA-LEU-methyl ester. Several parent to daughter ion (positive ion) transitions are noted on the figure.

of these peptides often contained cluster ions such as $(2M + H)^+$ or $(3M + H)^+$, in which M is the mass of the peptide; the daughter ion MS/MS spectra of these ions were characteristic in terms of the dominance of the process that regenerates the $(M + H)^+$ ion upon CID.

Desiderio and Sabbatini[795] used a similar experiment for the analysis of underivatized oligopeptides, including di-, tri-, and tetrapeptides, and the two pentapeptides leucine- and methionine-enkephalin. Ten micrograms of the peptide were required for deposition on the FD emitter. These authors make the point that "pure" peptide samples submitted to a mass spectrometry laboratory for analysis are usually contaminated to some extent with other peptides. The presence of the ions from these components in the mass spectrum will complicate the interpretation of the data in terms of the sequence of the target peptide. In these instances, the ability of daughter ion MS/MS to provide the sequence information for a mass-selected parent ion is quite helpful. Desiderio and co-workers have used daughter ion MS/MS in the study of several neuropeptides, including leucine enkephalin,[796] substance P,[797] and dynorphins.[798]

Use of MS/MS for sequence analysis of peptides and oligopeptides has been expanded by the development of FAB and secondary ion mass spectrometry.[799] In both methods, bombardment of a glycerol solution of a sample peptide provides a reasonably abundant, long-lasting signal for the protonated $(M + H)^+$ or deprotonated $(M - H)^-$ molecule of the peptide. Although the FAB spectra of smaller peptides contain fragment ions corresponding to characteristic and interpretable cleavages at the peptide linkages, in larger peptides the relative degree of fragmentation is reduced, and sequence information becomes increasingly difficult to deduce from the mass spectrum itself. Compounding this problem is the chemical noise that prevails in FAB mass spectra; known sequence ions can be identified despite the relatively poor signal-to-noise ratio, but unexpected sequence ions are not as readily apparent in the mass spectrum. The amelioration of chemical noise in a mass spectrum is a hallmark of the MS/MS experiment, and the ability to link parent ion/daughter ion pairs becomes a powerful sequencing tool.

Hunt et al.[800] reported the first use of CID-based daughter ion MS/MS in conjunction with fast atom bombardment and secondary ion mass spectrometry for the sequence analysis of oligopeptides. This work was an extension of earlier work on peptide sequencing by MS/MS,[801] but added the ability to directly ionize the peptides without the need for derivatization to increase volatility. In these experiments, the $(M + H)^+$ ion of the peptide was selected as the parent ion, and the low-energy daughter ion MS/MS spectrum obtained on a triple-quadrupole instrument. Fragment ions characteristic of both the N-terminus and the C-terminus of the peptide are observed in the daughter ion MS/MS spectrum (types A, B, and C, and X, Y, and Z). In some cases, the presence of hydrogen transfers may complicate the assignment of an ion as belonging to one series or the other, thus decreasing the confidence in the assignment of sequence. In these cases, Hunt et al.[800, 801] used derivatization reactions. The peptide solution was derivatized with a mixture of acetic anhydride and D_6-acetic anhydride to provide acetyl/D_3 acetyl derivatives at the N-terminus. The first quadrupole is adjusted to pass both the labeled and the unlabeled parent ions into the central collision quadrupole. The resultant daughter ion MS/MS spectrum contains a series of doublets 3 daltons apart corresponding to ions of the A series. This ability to "detune" selectively the quadrupole to pass both forms of the parent ion simultaneously is unique to this analyzer and also to the ion trap and the ion cyclotron resonance mass spectrometer. Rapid alternation between labeled derivatives of parent ions at different masses can be achieved with sector-based mass analyzers, providing an equivalent analytical result with a slight loss in sensitivity.

Parent ion MS/MS spectra have also been of considerable value in sequencing experiments since the first applications of SIMS and FAB to these problems. An early report by Barber et al.[802] used linked scanning to establish the ionic precursors to peptide fragments in tripeptides. A more comprehensive report soon followed[803] that carefully examined the mass spectra of angiotensin peptides; a decapeptide and an octapeptide were the two examples discussed. The protonated molecule is the dominant ion in positive ion FAB mass spectrum. Although some fragment ions were observed in the FAB mass spectrum itself, their relative abundances are typically 10% or less. In order to confirm and extend the interpretation of the mass spectrum, the unimolecular dissociations of each of the protonated molecules were examined. In the specific example cited, the $(M + H)^+$ ion at m/z 1296 of the dodecapeptide angiotensin was selected as the parent ion, and its unimolecular dissociations recorded. Cleavages corresponding to sequential steps in each of several discrete processes are observed. Significantly, fragmentation of the decapeptide down to a tripeptide was observed, with each sequential cleavage established by the observation of the appropriate daughter ion. Similar results were reported by Heerma et al.[804] in a daughter ion MS/MS study of a heptapeptide ALA-LEU-TRP(for)-ASN-PHE-ARG-ALA [TRP(for) is N-formyl tryptophan].

Cyclic peptides are especially difficult to sequence using classical methodology due to the fact that selective hydrolysis of such samples is difficult to achieve. MS/MS experiments with FAB ionization has been shown to provide the sequence of these cyclic peptides. Tomer et al.[805] used daughter ion MS/MS spectra obtained on a triple-sector mass spectrometer (EBE) to establish the mechanisms by which the protonated molecules $(M + H)^+$ of cyclic peptides dissociate in a CID experiment. This work was a necessary prerequisite for the application of this work to the structural elucidation of cyclic peptide toxins, described in the applications part of this section. A

Daughter ion MS/MS

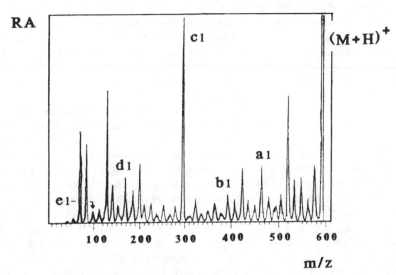

RA

c1

$(M+H)^+$

a1

d1

b1

e1-

100   200   300   400   500   600

m/z

Figure 6-32 ∎ Daughter ion MS/MS spectrum of the $(M+H)^+$ ion from a cyclic peptide cyclo-(LYS-PRO-dALA)$_2$. Losses a_1, b_1, c_1, d_1, and e_1 originate with initial protonation on the proline, and subsequent loss of LYS, LYS-ALA, LYS-ALA-PRO, LYS-ALA-PRO-LYS, and LYS-ALA-PRO-LYS-ALA, respectively.

pattern was established that is described by the initial protonation of an amide nitrogen, scission of the *N*-acyl bond, and subsequent fragmentation by loss of successive amino acid residues from the ring-opened acylium ion. The pattern of the fragmentation permitted the determination of the amino acid sequence of the cyclic peptide. The daughter ion MS/MS spectrum from the $(M+H)^+$ ion of a cyclic peptide cyclo-(LYS-PRO-d-ALA)$_2$ is shown in Figure 6-32. Scheme 6-2 lists the genesis of the daughter ions observed in the spectrum. In a typical case, 1 μL of a 1 mg/mL solution of the peptide in glycerol was sufficient to obtain the daughter ion

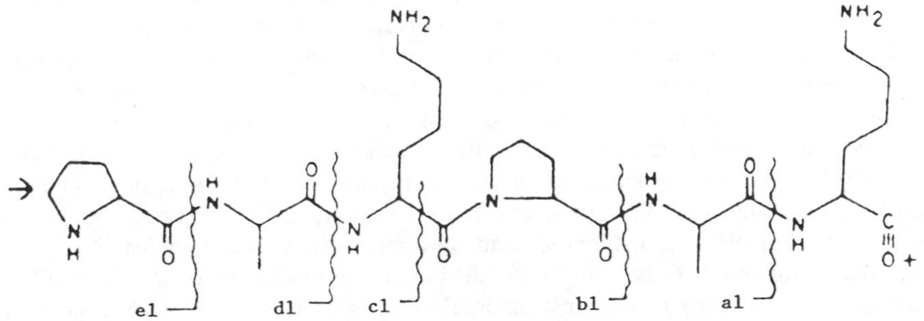

Scheme 2 ∎ Representation of the fragmentation of cyclo-(LYS-PRO-ALA)$_2$, assuming initial protonation at the indicated site on the proline followed by ring opening.

MS/MS spectra by averaging 20 scans obtained in a 20-s analysis time. Later work by Eckart et al.[806] extended the work to other cyclic peptides, including biologically active cyclic analogs of enkephaline and somatostatin. It was shown that unambiguous sequencing could not normally be obtained by interpretation of the daughter ion MS/MS spectrum of the protonated molecule alone. For confirmation of structure, the daughter ion MS/MS spectra of many of the fragment ions observed in the FAB mass spectrum are also required. While this is not difficult when the sample is in a high state of purity, complications may arise when the fragment ions can be formed by source fragmentations from several different peptide components.

Rigorous mechanistic studies of a set of peptides have been carried out to establish the general mechanisms of CID dissociation for these compounds. Lippstreu-Fisher and Gross[807] have used FAB ionization to create $(M + H)^+$ ions from a series of peptides containing from three to six amino acids. Daughter ion MS/MS spectra were obtained using high-energy collisions on an EBE instrument. Fragment ions that could be related to the sequence of the peptide were present in the FAB mass spectrum only with relatively low abundances and could not be used to unambiguously sequence the peptide. However, almost all of the daughter ions in the MS/MS spectrum of the $(M + H)^+$ ion could be assigned structures that result directly from cleavage of the peptide backbone. Study of the seven peptides revealed that 70% to 75% of the total ion current in the daughter ion MS/MS spectrum could be attributed to positive ions of three general types: B_j ions, in which the peptide bond is cleaved with charge retention on the N-terminal fragment, $(B_j - 28)$ ions , corresponding to the loss of 28 daltons (CO) from the B_j ions, and Y_j-type ions, in which the cleavage is at the peptide bond and the charge is retained on the C-terminal fragment; in each case, j is the number of amino acid residues in the fragment ion. A similar strategy (but with different forms of the daughter ions) was applied to the interpretation of the daughter ion MS/MS spectra obtained from protonated molecules of polyamino alcohols formed by reduction of the original peptides. The method was extended by application to simple mixtures of peptides and polyamino alcohols. These mixtures of peptides produced complicated FAB spectra, but the protonated molecules were easily the most abundant ions in the spectrum and could be easily identified. Sequence information via the MS/MS experiment could be obtained from as little as 80 to 100 ng of the peptide, with some changes in the MS/MS spectrum evident at these low levels due to the presence of several components in the beam selected as the parent ion. This result again underscores the importance of the purity of the sample used for sequencing, even when the protonated molecule is selected as the parent, and especially when the daughter ion MS/MS spectra of the lower mass fragment ions are measured.

Over the last few years, an increasing number of biologically active peptides have been isolated in which the N-terminus is blocked. In these cases, the direct application of the Edman degradation process is not possible. Furthermore, it has generally been the case that several other modifications to the backbone are present. Alkylation at the nitrogen is a common backbone modification encountered in these peptides. Daughter ion MS/MS analysis has been used with FAB ionization by Eckart et al.[808] to sequence a number of N-blocked, C-blocked, and backbone-N-alkylated peptides. Analysis of peptides could generally be achieved with sample concentrations of less than 0.1 μM and could be completed in an instrumental analysis time of 2 to 4 h. Measurements were made on a BEB instrument, with low-collision-energy CID in the third reaction region of the instrument. This was accomplished by selecting the parent ions through the first two sectors and passing them into a collision cell floated at a potential 400 V below the source potential. The daughter ions formed inside the cell are reaccelerated to 8000 V and analyzed by the final magnetic sector of the instrument. Because of the poor solubility of many of the peptides in a neat glycerol solution, the samples were first dissolved in 5 μL of a mixture of 33% hydrobromic acid in glacial acetic acid, and this solution was mixed with 5 μL of glycerol. The samples survived this relatively harsh treatment; acetylation was noted for some peptides that contain hydroxyl groups in the side chains of the constituent amino acids.

The interpretation of the mass spectra and daughter ion MS/MS spectra of these modified peptides can become quite complicated. The situation is no longer one in which the next residue is one of the 20 common amino acids; all possible points of alkylation and backbone modification must be considered. Eckart et al.[809] provide a concise summary of the interpretation of the spectral data in terms of these modifications, using common mass differences and exact mass measurements. A salient point of their discussion is that the initial correct identification of the protonated molecule of the peptide $(M + H)^+$ is critical for the subsequent daughter ion MS/MS experiment. The protocol essentially interprets the mass spectrum to indicate the presence of an N-alkyl substituent, calculates the mass of the expected protonated molecule, and then checks the spectrum to see if reasonable mass losses from this putative $(M + H)^+$ are observed. If internal consistency is achieved, the daughter ion MS/MS spectrum of the indicated protonated molecule is recorded. Eckart and Schwarz[809] note that N-alkylated amino acids may be present in isomeric forms. In the case of N-isobutylglycine and N-methylvaline, the daughter ion MS/MS spectra of the immonium ions formed in the FAB mass spectra of these two model compounds were clearly different. The fact that immonium ions of isomeric structures do not readily interconvert is consistent with the earlier work that used daughter ion MS/MS spectra of the immonium ions from leucine and isoleucine for their differentiation.[785-787]

The tert-butyloxycarbonyl (BOC) group is frequently used to protect amino groups in peptides and individual amino acids, and many peptides available commercially are sold in their BOC forms. Removal of the protecting group is straightforward and is achieved by dissolution of the peptide in a 25/75 mixture of concentrated hydrochloric acid/ethylacetate and evaporation to dryness at room temperature. Bathelt and Heerma[810] compared the utility of daughter ion MS/MS spectra in establishing the sequence of peptides analyzed by fast atom bombardment mass spectrometry in their BOC-protected or unprotected forms. In each case, the $(M + H)^+$ ion was selected as the parent ion. Figure 6-33 compares the daughter ion MS/MS spectra of the $(M + H)^+$ ions from BOC-LEU-SER-THR-OBz (top) and LEU-SER-THR-OBz (bottom), where OBz represents the benzyl ester group. As the figure shows, and in general for all six peptides studied, the daughter ion MS/MS spectra of the BOC forms of the peptides are more complex and contain less abundant sequence ions than do the daughter ion MS/MS spectra of the unprotected forms of the peptides. The ready fragmentation of the BOC group leads to a dilution of the ion current of the sequence ions into several different ion masses, and interpretation of the spectrum is therefore more difficult.

The discussion so far has been based on the interpretation of a mass spectrum or of a daughter ion MS/MS spectrum in terms of the predominant sequence ions that appear. There are several series of lower abundance fragment ions as well, corresponding to various losses of small groups from the amino acid residues. The mechanisms of

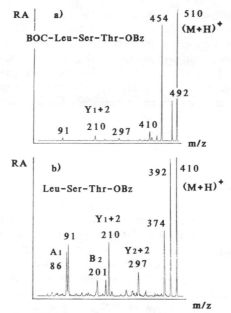

Figure 6-33 ■ Daughter ion MS/MS spectra of $(M + H)^+$ ions from BOC-LEU-SER-THR-OBz (top) and LEU-SER-THR-OBz (bottom), where OBz represents the benzyl ester group and BOC the butyloxycarbonyl protecting group.

these reactions have not in general been established and may in fact be more complicated than first apparent. Zwinselman et al.[811] have studied the mechanism of ammonia loss from the protonated molecule of arginine. Specific ^{15}N labeling shows that the $(M + H)^+$ ions formed in field desorption or metastable $(M + H)^+$ ions dissociate by loss of ammonia that contains the nitrogen of the guanidinyl group. In contrast, the $(M + H)^+$ ions formed by fast atom bombardment ionization lose ammonia that contains nitrogen found originally in both the guanidinyl group and the alpha-amino group. Metastable ions originally formed by FAB revert to loss of ammonia by a mechanism that involves only the guanidyl nitrogen. These differences may be attributed to changes in the site of protonation in the different ionization methods or in ions with different lifetimes. This point has also been expressed by Tomer et al.[805] in their discussion of the dissociations of cyclic peptides.

In another mechanistic study, Neumann and Derrick[812] studied the gas-phase chemistry of bradykinin. Field desorption was used to create $(M + H)^+$ ions from this nonapeptide, and CID was used to probe the chemistry through the daughter ion MS/MS spectra. A significant finding of this work is that there are a number of isomeric structures present for the lower mass sequence ions found in the positive ion FD spectrum, distinguished by their sites of protonation. Furthermore, the daughter ion MS/MS spectra of a number of doubly charged positive ions in the FD spectrum of bradykinin were recorded. Daughter ions formed by dissociation of a doubly charged ion include many ions also found in the FD spectrum itself. Careful interpretation of the mass spectra suggest that the structures, or the internal energies, of the ions formed by these two different paths may also be significantly different. Differences in daughter ion MS/MS spectra of ions of the same mass analyzed at two different FD emitter temperatures are also reproducibly different. The difference in internal energy was estimated to be very small in this instance, and so a difference in ion structure, and more specifically an isomerization of the original structure to a changed structure, was postulated. The origin of such a structural change may be clear in FD experiments in which the emitter temperature can be relatively high. Such changes may seem unlikely in FAB experiments carried out at room temperature, but the irradiation of the sample by the primary atom/ion beam may eventually lead to similar rearrangements in structure. Although the bulk of the glycerol matrix is at room temperature, the surface itself may be quite "hot."

Neumann and Derrick[219] have also estimated the energy transfer into peptide ions in the high-collision-energy CID process. The optimum pressure of the collision gas in the second reaction region of a BE instrument was defined as that pressure at which the summed abundance of the daughter ions is at a maximum for the particular parent ion selected. This pressure is different for the various parent ions selected and decreases as the mass of the parent ion increases. In all cases, the optimum collision gas pressure corresponds to an attenuation of the parent ion beam intensity by 60% to 70%. In general, translational energy losses in the collision process itself increase with increasing mass of the incident parent ions, ranging from 10 eV for ions of mass 450 daltons to almost 30 eV for ions above a mass of 1200 daltons. The value of q (see Chapter 3) for the amount of internal energy taken up by the ion can be estimated from the measured values of the translational energy loss. For the $(M + H)^+$ ion of bradykinin, a value of 21 eV is derived for a collision energy of 8 keV. Bricker and Russell[223] have made similar translational energy loss measurements on the ions derived from chlorophyll, but conclude that a significant fraction of this energy is transformed into excitation of the target gas atom. They suggest that the amount of internal energy deposited into a large polyatomic ion in a high-collision-energy experiment is more reasonably approximated as 10% to 25% of the translational energy loss value, and therefore the value will be about 2 to 5 eV of internal energy.

Mallis and Russell[813] have investigated the daughter ion MS/MS spectra of various forms of the parent ions of small peptides as a means to explore the energetics of CID excitation and dissociation mechanisms. The daughter ion MS/MS spectra of cationized molecular ions of peptides $(M + cation)^+$, where the cation is lithium, sodium, potassium, rubidium, or cesium, have been recorded. The predominant dissociation reactions induced upon collision are distinct. The daughter ions retain the alkali cation, which is thought to be preferentially bound between the more basic amino acids in the peptide, specifically histidine and arginine. This more localized charge tends to increase the structurally informative fragmentations. The absolute efficiency of the collision-induced dissociation process appears to be unchanged. Mallis and Russell[814] have extended the experiments to peptides in terms of the relative energetics of the dissociations to A-, B-, and C-type fragment ions from the N-terminal end of a peptide. At low internal energies of the parent ion, C-type daughter ions are observed almost exclusively, whereas at higher energies, all three forms of daughter ions are observed. It is also noted that under constant experimental conditions, the proportion of C-type ions in a daughter ion MS/MS spectrum increases as the mass of an $(M + H)^+$ ion increases. These data are interpreted to show that the amount of internal energy available for dissociating $(M + II)^+$ ions decreases as the mass of the parent ions increase.

The length of this section is testimonial to the intense interest in the mass-spectrometric sequencing of peptides. Some of the newest and most sophisticated MS/MS instrumentation is being devoted to this pursuit. Among these new instruments is a tandem quadrupole Fourier transform–ion cyclotron resonance mass spectrometer in which ions of interest are formed in a standard electron or chemical ionization source, selected by a mass-analyzing quadrupole, and passed into an ion cyclotron resonance cell in which they are stored, excited, and the daughter ions formed as a result of low-collision-energy CID mass-analyzed. Initial results with this instrument were described by Hunt et al.,[815] including positive ion mass spectra for renin substrate tetradecapeptide, bradykinin, gramicidin S, and melittin. For daughter ion MS/MS experiments, a stable population of the parent ions of these samples must be accumulated in the cell. Several thousand ions can be accommodated in the cell before ion overpopulation distorts the fields essential for mass selection and analysis; this same number of ions is also necessary to provide the initial parent ion population for daughter ion MS/MS experiments. Hunt et al.[816] have reported daughter ion MS/MS spectra of the $(M + H)^+$ ion of a tryptic peptide with 15 amino acid residues using photodissociation of the mass-selected parent ion at m/z 1772.9. Ten picomoles of the sample peptide were required, and secondary ion mass spectrometry was used to produce the $(M + H)^+$ ion. Ions of interest were passed through the quadrupoles into the ion cyclotron resonance cell, which was then pulsed with rf frequency to remove ions of below m/z 1600. The remaining ions were then exposed to a single pulse of 193-nm light from an argon fluoride excimer laser to cause the photodissociation. The daughter ions are mass-analyzed and the MS/MS spectrum stored. The daughter ion MS/MS spectra for 50 such laser pulses are summed to provide the spectrum shown in Figure 6-34. Recently, Hunt et al.[817] showed that the Q/ICR instrument was capable of recording mass spectra of peptides and small proteins in the mass range between 2000 and 13,000 daltons, thus providing molecular weight information for these samples. MS/MS experiments are not reported for these samples. Just as large peptides must be degraded prior to many FAB analyses, it will be necessary to dissociate such large ions in the ICR cell to smaller ionic fragments before sequencing can take place. The ability of the ion cyclotron resonance instrument to perform such multiple analyses and dissociations (MS^n) is encouraging.

Peptide sequencing based on CID in a Fourier transform–ion cyclotron resonance mass spectrometer has been demonstrated by Cody et al.[818] A few nanomoles of sample were admixed with KBr as a pellet and irradiated with the photon beam from a pulsed infrared (10.6-μm wavelength) laser. About 10^6 ions desorbed from the sample pellet make up the laser desorption mass spectrum. Specific parent ions are selected at a resolution of about 1000 and excited by CID (argon entrance through a pulsed-gas cell) to create daughter ions. Gramicidin D and gramicidin S were analyzed with this instrument, with the $(M + K)^+$ ions selected for the daughter ion MS/MS experiment. Since unit mass selection of these parent ions was not achieved, the daughter ions in the MS/MS spectrum appear as doublets. However, the sequence ions can still be determined with no difficulty, and 11 of the 15 sequence ions in gramicidin D can be assigned. The laser desorption mass spectrum of a commercial gramicidin D sample showed the presence of four components, including Val-1-gramicidin B, Val-1-gramicidin C, Val-1-gramicidin A, and Leu-1-gramicidin A.

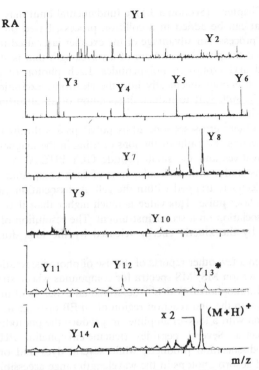

Figure 6-34 ■ Daughter ion MS/MS spectrum of the $(M + H)^+$ ion of a 15-residue tryptic peptide obtained using photodissociation of the parent ion. The sequence of the residue is PHE-GLN-GLU-THR-PHE-GLU-ASP-VAL-PHE-SER-ALA-SER-PRO-LXX-ARG, in which LXX represents undifferentiable leucine or isoleucine. The $(M + H)^+$ ion is at m/z 1772.9. The sequence of Y ions indicated on the figure are as follows: $Y_{14} = 1625.8$, the indicated $Y_{14}^{\wedge}$ ion is at m/z 1608.9, representing loss of ammonia from the Y_{14} ion; $Y_{13} = 1497.8$, the indicated Y_{13}^{*} ion represents loss of water from the Y_{13} ion; $Y_{12} = 1368.7$; $Y_{11} = 1267.6$; $Y_{10} = 1120.6$; $Y_9 = 991.5$; $Y_8 = 876.5$; $Y_7 = 777.4$; $Y_6 = 630.4$; $Y_5 = 543.3$; $Y_4 = 472.3$; $Y_3 = 385.2$; $Y_2 = 288.2$; $Y_1 = 175.1$.

Yang[819] has also used laser desorption in a Fourier transform–ion cyclotron resonance mass spectrometer to create $(M + H)^+$ and $(M − H)^-$ ions from BOC-protected peptides and CID daughter ion MS/MS experiments to determine the sequence of these ions. Water vapor was added to the cell at a very low pressure to act as the collision gas. Both positive and negative ion daughter ion MS/MS spectra could be recorded with equal ease, and the use of a constant background of water as the collision target avoided the need to pulse the collision gas into the cell.

As described previously in this chapter and in Section 3.3.2, an alternative to CID is photodissociation. As described previously, Hunt et al.[816] have used photodissociation for the analysis of a tryptic peptide. Photodissociation of peptides irradiated by UV laser light (193-nm wavelength) has also been demonstrated by Bowers et al.[109] using daughter and granddaughter ion MS/MS spectra measured in a Fourier transform–ion cyclotron resonance mass spectrometer. The goal of this work was to determine if photodissociation is more efficient than CID for the fragmentation of high mass ions.

As discussed in Chapter 3 (Section 3.3.1), a fundamental limit may be reached in terms of the energy that can be added in a collision process, based on the kinematics and dynamics of that process. An advantage of the excimer laser used in the photodissociation method is that the photons used are of a wavelength that is strongly absorbed by the carbonyl and aryl groups in polypeptides. Each photon of 193-nm wavelength deposits 6.42 eV of energy specifically into the electronic excitation of the ion. Even one photon may be sufficient to induce dissociation of an absorbing (and generally a low mass) parent ion.

A laser beam (2-cm^2 cross section, 20-ns pulse) passes through the ion cell of the instrument, intersecting the paths of the ions orbiting in the magnetic field. Bromobenzene and its derivatives and the small peptide GLY-PHE-ALA were used as model compounds in this study, with molecular ions of each compound formed by electron ionization and selectively trapped within the cell. A dissociation yield of up to 25% is reported for each laser pulse. This value is much higher than that reached with either CID or photodissociation on a sector instrument. The resolution of the mass measurement is lowered by the kinetic energy released by the ions during the process of dissociation.

There have been a few other reports of the use of photodissociation rather than CID to generate daughter ion MS/MS spectra for compounds whose structure incorporates the appropriate chromophores. Welch[820] described an experiment in which pulsed laser light was directed into the first reaction region of an EB instrument. Linked scans were used in conjunction with a lock-in amplifier to produce the photodissociation daughter ion MS/MS spectra. Several peptides (leucine-enkephalin, PHE-LEU-GLU-ILE) showed no photodissociation products, while several dyes and organometallic compounds that could absorb photons in the wavelength range accessible (454.5–514.5 nm) did produce daughter ions. The authors conclude that sufficient energy can be added by photons to induce dissociation even when ions with low average internal energies are examined.

Applications. The past five years of peptide sequencing by MS/MS has concentrated on the fundamental studies of proteins of known structure, establishing the characteristic fragmentations of common and modified amino acids, and placing a foundation under the experimental protocols that must be developed for the assignment of a sequence of an otherwise unknown peptide. At this point in the development of MS/MS, the fundamentals are in place, and as recent reports have shown, the rapid expansion of MS/MS into a standard method of peptide analysis is now taking place. This section reviews some applications of MS/MS to the sequence analysis of unknown peptides, beginning with the early examples of lower mass peptides and ending with the recent results of analysis of peptide parent ions with masses of several thousand daltons.

Gross et al.[821] described the use of daughter ion MS/MS and MS/MS/MS spectra in the elucidation of the structure of the fungal toxin from *Helminthosporium carbonum* (HC toxin), known to be a tetrapeptide consisting of two residues of alanine, a proline, and an unusual amino acid 2-amino-9,10-epoxy-8-oxodecanoic acid. Fast atom bombardment was used to create the $(M + H)^+$ ion at m/z 409. The daughter ion MS/MS spectrum of the cyclic peptide provides an ion at m/z 240 as the most abundant ion in the spectrum, and it is the sequence of this ion that was crucial in assigning the correct sequence. Accordingly, this daughter ion was formed in the first reaction region of an EBE instrument, passed through EB, and then the granddaughter ion MS/MS spectrum recorded by the final electric sector. The pattern of daughter ions was such that the sequence could be clearly established from among the limited number of possibilities present for the three-residue combination.

An analog of HC toxin described previously was identified using similar daughter ion MS/MS experiments by Kim et al.[822] In this analog, one of the alanine residues in the original cyclic structure is replaced by a glycine residue. The concentration of the analog toxin is about 5% that of the original HC toxin, and it had previously escaped detection. The substitution of the glycine for the alanine was readily apparent by comparison of the daughter ion MS/MS spectra of the $(M + H)^+$ ions from the two different tetrapeptides, as shown in Figure 6-35.

Eckart et al.[823] have described an MS/MS method for the sequence analysis of adipokinetic hormone I (AKH I), a peptide hormone from insects that mobilizes the release of lipids from body fat. The structure of this particular hormone was known from previous work, but this initial study emphasized that 5 μg of the pure sample could be sequenced with fast atom bombardment as the ionization method and daughter ion MS/MS analyses. Metastable ions that dissociate in the second reaction region of the instrument were studied in a BEB instrument. The complete sequence could be established by interpretation of the daughter ion MS/MS spectra of two fragment ions in the FAB mass spectrum that represent overlapping portions of the dodecapeptide. The same fast atom bombardment MS/MS experimental protocol was used to assign the sequence of the adipokinetic hormone from *Manduca sexta*.[824] The N-terminus is blocked in this nonapeptide, and so the direct Edman method is not possible. Two nanomoles of the substance, in this case isolated from tobacco hornworms, were required for the sequence analysis based on the metastable ion dissociations of the protonated molecule of the peptide at m/z 1008.

Witten et al.[825] have determined the amino acid sequences of two cockroach neuropeptides with a combination of exact mass measurements and metastable ion daughter ion MS/MS experiments. Most of the structure of the octapeptide could be deduced from the fast atom bombardment mass spectrum alone, especially with the help of the exact masses of the fragment ions, but the daughter ion MS/MS spectra provided additional confirmation of the putative structure. A linked-scan procedure was used to measure the daughter ion MS/MS spectrum; although the accuracy of the mass assignments in the MS/MS spectrum could be assigned only to ± 2 daltons, the differences in the mass between various amino acid residues in question were sufficiently large that this inaccuracy did not present a problem, although some difficulties might eventually surface based on the mass overlap problem. The same combination of fast atom bombardment and linked-scan daughter ion MS/MS scans was used to sequence the adipokinetic hormones from several different insect species, including *Schistocerca nitans, Schistocerca gregaria,* and *Locusta migratoria*.[826] All of the hormones are octapeptides with masses of the protonated molecules of about 900 daltons, depending on the exact amino acid composition, and the N-termini are blocked, thus preventing the direct application of the classical Edman degradation method.

Gade and Rinehart[827] used daughter ion MS/MS experiments to sequence the hypertrehalosaemic factor obtained from an Indian stick insect. Both linked-scan MS/MS scans on an EB instrument and daughter ion MS/MS scans on a BEEB instrument were used in this investigation of the decapeptide. The conclusion of the work was that the latter experiment, with its superior mass resolution of the daughter ions, is advantageous in these problems. In several cases, mass overlaps in the fast atom

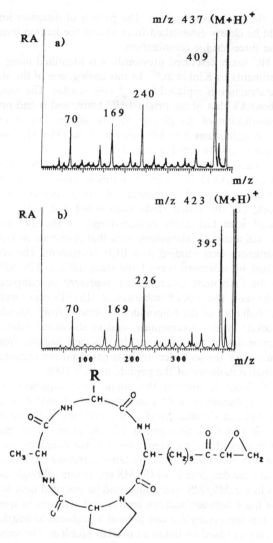

Figure 6-35 ■ Comparison of daughter ion MS/MS spectra of $(M + H)^+$ ions of two tetrapeptides differing in the substitution of glycine for alanine. The mass difference in the fragment ions highlights the substitution. For the structure shown, tetrapeptide (a) R = CH_3 and for (b) R = H. These tetrapeptides are forms of the HC toxin that attacks corn.

bombardment mass spectra were found, arising from the ^{13}C isotope peak of one sequence ion overlapping with a sequence ion generated in another process.

Krishnamurthy et al.[828] have used both high- and low-collision-energy daughter ion MS/MS spectra to study the structures of the cyclic peptides isolated from blue green algae. The peptide is toxic to the liver and has been implicated in both livestock and human illness. Cyclic peptides (dolastatins) were also studied using FAB MS/MS by

Eckart et al.[829] In this case, the source of the material is a sea hare resident in the Indian Ocean. The interest in the structures of these peptides arises from their demonstrated anticancer properties.[830]

Physalaemin-like peptides isolated from rabbit stomachs have also been studied by FAB MS/MS. The sequences of hepta- and octapeptides were established by interpretation of the daughter ion MS/MS spectra.[831] The MS/MS spectra of the methyl ester derivatives of the same series of peptides were also studied. This work describes the integration of a computer-aided interpretation system[832] into the analytical protocol. While most of the sequence ions are easily identified, others are not. The role of the computer is to provide a list of possible residue sequences consistent with all of the mass spectral data. In using the computer, the tendency to gravitate toward known or familiar sequences is reduced. Several other similar interpretive systems have been described for the sequencing of peptides.[833-836]

The range of applications of MS/MS to the sequencing of peptides continues to expand into all areas. Griffin et al.[837] have described MS/MS methodology that is used to sequence peptides derived from the enzyme aldose reductase, from glycoproteins from mites that are allergens, and from phosphorylated peptides. Aubagnac et al.[838] used daughter ion MS/MS techniques to monitor part of the reaction sequence in a solid-phase peptide synthesis scheme. Here the three aspects of speed of analysis, sensitivity, and specificity of MS/MS analysis are clearly needed. With sufficient experimental experience, MS/MS can be used to monitor the accuracy of standard or synthetic solutions of small peptides at the fmol level, as described by Desiderio and Dass.[839] The continuing growth of biotechnology and related fields carries a concomitant need for protein identification and peptide sequencing. The across-the-board applicability of the MS/MS experiment for such analyses, coupled with steadily decreasing costs in terms of instruments and time, portends a rapidly expanding use of MS/MS in peptide analysis.

6.8. Pharmaceutical Applications

An especially important analytical consideration in pharmaceutical analysis is the identification and quantitation of not only the major components of a drug mixture, but the minor components as well. In drug metabolism work, the toxicity of trace components is often of paramount importance. Approval or disapproval of drug use by regulatory agencies may hinge on the complete identification of all metabolites; the successful conversion of a candidate pharmaceutical into a commercial product depends in no small part on the capabilities of the analytical methods used for characterization. Ethical evaluation of such compound mixtures also requires that the most complete practical analysis be completed. The solution of such problems was one of the first areas in which MS/MS found application, based not only on its speed, sensitivity, and selectivity, but also on the ability of several MS/MS experiments, notably the parent ion and neutral loss MS/MS scans, to screen a complex mixture for compounds of appropriate chemical structure. An example is the use of MS/MS to identify potential metabolites of a candidate drug even at low concentrations, and these experiments will be discussed more fully in Section 6.8.3.

Since many drug products are proprietary, published reports of the use of MS/MS in pharmaceutical analyses are still relatively scarce. Nevertheless, the applications that can be described attest to a powerful analytical technique. Richter et al.[840] have described three areas of application for MS/MS: (1) qualitative work on known compounds, including the screening of formulations for active drug content, impurities, and synthetic markers (as in the investigations of patent violations); (2) structural analysis of new potential drugs and drug formulations; and (3) quantitative assay of drug metabolites in body fluids, as in pharmacokinetic studies. Of these, the latter has been most completely described in the open literature, and most of the former work remains proprietary. Examples of each of these areas of application have been described.[840] The present discussion subdivides pharmaceutical analysis by MS/MS into areas that deal primarily with drug structures (Section 6.8.1), analysis of drug residues in biological systems (Section 6.8.2), the determination of drug metabolites (Section 6.8.3), and, finally, the use of MS/MS in metabolic profiling and the diagnosis of diseases (Section 6.8.4).

6.8.1. Drug Assays and Drug Structures

Drug Assays. Quality control analysis of pharmaceutical preparations containing active drugs requires a fast and selective analysis of a targeted compound in a complex mixture of fillers and excipients, the composition of which is known. Given such a situation, an MS/MS method can be devised that minimizes matrix effects and provides a reliable means for the determination of the targeted compound. Duholke[841] described such an application of MS/MS in the direct analysis of prostaglandins and steroids in ointments and cream bases. For the steroid analysis, linked scans on an EB instrument were used to generate daughter ion MS/MS spectra of parent ions formed by chemical ionization. The mass spectrum of the topical cream studied produced ions that overlapped with the masses of the parent ions of the steroids of interest, and the daughter ion MS/MS spectrum also contained ions formed by the dissociations of these nonsteroid ions. However, certain daughter ions could be identified as characteristic of the steroid contained in the cream, and the steroid content could be quantitatively evaluated despite the presence of the matrix. The particular product investigated contained 0.025% of the steroid drug, but the amount of total sample introduced into the source of the mass spectrometer was not specified. Another assay of steroid content in cream by MS/MS has been more recently described.[842] In experiments similar in concept, McClusky et al.[843] described the use of daughter ion MS/MS spectra to identify salicylic acid, an impurity in aspirin, with direct introduction of the crushed aspirin tablet into the source of the mass spectrometer via the direct insertion probe.

MS/MS can also be used for assays of drug content in systems far removed from final products. Tondeur et al.[844] have described the use of daughter ion MS/MS in the identification and quantitation of the antibiotic toyocamycin directly in the fermentation broth in which it is produced. One microliter of the whole broth was mixed with 1 μL of glycerol. The daughter ion MS/MS spectrum of the $(M - H)^-$ ion of toyocamycin at m/z 290 was recorded. Detection limits in the whole broth for FAB mass spectra were 5 μg. The additional selectivity of MS/MS provided about 100-ng limits of detection when recording a complete daughter ion MS/MS spectrum. In the broths used to produce toyocamycin, usual concentrations are about 600 ng/μL.

Drug Structures. Pharmacologically active drugs are often of similar structure, and MS/MS experiments predicated on this relationship can be used to establish the structure of variants or to screen complex mixtures for the presence of several forms of the active drug. This structural similarity also catalyzes the use of MS/MS experiments in the study of drug metabolism, as outlined in Section 6.8.3.

Although FAB is now often used in the analysis of new drugs, the spectral interpretation can be thrown into question if there are several drugs present in a mixture in the glycerol solution, or if the irradiation of the glycerol solution generates artifact products that are closely related to the structure of the putative drug analyte. An example of the latter type has been described by McLafferty[845] in the FAB analysis of vitamin B-12. Daughter ion MS/MS experiments were used to characterize artifacts formed during continued irradiation of glycerol solutions of vitamin B-12. High-energy (10 keV) CID was used in an EBE mass spectrometer to create daughter ions from high mass ions in the vicinity of the molecular ion. Similarities in the daughter ion MS/MS spectra suggested that one of the artifact peaks is formed by intermolecular transfer of a cobalt atom from one molecule to the phosphate group of another, and that a few facile losses from the various analogs studied provide ions of a common structure.

One of the early applications of daughter ion MS/MS was to the analysis of barbiturates.[846] The electron ionization mass spectra of these compounds are in general not distinctive, but daughter ion MS/MS spectra of the $(M + H)^+$ or $(M + C_2H_5)^+$ ions formed in a chemical ionization source distinguished isomeric barbiturates such as amobarbital and pentobarbital. MS/MS experiments are now often used in conjunction with the analysis of compounds that cannot be made to pass through a gas chromatographic column. For instance, daughter ion MS/MS spectra obtained on a BE instrument have been used to study a series of related technetium and iron cationic complexes in use as organ-imaging agents.[847] Oxidation products of the original compounds can be characterized in mixtures, and isomeric forms of the complexes can be distinguished. The relative ion abundances in the daughter ion MS/MS spectra reflect the nature of the transition metal center and the relative size of the alkyl substituents attached to the chelating ligand.

Determination of the structure of antibiotics has long been of interest. David et al.[848] reported daughter ion and parent ion MS/MS spectra of anthracycline antibiotics. The dissociations of the protonated molecules formed in the chemical ionization source were characteristic of this structural class of antibiotics. Parent ion MS/MS spectra were used to elucidate the various dissociation pathways to the characteristic daughter ions. Barbalas et al.[849] have recorded the daughter ion MS/MS spectra of penicillins and other beta-lactams using high-energy CID on an EBE instrument. The electron ionization mass spectra of these compounds typically contain a large fragment ion at m/z 174 corresponding to $[C_{17}H_{12}NO_2S]^+$. The daughter ion MS/MS spectra of the m/z 174 ion in the electron ionization mass spectra of penicillins are distinct from those of the m/z 174 ion of identical empirical formula in the electron ionization mass spectra of other antibiotics such as the dihydrocephalosporins. Information about the original stereochemical configuration of the antibiotic molecule is lost in the dissociation to the ion at m/z 174, although interpretation of several of the dissociation pathways was aided by the use of isotope-resolved MS/MS, in which the daughter ion MS/MS spectrum of the ^{34}S-containing isotope was measured.

Cooper and Unger[850] have used daughter ion MS/MS spectra to differentiate isomeric forms of saframycin antibiotics. Exact mass measurements on the ions in the positive ion FAB mass spectrum were unable to establish the formula of two closely related antibiotics due to the very small difference in their masses (525.2703 and 525.2737, calculated resolution is 154,000). Figure 6-36 shows that the relative abun-

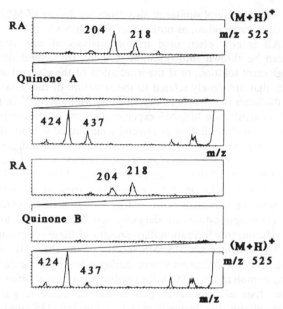

Figure 6-36 ■ Differentiation of isomeric saframycin antibiotics by daughter ion MS/MS spectra. The relative abundances of the ions at m/z 218 and m/z 204 are distinctive. The empirical formulas of both $(M + H)^+$ ions are $C_{28}H_{38}N_4O_6^+$.

dances of the daughter ions at m/z 218 and m/z 204 can be used to distinguish between the antibiotic structures.

Siegel et al.[851] have studied the daughter ion MS/MS spectra of polyether antibiotics, including monensin A, etheromycin, and several forms of maduramycin. Since these antibiotics are ionophores, the molecular ion current was carried by ions such as $(M + Na)^+$ and/or $(M + K)^+$ rather than the protonated molecule $(M + H)^+$. Sodium or potassium ions were present as ubiquitous contaminants in the sample or could be added as salts to the glycerol solution of the sample. If other monovalent ions such as cesium, rubidium, or silver were added, the appropriate cationized species were observed in the positive ion FAB mass spectrum. The polyether antibiotics studied were distinguished by either the dominance or the absence of an ion corresponding to loss of 62 daltons (concerted losses of carbon dioxide and water) from the cationized molecular ion. This loss also appeared in the metastable daughter ion MS/MS spectra and in the CID daughter ion MS/MS spectra. The presence of this ion was an indication that the carboxylic acid group near one end of the molecule was either free (in which case the loss was observed) or present as an ester or absent altogether (in which cases the loss was not observed). Daughter ion MS/MS analysis of the ion at $(M + \text{cation} - 62)^+$ was most useful in establishing the structure of the antibiotic and provided both a greater abundance and a greater diversity of daughter ions than did dissociation of the $(M + \text{cation})^+$ ion.

Holzmann et al.[852] have used negative ion fast atom bombardment and daughter ion MS/MS experiments to investigate the structures of drugs related to suramin, a trypanocidal (antiparasitic) drug used to treat river blindness. These drugs are particularly susceptible to thermal degradation and so had been resistent to analysis by electron or chemical ionization mass spectrometry. The presence of multiple sulfonate groups in the structure of the drug results in the production of a characteristic negative ion FAB mass spectrum. Since the structure also contains $-C=O-NH-$ peptide-like bonds, the patterns of dissociation discussed previously for the MS/MS analysis of peptides were applied with some modifications to the structural analysis of this class of compounds.

As more sophisticated instruments become available, MS/MS experiments are applied to increasingly complex problems. Roberts et al.[853] have used daughter ion MS/MS analysis for the structural elucidation of glycopeptide antibiotics using a four-sector BEEB instrument. Antibiotics in the vancomycin–ristocetin family consist of a peptide core, usually made up of unusual amino acids, attached to several carbohydrate structures. Masses of these molecules are usually in the 1400-to-2000-dalton range. The interpretation of MS/MS data provides the information necessary to deduce the nature of the carbohydrate units but also provides information about the nature of the peptide core itself. Figure 6-37 emphasizes the complexity of the

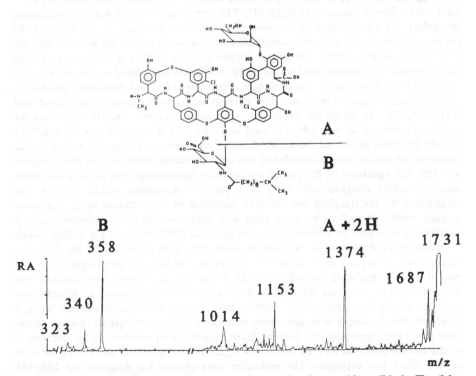

Figure 6-37 ■ Daughter ion MS/MS spectrum of a complex glycopeptide antibiotic. The (M + H)$^+$ ion is at m/z 1731. Loss of carbon dioxide from this ion produces the daughter ion at m/z 1687. Dissociation as indicated in the structure produces A and B. The ion at m/z 1374 is (A + 2H)$^+$, and that at m/z 358 is B$^+$.

structures under investigation as well as the range of ions recorded in the daughter ion MS/MS spectra.

Lay et al.[854] have described the differentiation of three classes of antihistamines by daughter ion MS/MS analysis. Positive ion FAB was used to create $(M + H)^+$ ions. Each class of antihistamines, namely ethylenediamine, ethanolamine, and propylamine derivatives, and their N-oxide analogs, provided neutral losses characteristic of each structural group.

6.8.2. Drug Residues

The sensitivity and selectivity of daughter ion MS/MS analyses have been put to use in a determination of sulfonamide drug residues in swine liver.[653] Routine tissue cleanup yields an extract that is directly analyzed with the direct insertion probe. The protonated molecular ion of the sulfonamide drug is produced by chemical ionization, selected by mass, and its daughter ion MS/MS spectrum recorded. Eighteen different drugs were analyzed, and several structure-characteristic daughter ions were identified. In some cases, isobaric isomer differentiation (sulfadoxine vs. sulfadimethoxine and sulfachlorpyridazine vs. sulfachlorpyrazine) was also demonstrated. Detection limits were reported as 0.1 ppm with 1.0 g of tissue sampled. In similar work, Finlay et al.[855] have used a hybrid geometry (EBQQ) MS/MS instrument to develop a rapid screening procedure for the analysis of residues of sulphonamide drugs in pig kidneys. In previous methods, the extensive sample cleanup necessary before the analysis was the rate-limiting step in sample throughput. Crude extracts of the pig kidney were placed on a moving belt (usually used for LC/MS interfaces) and passed sequentially into a chemical ionization source. The protonated molecules of all sulphonamides produce a characteristic daughter ion at m/z 156, corresponding to the common structural unit $NH_2C_6H_4SO_2^+$ in all of the drugs. For screening, a parent ion MS/MS scan for parents of the ion at m/z 156 was used and provided detection limits of about 0.1 mg/kg for most of the sulphonamide drugs. Confirmation of drugs identified in this screening procedure was accomplished through the measurement of the full daughter ion MS/MS spectrum of the parent ion and the requirement that each of the three most abundant daughter ions were of a relative abundance within 10% of that established by the daughter ion MS/MS spectrum of the standard drug. For more limited analyses involving no more than five sulphonamide drugs, multiple-reaction monitoring was used to establish detection limits of 0.05 mg/kg of pig kidney tissue.

Analyses of drug residues in food products are often prompted by the fact that regulations are in place detailing the use and permitted levels of drugs, specifically the use of growth-regulating steroids in cattle. Facino et al.[856] have used daughter ion MS/MS spectra to assay diethylstilbestrol, zeranol, and trenblone in plasma and various other tissues from cattle. Starting with 1.0-g samples of tissue and using a relatively simple sample homogenization, extraction, and centrifugation cleanup procedure (total time approximately 30 min), these targeted compounds could be determined in the tissue extracts at very low levels—14 pg for diethylstilbestrol and 150 pg for each of the other two estrogens. The molecular ions chosen for daughter ion MS/MS analysis were formed by electron ionization, which produces abundant molecular ions for these relatively simple compounds. The electron ionization mass spectra of the tissue extracts containing similar levels of the estrogens were very complex, and the presence of these compounds could not be discerned from those spectra. Although reported detection limits were twice as high in the MS/MS experiment with tissue extracts as with the pure samples, this sensitivity is obtained in the presence of all matrix components and is an excellent demonstration of the ability of the MS/MS experiment to discriminate against chemical noise. All three compounds could be determined in a single sample within an instrumental analysis time of 60 s, with quantitative assay over three orders of magnitude.

Similarly, an MS/MS-based assay for the antiparasitic drug ivermectin in cattle tissue has been developed by Tway et al.[857] Since this drug is active at very low levels, substantial sample cleanup was necessary before the MS/MS analysis could be completed without excessive interference from the matrix. This matrix interference persisted even through high-resolution mass measurements, suggesting that the sample cleanup procedure itself requires more refinement. The daughter ion MS/MS spectra of the parent ions of the two components of the commercial drug ivermectin each contains abundant ions used for selected reaction monitoring experiments. The lowest level of reliable detection was 5 to 10 ppb. Lower levels were compromised by the chemical noise that persisted even after a sample cleanup based on column fractionation. This level of detection is sufficient for assay of the drug even at the lowest levels of therapeutic use.

Analysis of drug levels in racehorse blood serum or urine is an exercise requiring high specificity, high sensitivity, a high speed of analysis, and a high number of samples. The identities of drugs for use in racehorses is established, and the behavior of standards can be readily established. "Disallowed" drugs are used for specific and common purposes, and the MS/MS behavior of these compounds can also be established ahead of time. Henion et al.[858] have described the use of LC/MS/MS for the analysis of sulfa drugs in racehorse urine and plasma. The use of atmospheric pressure ionization reduces the need for frequent source cleaning and seems also to reduce the level of matrix interference in the direct analysis of such samples without chromatographic separation. However, for large number of samples, analytical efficiency is best preserved if even a short chromatographic cleanup is incorporated into the analytical protocol. Sulfamethazine, sulfisoxazole, sulfadiazine, and sulfadimethoxine were studied, and the LC/MS/MS procedure provided an analytical assay of these drugs in the biological fluids at levels of from 0.5 to 10 μg/mL, corresponding to the usual levels of these drugs in equine plasma and urine. Further work by Henion et al.[859] explored the use of daughter ion MS/MS experiments in conjunction with the screening of samples by thin-layer chromatography (TLC). In these experiments, TLC was used to screen the biological samples. If the presence of a drug of interest was indicated, the sample was scraped from the TLC plate, eluted from the silica with solvent, and then analyzed by direct insertion probe MS/MS. Success for the analysis of drugs at trace levels was not possible with direct analysis of urine or blood serum by MS/MS or by analysis of the TLC material containing the sample. Only with sample elution was MS/MS successful in drug identification. These results are a consequence of the sample concentration inherent in the latter steps and missing from the first two, as described in Chapter 5. The chances of sample contamination do increase, however, with increased sample handling.

Brotherton and Yost[860] used daughter ion MS/MS to screen a wide variety of drugs in the blood serum of horses. It is claimed that the MS/MS method allows the simultaneous screening of as many as 50 drugs in the plasma in 5 min. Selected reaction monitoring is first used with an aliquot of the sample to indicate what drugs might be present, with full daughter ion MS/MS spectra recorded for the parent ions of the indicated drugs on a second aliquot of the sample. With 1 μL of blood serum introduced into the chemical ionization source via the direct insertion probe, detection limits of 1 to 10 ng were quoted for most of the drugs. With a simple extraction and cleanup of the blood serum, detection limits in the low-pg range could be obtained. The described procedure was capable of screening between 12 and 14 blood serum samples/h, assuming constant instrument performance. The analytical confidence in the analysis was greatly increased when the blood serum was brought through a simple deproteination procedure or after a more rigorous extraction of the drugs from the endogenous components. In particular, the ability of many of the drugs to bind with the proteins in the blood serum and changes in protein binding with time and sample treatment produced variations in the assayed levels. Changes in the level of free drug are also due to changes in the desorption profiles of the drug from the direct insertion probe.

Applications of MS/MS to the analysis of drugs in human plasma and urine samples have been reported. Haering et al.[861] measured ergotamine (used for the treatment of vascular headaches) in human plasma with daughter ion MS/MS on a triple-quadrupole mass spectrometer. Negative ion chemical ionization was used to produce the $(M - H)^-$ ion from a blood sample extracted in a relatively short procedure. Samples were introduced on the direct insertion probe. Single-reaction monitoring was used to extend the limit of detection to 2 pg/mL of serum with a reproducibility of 17.5% for nine samples. Control of the blank is the limiting factor in determination of the limits of detection, and the chemical noise in the blank is an explicit function of the care taken in the sample extraction and cleanup procedure. The detection limits in this method are approximately 50 times better than the previously reported procedure based on high-performance liquid chromatography. A radioimmunoassay provided equivalent sensitivity, but poorer selectivity and long-term precision. Figure 6-38 is a trace of the single-reaction monitoring experiment for a plasma sample from a volunteer, showing a signal for 20 pg/mL of ergotamine.

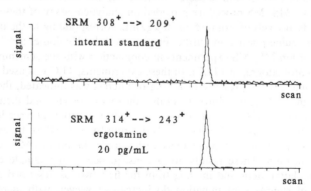

Figure 6-38 ■ Selected reaction monitoring experiment for analysis of ergotamine in plasma at a sample level of 20 pg/mL. The top trace is that for the internal standard dihydroergocryptine methanesulphonate and its reaction 308 → 209, and the bottom trace is the ergotamine reaction 314 → 243. In this case, the ion at m/z 314 is a fragment ion in the negative ion chemical ionization mass spectrum of ergotamine.

Finally, MS/MS experiments have been used to distinguish isomeric acylcarnitines in the urine samples of children with metabolic disorders by Gaskell et al.[862] Metabolic diseases such as Reye's syndrome and several acidemias change the distribution of carnitines excreted in the urine, and the quantitation of these compounds may evolve into a diagnostic tool. Since the carnitines produce abundant positive ions in fast atom bombardment ionization, this was the method used to create parent ions for daughter ion MS/MS spectra measured on a four-sector BEEB mass spectrometer. Daughter ion MS/MS spectra were characteristic of the isomeric structure of the carnitine. Butyrylcholine could be differentiated from isobutyrylcholine, and isomeric C_5- (valeryl, isovaleryl, and 2-methylbutyryl) and isomeric C_8- (n-octanoyl, valpropyl, and propylpentanyl) carnitines could be distinguished from each other. Although direct analysis of carnitines in urine was possible, better sensitivity was obtained when the carnitines were methylated to the preformed ionic compounds prior to FAB analysis and when the urine sample was lyophilized to remove endogenous interferences.

6.8.3. Drug Metabolites

Drug metabolism research often benefits from the functional group selectivity afforded by parent ion and neutral loss MS/MS scans. Since metabolites generally retain a large portion of the original drug structure, one or more of the ions present in the daughter ion MS/MS spectrum of the original drug should recur in the daughter ion MS/MS spectrum of each of the metabolites. Once these common ions are identified, they serve as characteristic markers for that drug structure, and the parent ion MS/MS experiment can be used to identify new metabolites or to quantitate metabolite levels with an isotopically labeled internal standard.

Perchalski et al.[863] described daughter ion, parent ion, and neutral loss MS/MS experiments implemented on a triple-quadrupole mass spectrometer for the identification and structural elucidation of drug metabolites. Chemical ionization mass spectrometry was used to create $(M + H)^+$ ions from blood serum samples that were analyzed either directly, after deproteination or after a simple extraction based on an acid/base fractionation. Samples were introduced into the source by distillation from the direct insertion probe, heated ballistically from 25° to 350° in 2 min. Monitoring for a specified parent ion/daughter ion pair was used for both daughter ion and parent ion MS/MS scan modes, and it was used to establish which compounds were initially present. A signal above background for any parent ion/daughter ion pair was a signal for the acquisition of a complete daughter ion MS/MS spectrum from the appropriate parent ion in a second sample. The daughter ion MS/MS spectrum so obtained was compared in its entirety against the standard spectrum stored in a computer library. Detection limits across the variety of drugs was generally in the 1- to 10-ng range, with lower detection limits possible with even rudimentary sample cleanup. Fractional distillation from the direct insertion probe allows the differentiation of drugs with the same nominal molecular weight by virtue of their evaporation profile, as well as the parent ion/daughter ion pair relationship established in the selected reaction monitoring experiment.

In contrast to the wide-ranging MS/MS procedure just described, MS/MS has also been used to focus on particular drugs. The compound 1-methyl-4-phenyl-1,2,3,6-tetrahydropyridine (MPTP) produces a Parkinsonian syndrome in humans and primates. It is thought that this compound is oxidized metabolically to 1-methyl-4-phenylpyridine, and that storage of this second compound in the primate brain amplifies its toxicity. Daughter ion MS/MS experiments completed by Johanessen et al.[864] confirmed the oxidation by comparison of the MS/MS spectrum obtained from a direct analysis of brain tissue with that of the standard compound.

An assay for the N-oxide metabolite of a cardiovascular drug benzazepine based on fast atom bombardment ionization and daughter ion MS/MS analysis has been developed by Straub and Levandoski.[865] The N-oxide metabolite is a zwitterion that thermally decomposes in electron and chemical ionization, but which produces a dominant protonated molecule $(M + H)^+$ in positive ion fast atom bombardment. The molecular weight of the metabolite can thus be clearly determined, but there is little fragmentation of this stable ion. Low-collision-energy CID on a triple-quadrupole instrument provides structurally important daughter ions of the parent ion at m/z 212. Multiple-reaction monitoring was used to follow the dissociation of the parent ion at m/z 212 to a daughter ion at m/z 180 (corresponding to loss of methanol), and simultaneously the dissociation of an isotopically labeled internal standard at m/z 217 to the daughter ion at m/z 185. Sensitivity down to levels of metabolite of approximately 100 pmol was demonstrated. A relatively extensive statistical evaluation of the MS/MS data was completed, including coefficients of variation for same-day and day-to-day analyses in three ranges of expected metabolite concentrations, reproducibilities for multiple runs, and comparison with a standard GC/MS method. In summary, precision ranged from 1% to 10%, and accuracy from 97% to 106% of the accepted (GC/MS) value. Of particular value in this work is that each assay requires less than 5 min, and extensive sample handling that had previously been shown to degrade the sample was avoided.

Off-line liquid chromatography was used by Lay et al.[866] to isolate glucoronide metabolites of doxylamine succinate and the antihistamine found in the commercial formulation Benedictin. The collected fractions were analyzed by FAB mass spectrometry. Exact mass measurements provided the empirical formula of these previously unknown metabolites, and daughter ion MS/MS spectra were interpreted, along with data from nuclear magnetic resonance, to provide their structures. Figure 6-39 shows the positive ion FAB mass spectrum of the combined LC fractions, with the masses of the metabolites marked (a), the FAB mass spectrum of an LC fraction containing only one metabolite (b), and the daughter ion MS/MS spectrum of the $(M + H)^+$ ion of that metabolite at m/z 449. Much of the structure of the metabolite could be established from the mass spectrum alone; the daughter ion MS/MS spectra were used to link clearly parent ions and structurally diagnostic daughter ions, removing doubts that the occurrence of such ions was due to the presence of impurities. Recently, Korfmacher et al.[867] have used thermospray MS/MS to extend their study of doxylamine metabolites; the advantage of this method is that the mass spectrometer is connected directly to the liquid chromatograph used to separate the components of the sample mixture.

Rudewicz and Straub[868] have used MS/MS in the analysis of conjugates of catecholamine. Conjugates are formed when the primary metabolites of a drug react with polar endogenous substances in the biological matrix such as amino acids, sulfates, or glucoronides. Means to study metabolic conjugates are limited and to date have mostly involved reformation of the metabolite structure and then inference back to the original conjugate structure. The particular sensitivity of FAB mass spectrometry

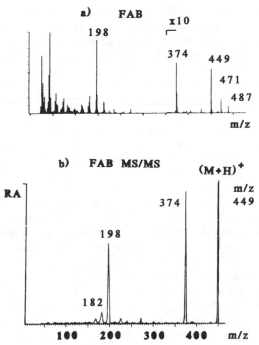

Figure 6-39 ■ (*a*) Positive ion FAB mass spectrum of a liquid chromatographic fraction containing one metabolite of doxylamine succinate. (*b*) Daughter ion MS/MS spectrum of (M + H)$^+$ ion at *m/z* 449, representing a single metabolite of doxylamine succinate.

to polar materials such as the conjugates makes their direct mass spectrometric analysis possible, and MS/MS is used to elucidate their structures. As Rudewicz and Straub[868] point out, CID of the molecular ions of the drug conjugates can often result in a daughter ion MS/MS spectrum in which the majority of the ion current is carried by a few stable ions characteristic of the conjugating group rather than the drug metabolite. Characteristic daughter ions or characteristic neutral losses established in CID studies of the standard drug may be of only minor relative abundance in the daughter ion MS/MS spectra of the conjugates. For example, alkyl sulfates produce a negative ion at *m/z* 97 (HSO$_4^-$) on CID; a parent ion scan for *m/z* 97 provides the distribution of alkyl sulfate conjugates in the sample. Aryl sulfate esters typically lose SO$_3$, and so a neutral loss scan for 80 daltons identifies conjugates of this type. O-Glucoronides undergo cleavage of the C—O glycosidic bond, with charge retention by the remainder of the drug structure; such structures are identified by a neutral loss scan for the glucoronide group with a mass of 176 daltons. MS/MS experiments that search for characteristic substructures of the administered drug are expected to identify primary metabolites. Other MS/MS scans based on the peculiar CID behavior of conjugates provide the distribution of the secondary metabolites.

Rudewicz and Straub used all of the MS/MS experiments just described to investigate the metabolic fate of ibopamine, a drug used to treat congestive heart failure. Urine samples from monkeys treated with the drug were analyzed directly by fast atom bombardment to provide a screening of the major metabolites. Liquid chromatographic separation was also coupled to MS/MS experiments via a thermospray ionization interface. FAB MS/MS provided a rapid screen of classes of conjugates, but the sensitivity of this analysis was 50 to 100 ng per identified component. Thermospray MS/MS provided better sensitivity (10–50 ng) in some cases, and the differentiation of isomers by retention time differences, but placed time constraints on the operation of the MS/MS instrument. In addition, FAB seemed less sensitive to small differences in source operating conditions.

The ability of MS/MS to analyze selected components in complex mixtures without pretreatment is an attractive feature, as just described, but the analysis of several such samples in succession can degrade the performance of the instrument. Furthermore, many drug metabolism studies make use of radiolabels that can be traced through an HPLC separation, providing an estimate of the total number of different metabolites and conjugates. The combination of the HPLC separation method with MS/MS therefore provides a confirmation that all of the metabolites have been identified. Typically, recent drug metabolism studies have been refined to include simple sample cleanup or chromatographic separations prior to the MS/MS analysis. Covey et al.[869] have used LC/MS/MS spectra to extend the understanding of phenylbutazone metabolism in the horse. Metabolism of this drug produces hydroxylated analogs, the two main isomers of which are distinguished by the position of the hydroxylation. Since the daughter ion MS/MS spectrum of each contained characteristic daughter ions reflecting the common structural element, their differentiation was straightforward. Additional selectivity was afforded by a difference in elution times from a short LC column. The first metabolite elutes at 25 s, the second at 45 s, each producing a parent ion at m/z 325. Since each of the metabolites has the same empirical formula, high-resolution mass spectrometry would be unable to distinguish between them based on exact mass measurement of the molecular ion. Since an ion at m/z 93 was the characteristic daughter ion for ions with the phenylbutazone structural nucleus, a simple LC separation was combined with parent ion MS/MS spectra of this selected daughter ion to establish the presence of several new metabolites of this drug, representing hydroxylation at still other sites in the parent drug molecule. These unknown metabolites are identified with total sample amounts of 10 to 50 ng. With the instrument setup with the basic MS/MS protocol, 60 samples of horse urine could be analyzed in 1 h. Such speeds are required for accurate pharmacokinetic studies; the authors are careful to point out that such performances are analytical "sprints" and the ability to analyze 1 sample/min does not translate into a sample analysis rate that can be maintained indefinitely.

Reviews of the use of MS/MS experiments in the determination of drug metabolites have been written by Yost et al.[870] and Straub et al.[871]

6.8.4. Diagnosis and Metabolic Profiling

As access to MS/MS instruments becomes more widespread, the results of MS/MS analyses are drawn closer to the formulation and solution of analytical and biochemical problems. A case in point is the diagnosis of metabolic disorders based on the outcome of MS/MS experiments and extension of the work based on GC/MS analysis that began as that technique became accepted in the medical community. In Section 6.8.2, the use of MS/MS in a structural study of carnitine and its derivatives was described.

Based on that work, Millington et al.[872] have used daughter ion MS/MS to assess quantitatively the distribution of carnitines in the urine of children suffering from propionic acidemia, methylmalonic aciduria, and Reye's syndrome. Aliquots of the urine were cleaned by passage through a Dowex ion exchange column and then analyzed by a combination of high-resolution fast atom bombardment mass spectrometry and daughter ion MS/MS. The combination of both techniques was required to unambiguously identify the structure of the carnitines present in the urine. The methodology is sufficiently well worked out so that diagnosis of a metabolic disorder can be suggested based on MS/MS analysis of amniotic fluid.[873]

Rinaldo et al.[874] have also studied the application of MS/MS to analysis of urines of patients suffering from methylamalonic acidurias. In this case, the identification and quantitation of methylamalonic acid in the urine was of interest. Direct exposure electron ionization was used to ionize the sample introduced on the direct insertion probe; the sample was a simple organic extract of the urine. The ions $(M + H)^+$ and $(M - OH)^+$ were abundant in the spectrum of a sample of the drug used as a standard; the daughter ion MS/MS spectra of these selected ions were measured. The urine from a healthy child was extracted, and the daughter ion MS/MS spectra of the ions at these same masses in the mass spectrum recorded. No correlation in the masses or abundances of the daughter ions with those established by the standards was noted. The urine of a child with the aciduria was also collected, extracted, and analyzed, and a match between the daughter ion mass spectra was found, establishing the presence of methylamalonic acid.

MS/MS experiments have also been used to study the distribution of carboxylic acids in urine. For instance, Rinaldo et al.[875] used negative ion chemical ionization to create $(M - H)^-$ ions from carboxylic acids present in urine and daughter ion MS/MS spectra to confirm their structures. No derivatization of the urine was required, as is necessary in the GC/MS technique normally used in the usual clinical assay. The only sample preparation was an organic solvent extraction of the acidified urine. Identifications of all of the C_4 to C_{10} dicarboxylic acids could be made, as well as oxalic, glyceric, and glyoxylic acids.

In earlier work, Hunt et al.[876] had demonstrated the use of MS/MS for the metabolic profiling of urinary acids. Molecular ions $(M - H)^-$ are formed by negative ion chemical ionization. Under the low-collision-energy CID conditions in a triple-quadrupole mass spectrometer, these parent ions characteristically lose 44 daltons (as loss of carbon dioxide) or 62 daltons (as loss of carbon dioxide and water). A neutral loss scan for losses of these masses then establishes the distribution of the carboxylic acids in the urine. Total sample preparation, instrument, and data analysis time for each urine sample was about 15 min. About 100 urinary acids could be detected in normal samples.

6.9. Applications to Continuous Flow Samples and Processes

Samples can readily be introduced into the source of a mass spectrometer in a stream or flow, rather than as discrete probe samples. Gas chromatography and liquid chromatography, of course, are examples of such methods. Flow injection analysis is a potential candidate for sample introduction, and the first such system is assuredly on the horizon. Mass spectrometric experiments based on a simple continuous flow sample introduction are essentially similar in execution and have already been described.

Brodbelt and Cooks[877] have described a simple interface between a reaction vessel and a triple-quadrupole MS/MS instrument that illustrates the ease with which MS/MS may be coupled with ongoing chemical processes. Part of the solution in a reaction vessel is continuously cycled through a short length of semipermeable capillary tubing, a short loop of which is routed into the high vacuum of the mass spectrometer not far from the ion source. Organic molecules that diffuse through the membrane can be detected in real time by the mass spectrometer. The authors estimate that components present in solution with concentrations of about 10^{-6} M can be detected. The system was used to follow the base-catalyzed methylation of cyclohexanone. In this particular experiment, parent ion MS/MS spectra were used to follow the appearance of each of the new methylated products formed in the reaction, since each dissociates upon collision to a common characteristic daughter ion. Problems with this interface that must be solved before its general application are the changes in permeability of the membrane with reaction solution temperature, the relatively long response time (1–5 min), and a susceptibility to memory effects, although this latter problem is reduced at higher temperatures.

However, the simplicity of the device portends an acceleration in the application of MS/MS to problems of process control, since it simplifies the error-prone and time-consuming sampling step. Maquestiau et al.[878] have used daughter ion MS/MS to establish the identities of products formed in a flash-vacuum pyrolysis reaction.

These applications of MS/MS evolved from novel methods for introduction of the sample into the mass spectrometer. The continuous flow/fast atom bombardment probe[879] is another recent development in sample introduction, and MS/MS will undoubtedly emerge in application to the study of reaction kinetics, as already demonstrated with mass spectrometry alone.[880-882] "Molecular microprobe" is a general description for a developing set of instruments that image organic surfaces, much as ion microprobes have been long used for generation of images based on atomic species. Analysis of organic ions by mass to charge ratios leads naturally into the use of MS/MS experiments to ascertain structure and reactivity. Constraints of absolute signal intensity in such experiments are an early concern, but new means of ion analysis and activation may make such experiments possible.

7

Conclusions and Outlook

7.1. Instrumentation in MS/MS

Many advances in MS/MS follow from the development of new instrumentation for mass analysis and the development of new means for ion activation to induce changes in ion mass, charge, or reactivity. Advances in each of these areas may seem disparate, but coalesce into instruments and experiments with significantly improved capabilities.

Two instrumental parameters of merit are the mass range and the resolution of the mass analyzer. The past five years have seen the mass ranges of sector instruments climb to 15,000 daltons, those of quadrupoles increase to 4000 daltons, those of Fourier transform–ion cyclotron resonance instruments to 16,000 daltons, and demonstration of a mass range in time of flight instruments to about 25,000 daltons. These limits may increase by factors of 2 or 3, but an order of magnitude increase is unlikely. For the purposes of MS/MS, the selection of higher mass ions for daughter ion analysis becomes more difficult as the width of the isotopic envelope increases. Resolution of mass analysis of 100,000 was available 10 years ago with sector instruments, and this value has not changed significantly. For lower mass ions, much higher mass resolution can be attained with Fourier transform–ion cyclotron resonance instruments. In MS/MS experiments, such instruments can now provide resolutions for parent ion selection of several thousands, and resolutions for daughter ion analysis of several tens of thousands, and in some instances, even greater resolution. Exact mass measurements in the determination of the empirical formulas of daughter ions will be a popular use of these instruments. New developments in the construction of MS/MS instruments can be expected. Hybrid instruments, first described only a few years ago, are already entering their second generation of commercial manufacture in EBQQ and BEQQ configurations. Hybrid instruments in which the quadrupoles precede the sectors have also been described. The refinement of hybrid instruments, ion-trapping instruments, and instruments using time of flight will continue in the next few years. These latter instruments represent conceptually new approaches to the acquisition of MS/MS data. As described in Chapter 2, the arrival of ions and neutrals at two separate detectors can be correlated, providing coincidence spectra that are the equivalent of a daughter ion MS/MS spectrum. Such instruments provide information about ion dissociations in a time window quite different from that normally accessed in experiments with sector instruments and similar to that studied in quadrupole mass spectrometers.

On the other end of the time scale for ion reactions, MS/MS experiments with ion-trapping instruments operate in a time domain that can be lengthened from tens of microseconds to seconds, and sometimes to minutes by the storage of ions in the cell of the mass spectrometer. Storage of ions for extended times and in specified locations makes possible new experiments in the activation of these ions, such as the photodissociation experiments described in the following paragraphs. It also allows a great range of variation in the time scale over which ion dissociation and ion/molecule association chemistry can be studied.

Future developments in instrumentation are anticipated in two essentially opposite directions. In the first instance, more complex and more costly instruments will be developed that offer improved performance in fundamental MS/MS experiments. Such instruments will also include the ability to pursue a number of approaches to the solution of difficult problems. The other anticipated direction of development is that of simple, less expensive MS/MS instruments, perhaps dedicated to a single type of application or analysis. These latter instruments will be totally automated. At the appropriate price, there will be a tremendous demand for these instruments, in a wide variety of applications areas, due to the sensitivity and selectivity of MS/MS.

Collisional activation has been a mainstay of MS/MS and is likely to remain the predominant activation method for the foreseeable future. The physics of the ion/molecule collision and the conversion of translational energy into internal energy (Section 3.3) are steadily being refined, although all the details of the process for polyatomic ions are not yet clear. Fundamental studies are needed to produce the sound framework for application of energy-resolved and angle-resolved MS/MS. Measurement of energy-resolved MS/MS data is already available in automated fashion on some commercial instruments, and such data have been used in empirical if not rigorous studies of ion structure.

A wider application of photodissociation, based on the expanded availability of high-output lasers of tunable wavelengths, is foreseen. Such experiments can be carried out on sector and quadrupole instruments (Section 3.3.2), but seem particularly well suited for ion activation in ion-trapping instruments, in which the ability to trap ions in specified orbits inside the cell is a distinct advantage. Surface-induced dissociation is simple in concept, but much remains to be learned about the underlying mechanisms. Accelerated development of surface-induced dissociation, and of photodissociation as well, will depend on demonstration of their applicability to high mass ions. Such ions, if they behave statistically, can accommodate within their many internal degrees of freedom a measure of energy that would cause the dissociation of smaller molecules. To cause the dissociation of higher mass ions, more energy may be required, and multiphoton absorption, or interaction of the ion with a solid surface, may provide the means to deliver that level of energy into the ion.

7.2. Outlook for Advanced Applications

The evolution of analytical MS/MS has progressed through the step at which the method leaves the province of the research laboratory and becomes part of the general analytical laboratory. The next developmental step is the transfer from the laboratory into the field. The future will find a much larger number of MS/MS instruments literally in the field for environmental monitoring and in production facilities for process assay and control. In the analytical laboratory, MS/MS will continue to expand, with MS/MS/MS experiments becoming more commonly used in the solution of difficult problems.

Chapter 6 provides examples of two present-day types of applications of MS/MS. The first type of application details capabilities well within the range of current MS/MS instrumentation and experiments, but extends the technique to a greater breadth of problems. Examples of this sort include the use of MS/MS in environmental applications, in natural products analyses, in food and flavor analyses, in the evaluation of industrial products and industrial processes, and in forensic chemistry. The second group of applications are those that involve a broad spectrum of developments in modern mass spectrometry, including new MS/MS experiments, new ionization methods, and new MS/MS instruments of substantially higher resolutions and mass ranges than those of the preceding generation. Examples include the sequence analysis of peptides and the structural analysis of biomolecules.

Expanded applications of MS/MS in such fields as environmental analysis, natural products studies, and food and flavor analysis depends explicitly upon access to instrumentation and on automatic operation of daughter ion, parent ion, and neutral loss scans. The lag in the development of general computer control of MS/MS instruments has hindered this expansion. However, several new instruments are available that provide an essentially automatic approach to MS/MS. The ion trap mass spectrometer is one such instrument (see Section 2.2.7). Within their limits of mass resolution and mass range, such instruments provide MS/MS capabilities to an expanded analytical audience. In addition, as long predicted, the cost of an MS/MS instrument has declined to a level at which the number of prospective users is also much larger. Increasingly, MS/MS will be a routine method of choice for problems significantly different from what has so far been accomplished. Unfortunately, such application may become steadily more remote from mass spectrometrists versed in the fundamentals and eager to collaborate. Finally, the results of such use is likely to be reported in unfamiliar forums. None of these expected changes relieves the responsibility of analytical chemists to evaluate such applications and to ensure their precision and accuracy; they only make such a task more challenging.

Current literature documents such applications of the MS/MS method. Applications to natural products are expanding as instrumentation becomes more widely available. Recent studies include the characterization of trichothecene mycotoxins by daughter ion MS/MS,[883,884] demonstrating the ability to quantitate the presence of these compounds in fermentation broths down to a limit of detection of 5 pg. Fast atom bombardment mass spectrometry was used with MS/MS experiments to confirm the presence of arsenobetaine and arsenocholine in seafood.[885] Daughter ion MS/MS experiments are also increasingly used in studies of organic synthetic and degradation reactions. Recent work includes the identification of compounds formed by flash vacuum pyrolysis of benotriazoles[886] and by the direct pyrolysis of polysulphides in the source of the mass spectrometer.[887]

Applications of the second type rely upon sophisticated instruments and meticulous experiments. Foremost among the examples are the analyses of biomolecules, and in particular, the sequencing of peptides based on the information in the daughter ion MS/MS spectrum. Such work involves the use of new ionization methods such as fast atom bombardment and secondary ion mass spectrometry, and control of the chemical and physical parameters that influence the sputtering of ions from liquid matrices.

Furthermore, the instruments used are more sophisticated to accommodate the larger mass ranges necessary for such analyses. Again, current literature is replete with increasingly sophisticated analyses in the biomolecular field. Photosystem proteins of spinach have been studied with FAB and daughter ion MS/MS.[888] The cyclotetrapeptide tentoxin has been sequenced by high-resolution mass spectrometry in conjunction with FAB MS/MS.[889] The structures of complex phospholipids were studied with daughter ion MS/MS in a successful method to identify the phosphorylated base.[890] An especially active area has been the use of MS/MS in the identification of drug metabolites,[891] including glucoronide metabolites of doxylamine,[892] glutathione conjugates of acetaminophen,[892] and metabolites of pyrrolizidine alkaloids.[893] Once identified, metabolites can be traced in complex biological samples as a function of time using sophisticated MS/MS techniques coupled with radiolabel methods.[894] MS/MS methods are also being more widely used for the analysis of sulphates in serum and urine samples.[895,896] The key to these applications, as described previously, is the greater availability of MS/MS instruments and the extensive computer control of MS/MS experiments.

7.3. Interpretation of MS/MS Data

Consider MS/MS data obtained from sector instruments (BE) as compared to those from triple-quadrupole instruments. For the solution to a given analytical problem, or for a particular ion structural study, experiments with either instrument may provide a solution. In contrast to the compilations of data that exist for electron ionization mass spectrometry, in which the instrumental particulars that apply to the collection of such data become secondary, the instruments and the experiments in MS/MS are as integral a part of the data as are the actual measurements of ion abundances and ion mass-to-charge ratios. Correlations between data measured with a sector instrument and with the triple-quadrupole instrument exist, but the differences are also a direct function of the instrument used to record the data. Even from instrument to instrument, considerable variation in measured data may be evident. Efforts to standardize the operating conditions in multiquadrupole MS/MS instruments are, however, making some progress.[898,899]

The interpretation of MS/MS data must therefore include a focus on the procedures of a specific laboratory, the characteristics of a particular instrument, and even the practices of a particular mass spectrometrist. This fact places even more emphasis on the fact that a community of mass spectrometrists must work together in sharing not only their data, but their experiments, and the accurate details of the function of their instruments. At present, the use of MS/MS still relies intimately on the the ability of the mass spectrometrist to formulate, complete, and interpret the results of the experiment.

As computer systems are more fully integrated with the control of the MS/MS instrument, the details and quality of the MS/MS data will come to depend more explicitly on the design and operation of the computer/mass spectrometer interface. This is especially so with computer-controlled triple-quadrupole mass spectrometers, in which the operating characteristics of each of the three quadrupoles are under direct computer control. Of all MS/MS instruments, the triple-quadrupole mass spectrometer is designed for the broadest analytical market. Invariable trade-offs in performance and flexibility are made by their manufacturers. Many applications of MS/MS can be completed without change of the default values of the operating parameters of the instrument. In some instances, these default values do not even enter into consideration until inadvertently changed, at which time a shift in analytical performance may become evident. To develop new experiments, or for the most demanding applications, mass spectrometrists cannot allow themselves to be drawn into situations in which their instrumental options are limited. To ensure a continuing ability to solve problems, to create new experiments, and to further the science of mass spectrometry, the spectrometrist must continue to master the technology of the instruments.

The final interpretation of MS/MS data, however, may become the province of a data system rather than the mass spectrometrist. Increasingly, the combined results of several MS/MS experiments are considered along with exact mass measurements to deduce structure or to assign unambiguously the presence of a specific compound in a mixture. For MS/MS alone, the experimental possibilities include daughter ion, parent ion, and neutral loss scans, energy-resolved MS/MS, isotope-resolved MS/MS, charge permutation reactions, ion/molecule association reactions, and sequential MS/MS experiments. It is increasingly clear that chemometric techniques, based on both heuristic algorithms and pattern recognition methods (see Section 5.3.3), will become increasingly involved in the interpretation of MS/MS data. The speed of implementation of such methods is limited primarily by software rather than hardware, and the increasing generality of chemometrics for the manipulation of generic analytical data is encouraging. It is unlikely that such advanced methods will arrive first via commercial manufacturers of instruments, since the lead time for such development is excessively long. It is more probable that computer-literate users will devise their own systems based on standard chemometric computer programs, as is already occurring.

7.4. Conclusions

In the decade since its introduction, MS/MS has justifiably earned a place among the analytical techniques used for complex mixture analysis and studies of fundamental ion chemistry. The number of instruments worldwide now exceeds several hundred, and the number of researchers devoted to the exploration of new applications and to the development of new experiments and new instrumentation appears to be steadily increasing. Development of MS/MS has been linked to the increased sophistication and reliability of the instrumentation upon which it is based. This includes advances in hardware such as electric and magnetic sectors used for mass analysis, but also the advent of more sophisticated electronics used for control of quadrupoles and Fourier transform–ion cyclotron resonance mass spectrometers, and by advances in data collection, processing and display, now demonstrating the ability to collect and manipulate data at extraordinary rates. The development of MS/MS has also been catalyzed by increasing interest in the behavior of gas-phase ions and the diversity of experiments that can be devised to force them to reveal their gas-phase chemical behavior.

A signature characteristic of MS/MS is that it has not been confined into specific areas of application, channeled into specific experiments, nor carried out only with specific, special-purpose instruments. MS/MS experiments, as discussed in Chapter 1, are based on changes in the mass, charge, or reactivity of an ion. To induce such changes, scientists have developed means of collisional activation, surface-induced dissociations, electron-induced dissociations, and dissociations induced by the absorption of photons. Both high- and low-collision-energy regimes have been explored, and the detailed changes in the MS/MS spectra as collision energy is varied have been studied. The effects of single collisions and of multiple collisions have been investigated. Use of a single activation step was followed by the introduction of experiments in which multiple separate activation reactions were used. Ion/molecule association reactions have been added to ion dissociations as a tool for study of the gas-phase chemistry of ions. The analysis of the ions so produced has taken place with electric and magnetic sectors, quadrupoles, time-of-flight instruments, Fourier transform–ion cyclotron resonance instruments, ion-trapping instruments, and combinations of each of these with the other. The diversity of MS/MS is such that it is difficult within a book such as this to describe adequately all aspects of its basis and application. Conversely, the heterogeneity of MS/MS ensures an extended period of development and refinement and expansion to encompass experiments and applications not yet foreseen.

References

1. Thomson, J. J. *Rays of Positive Electricity and Their Applications to Chemical Analysis.* Longmans: London, 1913.
2. Thomson, J. J. *Phil. Mag. VI* 1910, **18**, 824.
3. Aston, F. A. *Phil. Mag. VI* 1920, **39**, 449.
4. Dempster A. J. *Phys. Rev.* 1922, **20**, 631.
5. Hipple, J. A.; Condon, E. U. *Phys. Rev.* 1945, **68**, 54.
6. Barber, M.; Elliot, R. M. Presented at the ASTM E-14 Conference on Mass Spectrometry, Montreal, June 1964.
7. Hunt, W. W.; Jr.; Huffman, R. E.; Saari, J.; Wassel, G.; Betts, J. F.; Paufve, E. H.; Wyess, W.; Fluegge, R. A. *Rev. Sci. Instrum.* 1964, **35**, 88.
8. Ferguson, R. E.; McCulloh, K. E.; Rosenstock, H. M. *J. Chem. Phys.* 1965, **42**, 100.
9. Haddon, W. F.; McLafferty, F. W. *J. Am. Chem. Soc.* 1968, **90**, 4745.
10. Jennings, K. R. *Int. J. Mass Spectrom. Ion Phys.* 1968 **1**, 227.
11. Rosenstock, H. M.; Melton, C. E. *J. Chem. Phys.* 1957, **26**, 314.
12. Cooks, R. G.; Beynon, J. H.; Caprioli, R. M.; Lester, G. R. *Metastable Ions.* Elsevier: Amsterdam, 1973.
13. Danis, P. O.; Wesdemiotis, C.; McLafferty, F. W. *J. Am. Chem. Soc.* 1983, **105**, 7454.
14. Wade, A. P.; Enke, C. G.; Cooks, R. G. To be published.
15. Schwartz, J.; Vincenti, M.; Cooks, R. G.; Wade, A. P.; Enke, C. G., To be published
16. Whitten, W. B. *Int. J. Mass Spectrom. Ion Proc.* 1987, **77**, 165.
17. Dawson, P. H.; French, J. B.; Buckley, J. A.; Douglas, D. J.; Simmons, D., *Org. Mass Spectrom.* 1982, **17**, 205.
18. Stafford, G. C., Jr.; Kelly, P. E.; Syka, J. E. P.; Reynolds, W. E.; Todd, J. F. J. *Int. J. Mass Spectrom. Ion Proc.* 1984, **60**, 85.
19. Beynon, J. H.; Amy, J. W.; Baitinger, W. E. *J. Chem. Soc. Chem. Commun.* 1969, 723.
20. Beynon, J. H.; Caprioli, R. M.; Ast, T. *Org. Mass Spectrom.* 1971, **5**, 229.
21. Maurer, K. H.; Brunnee, C.; Kappus, G.; Habfast, K.; Schroder, U.; Schulze, P. Presented at the 19th Annual Conference on Mass Spectrometry and Allied Topics, Atlanta, GA, 1971.
22. Wachs, T.; Bente, P. F., III; McLafferty, F. W. *Int. J. Mass Spectrom. Ion Phys.* 1972, **9**, 333.
23. Beynon, J. H.; Cooks, R. G.; Amy, J. W.; Baitinger, W. E.; Ridley, T. Y. *Anal. Chem.* 1973, **45**, 1023A.

24. Silva, M. E. S. F.; Reed, R. I. Presented at the 21st Annual Conference on Mass Spectrometry and Allied Topics, San Francisco, CA, 1973.
25. White, F. A.; Rourke, F. M.; Sheffield, J. C. *Appl. Spectrosc.* 1958, **12**, 46.
26. Kondrat, R. W., Ph.D. Thesis, Purdue University, 1978.
27. Ast, T.; Bozorgzadeh, M. H.; Wiebers, J. L.; Beynon, J. H.; Brenton, A. G. *Org. Mass Spectrom.* 1979, **14**, 313.
28. Schaldach, B.; Grutzmacher, H.-F. *Org. Mass Spectrom.* 1980, **15**, 166.
29. Boyd, R. K.; Beynon, J. H. *Org. Mass Spectrom.* 1977, **12**, 163.
30. Bruins, A. P.; Jennings, K. R.; Evans, S. *Int. J. Mass Spectrom. Ion Phys.* 1978, **26**, 395.
31. Lacey, M. J.; Macdonald, C. G. *Org. Mass Spectrom.* 1977, **12**, 587.
32. Millington, D. S.; Smith, J. A. *Org. Mass Spectrom.* 1977, **12**, 264.
33. Haddon, W. F. *Org. Mass Spectrom.* 1980, **15**, 539.
34. Boyd, R. K.; Porter, C. J.; Beynon, J. H. *Org. Mass Spectrom.* 1981, **16**, 490.
35. Weston, A. F.; Jennings, K. R.; Evans, S.; Elliott, R. M. *Int. J. Mass Spectrom. Ion Phys.* 1976, **20**, 317.
36. Kemp, D. L.; Cooks, R. G.; Beynon, J. H. *Int. J. Mass Spectrom. Ion Phys.* 1976, **21**, 93.
37. Zakett, D.; Schoen, A. E.; Kondrat, R. W.; Cooks, R. G. *J. Am. Chem. Soc.* 1979, **101**, 678.
38. Shushan, B.; Boyd, R. K. *Int. J. Mass Spectrom. Ion Phys.* 1980, **34**, 37.
39. White, F. A.; Collins, T. L. *Appl. Spectrosc.* 1954, **8**, 169.
40. Van Asselt, N. P. F. B.; Maas, J. G.; Los, J. *Chem. Phys. Lett.* 1974, **24**, 555.
41. Carrington, A.; Milverton, D. R. J.; Sarre, P. J. *Mole. Phys.* 1976, **32**, 297.
42. Louter, G. J.; Boerboom, A. J. H.; Stalmeier, P. F. M.; Tuithof, H. H.; Kistemaker, J. *Int. J. Mass Spectrom. Ion Phys.* 1980, **33**, 335.
43. McLuckey, S. A.; Ouwerkerk, C. E. D.; Boerboom, A. J. H.; Kistemaker, P. G. *Int. J. Mass Spectrom. Ion Proc.* 1984, **59**, 85.
44. Stults, J. T.; Enke, C. G.; Holland, J. F. *Anal. Chem.* 1983, **55**, 1323.
45. Pinkston, J. D.; Rabb, M.; Watson, J. T.; Allison, J. *Rev. Sci. Instrum.* 1986, **57**, 583.
46. Danigel, H.; Jungclas, H.; Schmidt, L. *Int. J. Mass Spectrom. Ion Phys.* 1983, **52**, 223.
47. Begemann, W.; Dreihofer, S.; Meiwes-Broer, K. H.; Lutz, H. O. *J. Phys. D* 1986, **3**, 183.
48. Schey, K.; Cooks, R. G.; Grix, R.; Wollnik, H. *Int. J. Mass Spectrom. Ion Proc.*, 1987, **77**, 49.
49. White, F. A.; Forman, L. A. *Rev. Sci. Instrum.* 1967, **38**, 55.
50. Futrell, J. H.; Miller, C. D. *Rev. Sci. Instrum.* 1966, **37**, 1521.
51. Maquestiau, A.; Van Haverbeke, Y.; Flammang, R.; Abrassart, M.; Finet, D. *Bull. Soc. Chim. Belg.* 1978, **87**, 765.
52. McLafferty, F. W.; Todd, P. J.; McGilvery, D. C.; Baldwin, M. A. *J. Am. Chem. Soc.* 1980, **102**, 3360.
53. Vrscaj, V.; Kramer, V.; Medved, M.; Kralj, B.; Beynon, J. H.; Ast, T. *Int. J. Mass Spectrom. Ion Phys.* 1980, **33**, 409.
54. Russell, D. H.; Smith, D. H.; Warmack, R. J.; Bertram, L. K. *Int. J. Mass Spectrom. Ion Phys.* 1980, **35**, 381.
55. Gross, M. L.; Chess, E. K.; Lyon, P. A.; Crow, F. W.; Evans, S.; Tudge, H. *Int. J. Mass Spectrom. Ion Phys.* 1982, **42**, 243.
56. Rabrenovic, M.; Brenton, A. G.; Beynon, J. H. *Int. J. Mass Spectrom. Ion Phys.* 1983, **52**, 175.
57. Occolowitz, J. L. Presented at the 31st Annual Conference on Mass Spectrometry and Allied Topics, Boston, MA, 1983.
58. Beynon, J. H.; Harris, F. M.; Green, B. N.; Bateman, R. H. *Org. Mass Spectrom.* 1982, **17**, 55.
59. Bursey, M. M.; Hass, J. R. *J. Am. Chem. Soc.* 1985, **107**, 115.
60. Kunihiro, F.; Naito, M.; Naito, Y.; Kammei, Y.; Itagaki, Y. Presented at the 32nd Annual Conference on Mass Spectrometry and Allied Topics, San Antonio, TX, 1984.

61. Vestal, M. L.; Futrell, J. H. *Chem. Phys. Lett.* 1974, **28**, 559.

62. McGilvery, D. C.; Morrison, J. D. *Int. J. Mass Spectrom. Ion Phys.* 1978, **28**, 81.

63. Yost, R. A.; Enke, C. G. *J. Am. Chem. Soc.* 1978, **100**, 2274.

64. Yost, R. A.; Enke, C. G.; McGilvery, D. C.; Smith, D.; Morrison, J. D. *Int. J. Mass Spectrom. Ion Phys.* 1979, **30**, 127.

65. Hunt, D. F.; Shabanowitz, J.; Giordani, A. B. *Anal. Chem.* 1980, **52**, 386.

66. Zakett, D.; Cooks, R. G.; Fies, W. J. *Anal. Chim. Acta* 1980, **119**, 129.

67. Zakett, D.; Hemberger, P. H.; Cooks, R. G. *Anal. Chim. Acta* 1980, **119**, 149.

68. Morrison, J. D.; Stanney, K. A.; Tedder, J. Presented at the 34th Annual Conference on Mass Spectrometry and Allied Topics, Cincinnati, OH, 1986.

69. Beaugrand, C.; Devant, G.; Nermag, S. N.; Rolando, C.; Jaouen, D. Presented at the 34th Annual Conference on Mass Spectrometry and Allied Topics, Cincinnati, OH, 1986.

70. von Zahn, U.; Tartarczyk, H. *Phys. Lett.* 1964, **12**, 190.

71. Leventhal, J. J.; Friedmann, L. *J. Chem. Phys.* 1968, **49**, 1974.

72. Cosby, P. C.; Moran, T. F. *J. Chem. Phys.* 1970, **52**, 6157.

73. L'Hote, J. P.; Abbe, J.Ch.; Paulus, J. M.; Ingersheim, R. *Int. J. Mass Spectrom. Ion Phys.* 1971, **7**, 309.

74. Giardini-Guidoni, A.; Platania, R.; Zocchi, F. *Int. J. Mass Spectrom. Ion Phys.* 1974, **13**, 453.

75. Thomas, T. F.; Dale, F.; Paulson, J. F. *J. Chem. Phys.* 1977, **67**, 793.

76. Medley, S. S. *Rev. Sci. Instrum.* 1978, **49**, 698.

77. Armentrout, P.B; Beauchamp, J. L. *J. Am. Chem. Soc.* 1980, **102**, 1736.

78. Glish, G. L., Ph.D. Thesis, Purdue University, 1980.

79. Glish, G. L.; McLuckey, S. A.; Ridley, T. Y.; Cooks, R. G. *Int. J. Mass Spectrom. Ion Phys.* 1982, **41**, 157.

80. Schoen, A. E.; Amy, J. W.; Ciupek, J. D.; Cooks, R. G.; Dobberstein, P.; Jung, G. *Int. J. Mass Spectrom. Ion Proc.* 1985, **65**, 125.

81. Ciupek, J. D.; Amy, J. W.; Cooks, R. G.; Schoen, A. E. *Int. J. Mass Spectrom. Ion Proc.* 1985, **65**, 141.

82. Louris, J. N.; Wright, L. G.; Cooks, R. G.; Schoen, A. E. *Anal. Chem.* 1985, **57**, 2918.

83. Bateman, R. H.; Green, B. N.; Smith, D. C. Presented at the 30th Annual Conference on Mass Spectrometry and Allied Topics, Honolulu, HI, 1982.

84. Taylor, L. C. E.; Stradling, R. S.; Busch, K. L.; Qualls, W. C. Presented at the 32nd Annual Conference on Mass Spectrometry and Allied Topics, San Antonio, TX, 1984.

85. DiDonato, G. C.; Busch, K. L. *Anal. Chem.* 1986, **58**, 229.

86. Harris, F. M.; Keenan, G. A.; Bolton, P. D.; Davies, S. B.; Singh, S.; Beynon, J. H. *Int. J. Mass Spectrom. Ion Proc.* 1984, **58**, 273.

87. Young, S. E.; Butrill, S. E., Jr. Presented at the 31st Annual Conference on Mass Spectrometry and Allied Topics, Boston, MA, 1983.

88. Glish, G. L.; Goeringer, D. E. *Anal. Chem.* 1984, **56**, 2291.

89. Glish, G. L.; McLuckey, S. A.; McKown, H. S. *Anal. Instrum.* 1987, **16**, 191.

90. Mamyrin, B. A.; Darataev, V. I.; Shmikk, D. V.; Zagulin, V. A. *Sov. Phys.-JETP* 1973, **37**, 45.

91. Rockwood, A. L. Presented at the 34th Annual Conference on Mass Spectrometry and Allied Topics, Cincinnati, OH, 1986.

92. Muga, M. L. Presented at the 32nd Annual Conference on Mass Spectrometry and Allied Topics, San Antonio, TX, 1984.

93. Glish, G. L.; McLuckey, S. A.; McKown, H. S. To be published.

94. Glish, G. L.; McLuckey, S. A.; McBay, E. H.; Bertram, L. K. *Int. J. Mass Spectrom. Ion Proc.* 1986, **70**, 321.

95. Bricker, D. L.; Adams, T. A., Jr.; Russell, D. H. *Anal. Chem.* 1983, **55**, 2417.

96. McIver, R. T., Jr. Presented at the 29th Annual Conference on Mass Spectrometry and Allied Topics, Minneapolis, MN, 1981.
97. Cody, R. B.; Freiser, B. S. *Int. J. Mass Spectrom. Ion Phys.* 1982, **41**, 199.
98. Cody, R. B.; Burnier, R. C.; Cassady, C. J.; Freiser, B. S. *Anal. Chem.* 1982, **54**, 2225.
99. Cody, R. B.; Freiser, B. S. *Anal. Chem.* 1982, **54**, 1431.
100. Jacobson, D. B.; Freiser, B. S. *J. Am. Chem. Soc.* 1984, **106**, 4623.
101. Carlin, T. J.; Freiser, B. S. *Anal. Chem.* 1983 **55**, 571.
102. Ghaderi, S.; Littlejohn, D. P. Presented at the 33rd Annual Conference on Mass Spectrometry and Allied Topics, San Diego, CA, 1985.
103. McIver, R. T., Jr.; Hunter, R. L.; Bowers, W. D. *Int. J. Mass Spectrom. Ion Proc.* 1985, **64**, 67.
104. Wise, M. B. *Anal. Chem.* 1987, **59**, 2289.
105. Marshall, A. G.; Wang, T-C. L.; Ricca, T. L. *J. Am. Chem. Soc.* 1985, **107**, 7893.
106. Dunbar, R. C. *Gas Phase Ion Chemistry*, Bowers, M. T., Ed. Academic: New York, 1979, Volume 2.
107. Bowers, W. D.; Delbert, S.; Hunter, R. L.; McIver, R. T., Jr. *J. Am. Chem. Soc.* 1984, **106**, 7288.
108. Cassady, C. J.; Freiser, B. S. *J. Am. Chem. Soc.* 1984, **106**, 6176.
109. Bowers, W. D.; Delbert, S.; McIver, R. T., Jr. *Anal. Chem.* 1986, **58**, 969.
110. Cody, R. B.; Freiser, B. S. *Anal. Chem.* 1979, **51**, 547.
111. Cody, R. B.; Freiser, B. S. *Anal. Chem.* 1987, **59**, 1054.
112. Syka, J. E. P.; Fies, W. J. Presented at the 35th Annual Conference on Mass Spectrometry and Allied Topics, Denver, CO, 1987.
113. Louris, J. N.; Cooks, R. G.; Syka, J. E. P.; Kelley, P. E.; Stafford, G. C., Jr.; Todd, J. F. J. *Anal. Chem.* 1987, **59**, 1677.
114. Beynon, J. H.; Cooks, R. G.; Keough, T. *Int. J. Mass Spectrom. Ion Phys.* 1974, **14**, 437.
115. Glish, G. L.; Todd, P. J. *Anal. Chem.* 1982, **54**, 842.
116. van Asslet, N. P. F. B.; Maas, J. G.; Los, J. *Chem. Phys.* 1974, **5**, 429.
117. Huber, B. A.; Miller, T. M.; Cosby, P. C.; Zeman, H. D.; Leon, R. L.; Moseley, J. T.; Peterson, J. R. *Rev. Sci. Instrum.* 1977, **48**, 1306.
118. Mukhtar, E. S.; Griffiths, I. W.; Harris, F. M.; Beynon, J. H. *Org. Mass Spectrom.* 1980, **15**, 51.
119. Harris, F. M.; Mukhtar, E. S.; Griffiths, I. W.; Beynon, J. H. *Proc. Roy. Soc. London A* 1981, **374**, 461.
120. Mukhtar, E. S.; Griffiths, I. W.; Harris, F. M.; Beynon, J. H. *Int. J. Mass Spectrom. Ion Phys.* 1981, **37**, 159.
121. Krailler, R. E.; Russell, D. H. *Anal. Chem.* 1985, **57**, 1211.
122. Tajima, S.; Tobita, S.; Ogino, K.; Niwa, Y. *Org. Mass Spectrom.* 1986, **21**, 236.
123. Schulz, P. A.; Gregory, D. C.; Meyer, F. W.; Phaneuf, R. A. *J. Chem. Phys.* 1986, **85**, 3386.
124. Vekey, K.; Brenton, A. G.; Beynon, J. H. *Int. J. Mass Spectrom. Ion Proc.* 1986, **70**, 277.
125. Miller, P. E.; Denton, M. B. *Int. J. Mass Spectrom. Ion Proc.* 1986, **72**, 223.
126. Mabud, Md. A.; Dekrey, M. J.; Cooks, R. G. *Int. J. Mass Spectrom. Ion Proc.* 1985, **67**, 285.
127. Forst, W. *Theory of Unimolecular Reactions*. Academic: New York, 1973.
128. Robinson, P. J.; Holbrook, K. A. *Unimolecular Reactions* Wiley: London, 1972.
129. Beynon, J. H.; Gilbert, J. R. *Application of Transition State Theory to Unimolecular Reactions, An Introduction*. Wiley: New York, 1984.
130. Wahrhaftig, A. L. In *Gaseous Ion Chemistry and Mass Spectrometry*, Futrell, J. H., Ed. Wiley: New York, 1986.
131. Lorquet, J. C.; Barbier, C.; Leyh-Nihant, B. *Adv. Mass Spectrom.* 1986, **10A**, 71.
132. Meisels, G. G.; Verboom, G. M. L.; Weiss, M. J.; Hsieh, T.C. *Adv. Mass Spectrom.* 1980, **8A**, 104.

133. Klots, C. E. In *Kinetics of Ion-Molecule Reactions.* Ausloos, P., Ed. Plenum: New York, 1979.
134. Glish, G. L.; Goeringer, D. E.; McLuckey, S. A. Presented at the 33rd Annual Conference on Mass Spectrometry and Allied Topics, San Diego, CA, 1985.
135. Levsen, K. *Fundamental Aspects of Organic Mass Spectrometry.* Springer-Verlag: Weinheim, 1978.
136. Beckey, H. D. *Principles of Field Ionization and Field Desorption Mass Spectrometry.* Pergamon: Oxford, 1977.
137. Nibbering. N. M.M. *Mass Spectrom. Rev.* 1984, **3**, 445.
138. Lifshitz, C.; Gefen, S. *Org. Mass Spectrom.* 1984, **19**, 197.
139. Lifshitz, C. *Mass Spectrom. Rev.* 1982, **1**, 309.
140. Aviyente, V.; Shaked, M.; Feinmesser, A.; Gefen, S.; Lifshitz, C. *Int. J. Mass Spectrom. Ion Proc.* 1986, **70**, 67.
141. Lehman, T. A.; Bursey, M. M. *Ion Cyclotron Resonance Spectrometry.* Wiley: New York, 1976.
142. Lifshitz, C.; Gotchiguian, P.; Roller, R. *Chem. Phys. Lett.* 1983, **95**, 106.
143. Gefen, S.; Lifshitz, C. *Int. J. Mass Spectrom. Ion Proc.* 1984, **58**, 251.
144. Lifshitz, C.; Gefen, S.; Arakawa, R. *J. Phys. Chem.* 1984, **88**, 4242.
145. Morgan, R. P.; Brenton, A. G.; Beynon, J. H. *Int. J. Mass Spectrom. Ion Phys.* 1979, **29**, 195.
146. Jennings, K. R. In *Gas Phase Ion Chemistry.* Bowers, M. T., Ed. Academic: New York, 1979, Volume 2.
147. Los, J.; Govers, T. R. In *Collision Spectroscopy.* Cooks, R. G.; Ed., Plenum: New York, 1978.
148. Holmes. J. L.; Terlouw, J. K. *Org. Mass Spectrom.* 1980, **15**, 383.
149. Fraefel, A.; Seibl, L. *Mass Spectrom. Rev.* 1985, **4**, 151.
150. Haney, M. A.; Franklin, J. L. *J. Chem. Phys.* 1968, **48**, 4093.
151. Klots, C. E. *J. Chem. Phys.* 1973, **58**, 5364.
152. Burgers, P. C.; Holmes, J. L. *Int. J. Mass Spectrom. Ion Proc.* 1984, **58**, 15.
153. Beynon, J. H. In *Gas Phase Ion Chemistry*, Bowers, M. T., Ed. Academic: New York, 1979, Volume 2.
154. Beynon, J. H.; Bertrand, M.; Cooks, R. G. *J. Am. Chem. Soc.* 1973, **95**, 1739.
155. Day, R. J.; Krause, D. A.; Jorgensen, W. L.; Cooks, R. G. *Int. J. Mass Spectrom. Ion Phys.* 1979, **30**, 83.
156. Beynon, J. H. *Proc. Roy. Soc. London A* 1981, **378**, 1.
157. Occolowitz, J. L. *J. Am. Chem. Soc.* 1969, **91**, 5202.
158. Yeo, A. N. H.; Williams, D. H. *J. Am. Chem. Soc.* 1971, **93**, 395.
159. Tsang, C. W.; Harrison, A. G. *Org. Mass Spectrom.* 1973, **7**, 1377.
160. Aston, F. W. *Proc. Cambridge Philos. Soc.* 1919, **19**, 317.
161. Kupriyanov, S. E.; Perov, A. A. Russ. *J. Phys. Chem.* 1965, **39**, 871.
162. McLafferty, F. W.; Bente, P. F.; Kornfeld, R.; Tsai, S.-C.; Howe, I. *J. Am. Chem. Soc.* 1973, **95**, 2120.
163. McLafferty, F. W.; Kornfeld, R.; Haddon, W. F.; Levsen, K.; Sakai, I.; Bente, P. F.; Tsai, S.-C.; Schuddemage, H. D. R. *J. Am. Chem. Soc.* 1973, **95**, 3886.
164. Yost, R. A.; Enke, C. G. *Anal. Chem.* 1979, **51**, 1251A.
165. Laramee, J. A.; Cameron, D.; Cooks, R. G. *J. Am. Chem. Soc.* 1981, **103**, 12.
166. Kim, M. S.; McLafferty, F. W. *J. Am. Chem. Soc.* 1978, **100**, 3279.
167. van Tilborg, M. W. E. M.; van Thuijl, J. *Org. Mass Spectrom.* 1983, **18**, 331.
168. van Tilborg, M. W. E. M.; van Thuijl, J. *Org. Mass Spectrom.* 1984, **19**, 217.
169. van Tilborg, M. W. E. M.; van Thuijl, J. *Org. Mass Spectrom.* 1984, **19**, 569.
170. Cooks, R. G., Ed. *Collision Spectroscopy.* Cooks, R. G., Ed. Plenum: New York, 1979.
171. Levsen, K.; Schwarz, H. *Mass Spectrom. Rev.* 1983, **2**, 77.
172. Douglas, D. J. *J. Phys. Chem.* 1982, **86**, 185.

173. Child, M. S. *Molecular Collision Theory*. Academic: London, 1974.
174. Chantry, P. J. *J. Chem. Phys.* 1971, **55**, 2746.
175. Vestal, M. V. In *Fundamental Processes in Radiation Chemistry*, Ausloos, P., Ed. Interscience: New York, 1968.
176. Klots, C. E. *J. Chem. Phys.* 1976, **64**, 1976.
177. Boyd, R. K.; Kingston, E. E.; Brenton, A. G.; Beynon, J. H. *Proc. Roy. Soc. London A* 1984, **392**, 59.
178. Waddell, D. S.; Boyd, R. K.; Brenton, A. G.; Beynon, J. H. *Int. J. Mass Spectrom. Ion Proc.* 1986, **68**, 71.
179. Ewald, H.; Seibt, W. In *Recent Developments in Mass Spectroscopy*, Ogata, K.; Hawakawa, T., Eds. University Park Press: Baltimore, 1970.
180. Gentry, W. R. In *Gas Phase Ion Chemistry*. Bowers, M. T., Ed. Academic: New York, 1979, Volume 2.
181. Toennies, J. P. *Annu. Rev. Phys. Chem.* 1976, **27**, 225.
182. Appell, J. In *Collision Spectroscopy*. Cooks, R. G., Ed. Plenum: New York, 1978.
183. Beynon, J. H. *Adv. Mass Spectrom.* 1986, **10A**, 437.
184. Wachs, T.; McLafferty, F. W. *Int. J. Mass Spectrom. Ion Phys.* 1977, **23**, 243.
185. Ast, T.; Porter, C. J.; Proctor, C. J.; Beynon, J. H. *Bull. Soc. Chim. Beograd.* 1981, **46**, 135.
186. Proctor, C. J.; Porter, C. J.; Brenton, A. G.; Beynon, J. H. *Int. J. Mass Spectrom. Ion Phys.* 1981, **39**, 9.
187. Szulejko, J. E.; Howe, I.; Beynon, J. H. *Int. J. Mass Spectrom. Ion Phys.* 1981, **37**, 27.
188. Ast, T.; Beynon, J. H.; Cooks, R. G. *J. Am. Chem. Soc.* 1972, **94**, 1004.
189. Cooks, R. G.; Hendricks, L.; Beynon, J. H. *Org. Mass Spectrom.* 1975, **10**, 625.
190. Dawson, P. H. *Int. J. Mass Spectrom. Ion Phys.* 1982, **43**, 195.
191. Dawson, P. H.; French, J. B.; Buckley, J. A.; Douglas, D. J.; Simmons, D. *Org. Mass Spectrom.* 1982, **17**, 212.
192. Dawson, P. H. *Int. J. Mass Spectrom. Ion Phys.* 1983, **50**, 287.
193. Dawson, P. H.; Sun, W.-F. *Int. J. Mass Spectrom. Ion Phys.* 1982, **44**, 51.
194. Dawson, P. H.; Sun, W.-F. *Int. J. Mass Spectrom. Ion Proc.* 1984, **61**, 123.
195. Dawson, P. H. *Int. J. Mass Spectrom. Ion Proc.* 1985, **66**, 151.
196. Shirts, R. B. In *Gaseous Ion Chemistry and Mass Spectrometry*, Futrell, J. H., Ed. Wiley: New York, 1986.
197. Levine, R. D.; Bernstein, R. B. *Molecular Reaction Dynamics*. Oxford University Press: New York, 1974.
198. Mahan, B. H. *Acc. Chem. Res.* 1970, **3**, 393.
199. Durup, J. In *Recent Developments in Mass Spectroscopy*. Ogata, K.; Hawakawa, T., Eds. University Park Press: Baltimore, 1970.
200. McLafferty, F. W. *Phil. Trans. Roy. Soc. London A* 1979, **293**, 93.
201. Levsen, K.; Schwarz, H. *Angew. Chem. Int. Ed. Engl.* 1976, **15**, 509.
202. Herman, Z.; Futrell, J. H.; Friedrich, B. *Int. J. Mass Spectrom. Ion Proc.* 1984, **58**, 181.
203. Massey, H. S. W. *Rep. Prog. Phys.* 1949, **12**, 3279.
204. Hasted, J. B.; Lee, A. R. *Proc. Phys. Soc.* 1962, **79**, 702.
205. Massey, H. S. W.; Burhop, E. H. S. *Electronic and Ionic Impact Phenomena*. Oxford University Press: Oxford, 1952.
206. Todd, P. J.; McLafferty, F. W. In *Tandem Mass Spectrometry*. McLafferty, F. W., Ed. Wiley: New York, 1983.
207. Jackson, J. D. *Classical Electrodynamics*. Wiley: New York, 1975.
208. Christophorou, L. G. *Atomic and Molecular Radiation Physics*. Wiley: London, 1971.
209. Russek, A. *Physica* 1970, **48**, 165.
210. Mahan, B. H. *J. Chem. Phys.* 1970, **52**, 5221.
211. Hasted, J. B. *Physics of Atomic Collisions*, Second Edition. Elsevier: New York, 1972.
212. Eastes, W.; Toennies, J. P. *J. Chem. Phys.* 1979, **70**, 1644.

213. Eastes, W.; Toennies, J. P. *J. Chem. Phys.* 1979, **70**, 1652.

214. Greene, R. F.; Hall, R. B.; Sondergaard, N. A. *J. Chem. Phys.* 1977, **66**, 3171.

215. Singh, S.; Harris, F. M.; Boyd, R. K.; Beynon, J. H. *Int. J. Mass Spectrom. Ion Proc.* 1985, **66**, 131.

216. Fernandez, S. M.; Erikson, F. J.; Bray, A. V.; Pollack, E. *Phys. Rev. A* 1975, **12**, 1252.

217. Anderson, N.; Vedder, M.; Russek, A.; Pollack, E. *Phys Rev. A* 1980, **21**, 782.

218. van der Zande, W. J.; de Bruijn, D. P.; Los, J.; Kistemaker, P. G.; McLuckey, S. A. *Int. J. Mass Spectrom. Ion Proc.* 1985, **67**, 161.

219. Neumann, G. M.; Derrick, P. J. *Org. Mass Spectrom.* 1984, **19**, 165.

220. Neumann, G. M.; Sheil, M. M.; Derrick, P. J. *Z. Naturforsch.* 1984, **39a**, 584.

221. Gilbert, R. G.; Sheil, M. M.; Derrick, P. J. *Org. Mass Spectrom.* 1985, **20**, 431.

222. Griffiths, I. W.; Mukhtar, E. S.; March, R. E.; Harris, F. M.; Beynon, J. H. *Int. J. Mass Spectrom. Ion Phys.* 1981, **39**, 125.

223. Bricker, D. L.; Russell, D. H. *Anal. Chem.* 1986, **108**, 6174.

224. Laramee, J. A.; Carmody, J. J.; Cooks, R. G. *Int. J. Mass Spectrom. Ion Phys.* 1979, **31**, 333.

225. Laramee, J. A.; Hemberger, P. H.; Cooks, R. G. *J. Am. Chem. Soc.* 1979, **101**, 6460.

226. Fedor, D. M.; Cooks, R. G. *Anal. Chem.* 1980, **52**, 679.

227. Hubik, A. R.; Hemberger, P. H.; Laramee, J. A.; Cooks, R. G. *J. Am. Chem. Soc.* 1980, **102**, 3997.

228. McLuckey, S. A.; Cooks, R. G. In *Tandem Mass Spectrometry*, McLafferty, F. W., Ed. Wiley: New York, 1983.

229. Verma, S.; Ciupek, J. D.; Cooks, R. G.; Schoen, A. E.; Dobberstein, P. *Int. J. Mass Spectrom. Ion Proc.* 1983, **52**, 311.

230. Lindholm, E. In *Ion–Molecule Reactions*, Franklin, J. L., Ed. Plenum: New York, 1972.

231. Baer, T. In *Gas Phase Ion Chemistry*, Bowers, M. T., Ed. Academic: New York, 1979, Volume 2.

232. Dannacher, J. *Org. Mass Spectrom.* 1984, **19**, 253.

233. Hemberger, P. H.; Laramee, J. A.; Hubik, A. R.; Cooks, R. G. *J. Phys. Chem.* 1981, 85, 2335.

234. Todd, P. J.; Warmack, R. J.; McBay, E. H. *Int. J. Mass Spectrom. Ion Phys.* 1983, **50**, 299.

235. van der Zande, W. J.; McLuckey, S. A.; de Bruijn, D. P.; Los, J.; Kistemaker, P. G. *Adv. Mass Spectrom.* 1986, **10B**, 1123.

236. Boyd, R. K.; Kingston, E. E.; Brenton, A. G.; Beynon, J. H. *Proc. Roy. Soc. London A* 1984, **392**, 89.

237. Yost, R. A.; Enke, C. G. In *Tandem Mass Spectrometry*, McLafferty, F. W., Ed. Wiley: New York, 1983.

238. McLuckey, S. A.; Glish, G. L.; Cooks, R. G. *Int. J. Mass Spectrom. Ion Phys.* 1981, **39**, 219.

239. Fetterolf, D. D.; Yost, R. A. *Int. J. Mass Spectrom. Ion Phys.* 1982, **44**, 37.

240. McLuckey, S. A.; Sallans, L.; Cody, R. B.; Burnier, R. C.; Verma, S.; Freiser, B. S.; Cooks, R. G. *Int. J. Mass Spectrom. Ion Phys.* 1982, **44**, 215.

241. Nacson, S.; Harrison, A. G. *Int. J. Mass Spectrom. Ion Proc.* 1985, **63**, 325.

242. Zwinselman, J. J.; Nacson, S.; Harrison, A. G. *Int. J. Mass Spectrom. Ion Proc.* 1985, **67**, 93.

243. Dawson, P. H.; Sun, W.-F. *Int. J. Mass Spectrom. Ion Proc.* 1986, **70**, 97.

244. Harrison, A. G.; Lin, M. S. *Int. J. Mass Spectrom. Ion Phys.* 1983, **51**, 353.

245. Ellenbroek, T.; Toennies, J. P. *Chem. Phys.* 1982, **71**, 309.

246. Kenttämaa, H. I.; Cooks, R. G. *Int. J. Mass Spectrom. Ion Proc.* 1985, **64**, 79.

247. Wysocki, V. H.; Kenttämaa, H. I.; Cooks, R. G. *Int. J. Mass Spectrom. Ion Proc.* 1987, **75**, 181.

248. DeKrey, M. J.; Kenttämaa, H. I.; Cooks, R. G. *Org. Mass Spectrom.* 1986, **21**, 193.

249. McLuckey, S. A. *Org. Mass Spectrom.* 1984, **19**, 545.

250. Boyd, R. K.; Bott, B. A.; Beynon, J. H.; Harvan, D. J.; Hass, J. R. *Int. J. Mass Spectrom. Ion Proc.* 1985, **66**, 253.
251. Ciupek, J. D.; Zakett, D.; Cooks, R. G.; Wood, K. V. *Anal. Chem.* 1982, **54**, 2215.
252. Ouwerkerk, C. E. D.; McLuckey, S. A.; Kistemaker, P. G.; Boerboom, A. J. H. *Int. J. Mass Spectrom. Ion Proc.* 1984, **56**, 11.
253. Todd, P. J.; McLafferty, F. W. *Int. J. Mass Spectrom. Ion Phys.* 1981, **38**, 371.
254. Kim, M. S. *Int. J. Mass Spectrom. Ion Phys.* 1983, **50**, 189.
255. Kim, M. S. *Int. J. Mass Spectrom. Ion Phys.* 1983, **51**, 279.
256. Dawson, P. H.; Douglas, D. J. In *Tandem Mass Spectrometry*, McLafferty, F. W., Ed. Wiley: New York, 1983.
257. Dunbar, R. C. In *Gas Phase Ion Chemistry*, Bowers, M. T., Ed. Academic: New York, 1984, Volume 3.
258. Dunbar, R. C. In *Molecular Ions: Spectroscopy, Structure and Chemistry*, Miller, T. A., Ed. North-Holland: Amsterdam, 1983.
259. Thorne, L. R.; Beauchamp, J. L. In *Gas Phase Ion Chemistry*, Bowers, M. T., Ed. Academic: New York, 1984, Volume 3.
260. Bomse, D. S.; Woodin, R. L.; Beauchamp, J. L. *J. Am. Chem. Soc.* 1979, **101**, 5503.
261. Louris, J. N.; Brodbelt, J. S.; Cooks, R. G. *Int. J. Mass Spectrom. Ion Proc.* 1987, **75**, 345.
262. Wagner-Redeker, W.; Levsen, K. *Org. Mass Spectrom.* 1981, **16**, 538.
263. Welch, M. J.; Sams, R.; White, E., V *Anal. Chem.* 1986, **58**, 890.
264. Jarrold, M. F.; Illies, A. J.; Bowers, M. T. *J. Chem. Phys.* 1983, **79**, 6086.
265. Morgan, T. G.; Kingston, E. E.; Harris, F. M.; Beynon, J. H.; *Org. Mass Spectrom.* 1982, **17**, 594.
266. Griffiths, I. W.; Harris, F. M.; Mukhtar, E. S.; Beynon, J. H. *Int. J. Mass Spectrom. Ion Phys.* 1981, **38**, 127.
267. Griffiths, I. W.; Mukhtar, E. S.; Harris, F. M.; Beynon, J. H. *Int. J. Mass Spectrom. Ion Phys.* 1981, **38**, 333.
268. Mukhtar, E. S.; Griffiths, I. W.; March, R. E.; Harris, F. M.; Beynon, J. H. *Int. J. Mass Spectrom. Ion Phys.* 1981, **41**, 61.
269. Griffiths, I. W.; Mukhtar, E. S.; Harris, F. M.; Beynon, J. H. *Int. J. Mass Spectrom. Ion Phys.* 1982, **43**, 283.
270. Fukuda, E. K.; Campana, J. E. *Anal. Chem.* 1985, **57**, 949.
271. Fukuda, E. K.; Campana, J. E. *Int. J. Mass Spectrom. Ion Proc.* 1985, **65**, 321.
272. Kingston, E. E.; Morgan, T. G.; Harris, F. M.; Beynon, J. H. *Int. J. Mass Spectrom. Ion Phys.* 1982, **43**, 261.
273. Kingston, E. E.; Morgan, T. G.; Harris, F. M.; Beynon, J. H. *Int. J. Mass Spectrom. Ion Phys.* 1983, **47**, 73.
274. Morgan, T. G.; March, R. E.; Harris, F. M.; Beynon, J. H. *Int. J. Mass Spectrom. Ion Proc.* 1984, **61**, 41.
275. Kim, M. S.; Morgan, T. G.; Kingston, E. E.; Harris, F. M. *Org. Mass Spectrom.* 1983, **18**, 582.
276. Harris, F. M.; Beynon, J. H. In *Gas Phase Ion Chemistry*, Bowers, M. T., Ed. Academic: New York, 1984, Volume 3.
277. van Velzen, P. N. T.; van der Hart, W. J.; van der Greef, J.; Nibbering, N. M.M.; Gross, M. L. *J. Am. Chem. Soc.* 1982, **104**, 1208.
278. Freiser, B. S. *Int. J. Mass Spectrom. Ion Phys.* 1978, **26**, 39.
279. Fedor, D. W.; Cody, R. B.; Burinsky, D. J.; Freiser, B. S.; Cooks, R. G. *Int. J. Mass Spectrom. Ion Phys.* 1979, **31**, 27.
280. Cooks, R. G.; Terwilliger, D. T.; Ast, T.; Beynon, J. H.; Keough, T. *J. Am. Chem. Soc.* 1975, **97**, 1583.
281. Cooks, R. G.; Ast, T.; Beynon, J. H. *Int. J. Mass Spectrom. Ion Phys.* 1975, **16**, 348.
282. Gandy, R. M.; Ampulski, R.; Prusaczyk, J.; Johnsen, R. H. *Int. J. Mass Spectrom. Ion Phys.* 1977, **24**, 363.

283. DeKrey, M. J.; Mabud, Md. A.; Cooks, R. G.; Syka, J. E. P. *Int. J. Mass Spectrom. Ion Proc.* 1985, **67**, 295.
284. Schey, K. L.; Cooks, R. G.; Kraft, A.; Grix, A.; Wollnik, H. Presented at the 36th Annual Conference on Mass Spectrometry and Allied Topics, San Francisco, CA, 1988.
285. McDaniel, E. W.; Cermak, V.; Dalgarno, A.; Ferguson, E. E.; Friedman, L. *Ion–Molecule Reactions*, Wiley: New York, 1970.
286. Dempster, A. J. *Phil. Mag.* 1916, **31**, 438.
287. Smith, D.; Adams, N. G. In *Gas Phase Ion Chemistry*, Bowers, M. T., Ed. Academic: New York, 1979, Volume 1.
288. Ferguson, E. E.; Fehsenfeld, F. C.; Albritton, D. L. In *Gas Phase Ion Chemistry*, Bowers, M. T., Ed. Academic: New York, 1979, Volume 1.
289. Bowers, M. T., Ed. *Gas Phase Ion Chemistry.* Academic: New York, 1979, Volume 1.
290. Franklin, J. L., Ed. *Ion–Molecule Reactions*. Plenum: New York, 1979.
291. Jennings, K. R. *Adv. Mass Spectrom.* 1986, **10A**, 303.
292. Talrose, V. L. *Adv. Mass Spectrom.* 1980, **8A**, 147.
293. Hseih, E. T.-Y.; Castleman, A. W. *Int. J. Mass Spectrom. Ion Phys.* 1981, **40**, 295.
294. Ausloos, P. J., Ed. *"Kinetics of Ion–Molecule Reactions." NATO Advanced Study Institutes Series B* Volume 40. Plenum: New York, 1979.
295. Ausloos, P. J., Ed. *"Interaction Between Ions and Molecules." NATO Advanced Study Institutes Series B*, Volume 6. Plenum: New York, 1975.
296. Field, F. H. In *Ion–Molecule Reactions*, Franklin, J. L., Ed. Plenum: New York, 1972.
297. George, T. F.; Suplinskas, R. J. *J. Chem. Phys.* 1969, **51**, 3666.
298. Chesnavich, W. J.; Bass, L.; Su, T.; Bowers, M. T. *J. Chem. Phys.* 1981, **74**, 2228.
299. Ono, Y.; Linn, S. H.; Tzeng, W.-B.; Ng, C. Y. *J. Chem. Phys.* 1984, **80**, 1482.
300. Meisels, G. G.; Verboom, G. M. L.; Weiss, M. J.; Hseih, T. C. *J. Am. Chem. Soc.* 1979, **101**, 7189.
301. Ceyer, S. T.; Tiedemann, P. W.; Ng, C. Y.; Mahan, B. H.; Lee, T. Y. *J. Chem. Phys.* 1979, **70**, 2138.
302. Henchman, M. In *Ion–Molecule Reactions*, Franklin, J. L., Ed. Plenum: New York, 1972.
303. Li, Y.; Herman, J. A.; Harrison, A. G. *Can. J. Chem.* 1981, **59**, 1753.
304. Futrell, J. H.; Tiernan, T. O. In *Ion–Molecule Reactions*, Franklin, J. L., Ed. Plenum: New York, 1972.
305. Busch, K. L.; Kruger, T. L.; Cooks, R. G. *Anal. Chim. Acta* 1981, **119**, 153.
306. Beynon, J. H.; Caprioli, R. M.; Baitinger, W. E.; Amy, J. W. *Org. Mass Spectrom.* 1970, **3**, 455.
307. Mathur, B. P.; Abbey, L. E.; Burgess, E. M.; Moran, T. F. *Org. Mass Spectrom.* 1980, **15**, 312.
308. Mathur, B. P.; Burgess, E. M.; Bostwick, D. E.; Moran, T. F. *Org. Mass Spectrom.* 1981, **16**, 92.
309. Jones, B. E.; Abbey, L. E.; Chatham, H. L.; Hanner, A. W.; Teleshefsky, L. A.; Burgess, E. M.; Moran, T. F. *Org. Mass Spectrom.* 1982, **17**, 10.
310. Appling, J. R.; Jones, B. E.; Abbey, L. E.; Bostwick, D. E.; Moran, T. F. *Org. Mass Spectrom.* 1983, **18**, 282.
311. Appling, J. R.; Musier, K. M.; Moran, T. F. *Org. Mass Spectrom.* 1984, **19**, 412.
312. Appling, J. R.; Burdick, G. W.; Moran, T. F. *Org. Mass Spectrom.* 1985, **20**, 343.
313. Guenat, C.; Maquin, F.; Stahl, D. *Int. J. Mass Spectrom. Ion Proc.* 1984, **59**, 121.
314. Mabud, Md. A.; DeKrey, M. J.; Cooks, R. G.; Ast, T. *Int. J. Mass Spectrom. Ion Proc.* 1986, **69**, 277.
315. Kenttämaa, H. I.; Wood, K. V.; Busch, K. L.; Cooks, R. G. *Org. Mass Spectrom.* 1983, **18**, 561.
316. Howe, I.; Bowie, J. H.; Szulejko, J. E.; Beynon, J. H. *J. Chem. Soc. Chem. Commun.* 1979, 983.
317. Howe, I.; Bowie, J. H.; Szulejko, J. E.; Beynon, J. H. *Int. J. Mass Spectrom. Ion Phys.* 1980, **34**, 99.

318. Bowie, J. H.; Blumenthal, T. *J. Am. Chem. Soc.* 1975, **97**, 2959.
319. McClusky, G. A.; Kondrat, R. W.; Cooks, R. G. *J. Am. Chem. Soc.* 1978, **100**, 6045.
320. Zakett, D.; Ciupek, J. D.; Cooks, R. G. *Anal. Chem.* 1981, **53**, 723.
321. Shushan, B.; Douglas, D. J. *Org. Mass Spectrom.* 1982, **17**, 198.
322. Keough, T.; Beynon, J. H.; Cooks, R. G. *J. Am. Chem. Soc.* 1973, **95**, 1695.
323. Lee, A. R.; Jonathan, P.; Brenton, A. G.; Beynon, J. H. *Int. J. Mass Spectrom. Ion Proc.* 1987, **75**, 329.
324. Feng, R.; Wesdemiotis, C.; Zhang, M.-Y.; Drinkwater, D. E.; McLafferty, F. W. Presented at the 35th Annual Conference on Mass Spectrometry and Allied Topics, Denver, CO, 1987.
325. Beynon, J. H. *Anal. Chem.* 1970, **42**, 97A.
326. Cooks, R. G.; Beynon, J. H.; Ast, T. *J. Am. Chem. Soc.* 1972, **94**, 1004.
327. Ast, T.; Beynon, J. H.; Cooks, R. G. *J. Am. Chem. Soc.* 1972, **94**, 6611.
328. Cooks, R. G.; Ast, T.; Beynon, J. H. *Int. J. Mass Spectrom. Ion Phys.* 1973, **11**, 490.
329. Ast, T. *Adv. Mass Spectrom.* 1986, **10A**, 471.
330. Maquin, F.; Stahl, D.; Sawaryn, A.; von R. Schleyer, P.; Koch, W.; Frenking, G.; Schwarz, H. *J. Chem Soc. Chem. Commun.* 1984, 504.
331. Stahl, D.; Maquin, F. *Org. Mass Spectrom.* 1984, **19**, 202.
332. Holmes, J. L. *Org. Mass Spectrom.* 1985, **20**, 169.
333. Stahl, D.; Maquin, F.; Gaumann, T.; Schwarz, H.; Carrupt, P. A.; Vogel, P. *J. Am. Chem. Soc.* 1985, **107**, 5049.
334. McLafferty, F. W.; Barbalas, M. P.; Turecek, F. *J. Am. Chem. Soc.* 1983, **105**, 1.
335. Barbalas, M. P.; Turecek, F.; McLafferty, F. W. *Org. Mass Spectrom.* 1982, **17**, 595.
336. Gellene, G. I.; Porter, R. F. *Int. J. Mass Spectrom. Ion Proc.* 1985, **64**, 55.
337. Cooks, R. G.; Beynon, J. H.; Litton, J. F. *Org. Mass Spectrom.* 1975, **10**, 503.
338. Bowen, R. D.; Barbalas, M. P.; Pagano, F. R.; Todd, P. J.; McLafferty, F. W. *Org. Mass Spectrom.* 1980, **15**, 51.
339. Dass, C.; Peake, D. A.; Gross, M. L. *Org. Mass Spectrom.* 1986, **21**, 34.
340. Jarrold, M. F.; Illies, A. J.; Liu, S.; Bowers, M. T. *Org. Mass Spectrom.* 1983, **18**, 388.
341. Kingston, E. E.; Brenton, A. G.; Beynon, J. H.; Flammang, R.; Maquestiau, A. *Int. J. Mass Spectrom. Ion Proc.* 1984, **62**, 317.
342. Kingston, E. E.; Beynon, J. H.; Ast, T.; Flammang, R.; Maquestiau, A. *Org. Mass Spectrom.* 1985, **20**, 546.
343. Danis, P. O.; Feng, R.; McLafferty, F. W. *Anal. Chem.* 1986, **58**, 355.
344. McLafferty, F. W. *Adv. Mass Spectrom.* 1986, **10A**, 493.
345. Holmes, J. L.; Mommers, A. A.; Terlouw, J. K.; Hop, C. E. C. A. *Int. J. Mass Spectrom. Ion Proc.* 1986, **68**, 249.
346. Wesdemiotis, C.; Danis, P. O.; Feng, R.; Tso, J.; McLafferty, F. W. *J. Am. Chem. Soc.* 1985, **107**, 8059.
347. Danis, P. O.; Feng, R.; McLafferty, F. W. *Anal. Chem.* 1986, **58**, 348.
348. Howe, I. In *Mass Spectrometry: A Specialist Periodical Report*, *Volume* 7. Johnstone, R. A. W., Ed. Burlington House: London, 1984, Chapter 4.
349. Baldwin, M. A. In *Mass Spectrometry: A Specialist Periodical Report*, *Volume* 8. Johnstone, R. A. W., Ed. Burlington House: London, 1985, Chapter 2.
350. Clark, T. *A Handbook of Computational Chemistry*. Wiley, New York, 1985.
351. Traeger, J. C.; McLoughlin, R. G. *Int. J. Mass Spectrom. Ion Phys.* 1978, **27**, 319.
352. Lossing, F. P.; Traeger, J. C. *Int. J. Mass Spectrom. Ion Phys.* 1976, **19**, 9.
353. Morgan, R. P.; Derrick, P. J. *Org. Mass Spectrom.* 1975, **10**, 563.
354. Holmes, J. L.; Weese, G. M.; Blair, A. S.; Terlouw, J. K. *Org. Mass Spectrom.* 1977, **12**, 424.
355. Rosenstock, H. M.; Dibeler, V. H.; Harllee, F. N. *J. Chem. Phys.* 1964, **40**, 591.
356. Shannon, T. S.; McLafferty, F. W. *J. Am. Chem. Soc.* 1966, **88**, 5021.
357. Levsen, K.; McLafferty, F. W. *J. Am. Chem. Soc.* 1974, **96**, 139.
358. Jones, E. G.; Bauman, L. E.; Beynon, J. H.; Cooks, R. G. *Org. Mass Spectrom.* 1973, **7**, 185.

359. Levsen, K.; Beckey, H. D. *Org. Mass Spectrom.* 1974, **9**, 570.
360. McLafferty, F. W.; Hirota, A.; Barbalas, M. P.; Pegues, R. F. *Int. J. Mass Spectrom. Ion Phys.* 1980, **35**, 299.
361. McLafferty, F. W. *Acc. Chem. Res.* 1980, **13**, 33.
362. Hvistendahl, G.; Williams, D. H. *J. Am. Chem. Soc.* 1975, **97**, 3097.
363. Porter, C. J.; Morgan, R. P.; Beynon, J. H. *Int. J. Mass Spectrom. Ion Phys.* 1978, **28**, 326.
364. Van Koppen, P. A. M.; Illies, A. J.; Kirchner, N. J.; Bowers, M. T. *Org. Mass Spectrom.* 1982, **17**, 399.
365. Kemp. D. L.; Beynon, J. H.; Cooks, R. G. *Org. Mass Spectrom.* 1976, **11**, 857.
366. Burgers, P. C.; Terlouw, J. K.; Holmes, J. L. *Org. Mass Spectrom.* 1982, **17**, 369.
367. Holmes, J. L.; Mommers, A. A.; Szulejko, J. E.; Terlouw, J. K.; *J. Chem. Soc. Chem. Commun.* 1984, 165.
368. Holmes, J. L.; Terlouw, J. K.; Burgers, P. C.; Rye, R. T. B. *Org. Mass Spectrom.* 1980, **15**, 149.
369. Wagner-Redeker, W.; Levsen, K.; Schwarz, H.; Zummack, W. *Org. Mass Spectrom.* 1981, **16**, 361.
370. Bomse, D. S.; Beauchamp, J. L. *J. Am. Chem. Soc.* 1981, **103**, 3292.
371. Watson, C. H.; Baykut, G.; Eyler, J. R. *Anal. Chem.* 1987, **59**, 1133.
372. Tomer, K. B. *Org. Mass Spectrom.* 1974, **9**, 686.
373. Nibbering, N. M.M. *Tetrahedron* 1973, **29**, 385.
374. McIver, R. T., Jr. *Org. Mass Spectrom.* 1975, **10**, 396.
375. Beauchamp, J. L.; Dunbar, R. C. *J. Am. Chem. Soc.* 1970, **92**, 1477.
376. Staley, R. H.; Corderman, R. R.; Foster, M. S.; Beauchamp, J. L. *J. Am. Chem. Soc.* 1974, **96**, 1260.
377. Jalonen, J. *J. Chem. Soc. Chem. Commun.* 1985, 872.
378. Blair, A. S.; Harrison, A. G. *Can. J. Chem.* 1973, **51**, 703.
379. Gross, M. L.; McLafferty, F. W. *J. Am. Chem. Soc.* 1971, **93**, 1267.
380. Gross, M. L.; Russell, D. H.; Aerni, R. J.; Bronczyk, S. A. *J. Am. Chem. Soc.* 1977, **99**, 3603.
381. McEwen, C. N.; Rudat, M. A. In *Tandem Mass Spectrometry*, McLafferty, F. W., Ed. Wiley: New York, 1983.
382. McEwen, C. N.; Rudat, M. A. *J. Am. Chem. Soc.* 1979, **101**, 6470.
383. McEwen, C. N.; Rudat, M. A. *J. Am. Chem. Soc.* 1981, **103**, 4343.
384. Rudat, M. A.; McEwen, C. N. *J. Am. Chem. Soc.* 1981, **103**, 4349.
385. McEwen, C. N. *Mass Spec. Rev.* 1986, **5**, 521.
386. Burgers, P. C.; Holmes, J. L.; Mommers, A. A.; Terlouw, J. K. *Chem. Phys. Lett.* 1983, **102**, 1.
387. Burgers, P. C.; Holmes, J. L.; Mommers, A. A.; Szulejko, J. E.; Terlouw, J. K. *Org. Mass Spectrom.* 1984, **19**, 442.
388. Terlouw, J. K.; Holmes, J. L.; Burgers, P. C. *Int. J. Mass Spectrom. Ion Proc.* 1985, **66**, 239.
389. Porter, C. J.; Beynon, J. H.; Ast, T. *Org. Mass Spectrom.* 1981, **16**, 101.
390. Glish, G. L.; Cooks, R. G. *J. Am. Chem. Soc.* 1978, **100**, 6720.
391. Burinsky, D. J.; Cooks, R. G. *Org. Mass Spectrom.* 1983, **18**, 410.
392. Rose, M. E. *Org. Mass Spectrom.* 1981, **16**, 323.
393. Boyd, R. K.; Beynon, J. H. *Int. J. Mass Spectrom. Ion Phys.* 1977, **23**, 163.
394. Williams, D. H.; Hvistendahl, G. *J. Am. Chem. Soc.* 1974, **96**, 6753.
395. Williams, D. H.; Hvistendahl, G. *J. Am. Chem. Soc.* 1974, **96**, 6755.
396. Bowen, R. D.; Williams, D. H. *Int. J. Mass Spectrom. Ion Phys.* 1979, **29**, 47.
397. Howe, I.; Williams, D. H. *J. Chem. Soc. Chem. Commun.* 1971, 1195.
398. Porter, C. J.; Kralj, B.; Brenton, A. G.; Beynon, J. H. *Org. Mass Spectrom.* 1980, **15**, 619.
399. Miller, D. L.; Gross, M. L. *J. Am. Chem. Soc.* 1983, **105**, 3783.
400. Bone, L. I.; Futrell, J. H. *J. Chem. Phys.* 1967, **46**, 4084.

401. Jacobson, D. B.; Freiser, B. S. *J. Am. Chem. Soc.* 1983, **105**, 36.
402. O'Lear, J. R.; Wright, L. G.; Louris, J. N.; Cooks, R. G. *Org. Mass Spectrom.* 1987, **22**, 348.
403. Rabrenovic, M.; Proctor, C. J.; Ast, T.; Herbert, C. G.; Brenton, A. G.; Beynon, J. H. *J. Phys. Chem.* 1983, **87**, 3305.
404. Proctor, C. J.; Porter, C. J.; Ast, T.; Bolton, P. D.; Beynon, J. H. *Org. Mass Spectrom.* 1981, **16**, 454.
405. Stahl, D.; Maquin, F. *Chem. Phys. Lett.* 1984, **108**, 613.
406. Rabrenovic, M.; Beynon, J. H. *Int. J. Mass Spectrom. Ion Proc.* 1984, **56**, 85.
407. Kiser, R. W. *J. Chem. Phys.* 1962, **36**, 2964.
408. Dorman, F. H.; Morrison, J. D.; Nicholson, A. J. C. *J. Chem. Phys.* 1959, **31**, 1335.
409. Ast, T.; Porter, C. J.; Proctor, C. J.; Beynon, J. H. *Chem. Phys. Lett.* 1981, **78**, 439.
410. Hanner, A. W.; Hanner, T. F. *Org. Mass Spectrom.* 1981, **16**, 512.
411. Siegbahn, E. M. *Chem. Phys.* 1982, **66**, 443.
412. Pople, J. A.; Tidor, B.; von Schleyer, P. R. *Chem. Phys. Lett.* 1982, **88**, 533.
413. Bordas-Nagy, J.; Holmes, J. L.; Mommers, A. A. *Org. Mass Spectrom.* 1986, **21**, 629.
414. Illies, A. J.; Bowers, M. T. *Chem. Phys.* 1982, **66**, 281.
415. Moore, J. H., Jr.; Doering, J. P. *J. Chem. Phys.* 1970, **52**, 1692.
416. Moore, J. H., Jr. *Phys. Rev. A* 1974, **10**, 724.
417. Herrero, F. A.; Doering, J. P. *Phys. Rev. Lett.* 1972, **29**, 609.
418. Eastes, W.; Ross, U.; Toennies, J. P. *J. Chem. Phys.* 1979, **70**, 1652.
419. Annis, B. K.; Stockdale, J. A. D. *J. Chem. Phys.* 1981, **74**, 297.
420. Stockdale, J. A. D.; Annis, B. K. *Chem. Phys. Lett.* 1981, **82**, 451.
421. Annis, B. K.; Stockdale, J. A. D. *Chem. Phys. Lett.* 1980, **74**, 365.
422. Burgers, P. C.; Holmes. J. L. *Org. Mass Spectrom.* 1982, **17**, 123.
423. Burgers, P. C.; Mommers, A. A.; Holmes, J. L. *J. Am. Chem. Soc.* 1983, **105**, 5976.
424. Burgers, P. C.; Holmes, J. L. *Org. Mass Spectrom.* 1984, **19**, 452.
425. Holmes, J. L.; Lossing, F. P.; Terlouw, J. K.; Burgers, P. C. *Can. J. Chem.* 1983, **61**, 2305.
426. Gilbert, J. R.; Stace, A. J. *Int. J. Mass Spectrom. Ion Phys.* 1974, **15**, 311.
427. Kebarle, P. *Ann. Rev. Phys. Chem.* 1977, **28**, 445.
428. Aue, D. H.; Webb, H. M.; Bowers, M. T. *J. Am. Chem. Soc.* 1976, **98**, 311.
429. Wolf, J. F.; Staley, R. H.; Koppel, I.; Taagepera, M.; McIver, R.T, Jr.; Beauchamp, J. L.; Taft, R. W. *J. Am. Chem. Soc.* 1977, **99**, 5417.
430. Yamdagni, R.; Kebarle, P. *J. Am. Chem. Soc.* 1973, **95**, 3504.
431. Lau, Y. K.; Ikuta, S.; Kebarle, P. *J. Am. Chem. Soc.* 1982, **104**, 1463.
432. Wren, A.; Gilbert, P.; Bowers, M. T. *Rev. Sci. Instrum.* 1978, **49**, 531.
433. Walder, R.; Franklin, J. L. *Int. J. Mass Spectrom. Ion Phys.* 1980, **36**, 85.
434. Aue, D. H.; Bowers, M. T. In *Gas Phase Ion Chemistry*, Bowers, M. T., Ed. Academic: New York, 1979, Volume 2.
435. Bursey, M. M.; Harvan, D. J.; Hass, J. R.; Becker, E. I.; Arison, B. H. *Org. Mass Spectrom.* 1984, **19**, 160.
436. McLuckey, S. A.; Cameron, D.; Cooks, R. G. *J. Am. Chem. Soc.* 1981, **103**, 1313.
437. Boand, G.; Houriet, R.; Gaumann, T. *J. Am. Chem. Soc.* 1983, **105**, 2203.
438. Burinsky, D. J.; Fukuda, E. K.; Campana, J. E. *J. Am. Chem. Soc.* 1984, **106**, 2770.
439. McLuckey, S. A.; Schoen, A. E.; Cooks, R. G. *J. Am. Chem. Soc.* 1982, **104**, 848.
440. Chen, J. H.; Hays, J. D.; Dunbar, R. C. *J. Phys. Chem.* 1984, 4759.
441. Welch, M. J.; Pereles, D. J.; White, E., V *Org. Mass Spectrom.* 1985, **20**, 425.
442. McLuckey, S. A.; Verma, S.; Cooks, R. G.; Farncombe, M. J.; Mason, R. S.; Jennings, K. R. *Int. J. Mass Spectrom. Ion Phys.* 1983, **48**, 423.
443. Boyd, R. K.; Harris, F. M.; Beynon, J. H. *Int. J. Mass Spectrom. Ion Proc.* 1985, **66**, 185.
444. Bicking, C. A. In *Treatise on Analytical Chemistry*, Part I, Volume **1**, Second edition, Kolthoff, I. M.; Elving, P. J., Eds. Wiley: New York, 1978.
445. Gill, R.; Moffatt, A. C. *Anal. Proc.* 1982, **19**, 170.

446. Poole, C. F.; Schuette, S. A. *J. High Res. Chrom. Chrom. Comm.* 1983, **6**, 526.

447. Ende, M.; Spiteller, G. *Mass Spectrom. Rev.* 1982, **1**, 29.

448. Drozd, J. "Chemical Derivatization in Gas Chromatography." In *J. Chrom. Library*, Volume 19. Elsevier: New York, 1981.

449. Hunt, D. F.; Stafford, G. C., Jr.; Crow, F. W.; Russell, J. W. *Anal. Chem.* 1976, **48**, 2098.

450. Johnson, G. S.; Ruliffson, W. S.; Cooks, R. G. *Carbohydrate Res.* 1971, **18**, 233.

451. Zakett, D.; Cooks, R. G. *Analysis of Synfuels by MS/MS*, Blaustein, B. D.; Bockrath, B. C.; Friedman, S., Eds., New Approaches in Coal Chemistry, ACS Symposium Series, Number 169, American Chemical Society: Washington, D. C., 1981.

452. Busch, K. L.; Unger, S. E.; Vincze, A.; Cooks, R. G.; Keough, T. *J. Am. Chem. Soc.* 1982, **104**, 1507.

453. DiDonato, G. C.; Busch, K. L. *Anal. Chim. Acta* 1985, **171**, 233.

454. Kidwell, D. A.; Ross, M. M.; Colton, R. J. *J. Am. Chem. Soc.* 1984, **106**, 2219.

455. DiDonato, G. C.; Busch, K. L. *Biomed. Mass Spectrom.* 1985, **12**, 354.

456. Ross, M. M.; Kidwell, D. A.; Colton, R. J. *Int. J. Mass Spectrom. Ion Proc.* 1985, **63**, 141.

457. Ligon, W. V.; Dorn, S. B. *Int. J. Mass Spectrom. Ion Proc.* 1984, **57**, 75.

458. Ligon, W. V.; Dorn, S. B. *Anal. Chem.* 1986, **58**, 1889.

459. Flurer, R. A.; Busch, K. L. *Abstracts of Papers*, 1986 Pittsburgh Conference and Exposition on Analytical Chemistry and Applied Spectroscopy, March 1986, Atlantic City, NJ, Paper 864.

460. Busch, K. L.; DiDonato, G. C. Presented at the 34th Annual Conference on Mass Spectrometry and Allied Topics, Cincinnati, OH, 1986.

461. Hunt, D. F. *Int. J. Mass Spectrom. Ion Phys.* 1982, **45**, 111.

462. Field, F. H. *Adv. Mass Spectrom.* 1986, **10**, 271.

463. Mark, T. D. *Int. J. Mass Spectrom. Ion Phys.* 1982, **45**, 125.

464. Mark, T. D.; Dunn, G. H. *Electron Impact Ionization*. Springer-Verlag: New York, 1985.

465. Harrison, A. G. *Chemical Ionization Mass Spectrometry*. CRC Press: Boca Raton, FL, 1983.

466. Hunt, D. F. *Adv. Mass Spectrom.* 1974, **6**, 517.

467. Mitchum, R. K.; Korfmacher, W. A. *Anal. Chem.* 1983, **55**, 1485A.

468. Barber, M.; Bordoli, R. J.; Elliott, G. J.; Sedgwick, R. D.; Tyler, A. N. *Anal. Chem.* 1982, **54**, 645A.

469. Barber, M.; Bordoli, R. J.; Sedgwick, R. D.; Tyler, A. N. *Nature* 1981, **293**, 270.

470. Day, R. J.; Unger, S. E.; Cooks, R. G. *Anal. Chem.* 1980, **52**, 557A.

471. Benninghoven, A.; Sichtermann, W. *Org. Mass Spectrom.* 1977, **12**, 595.

472. Macfarlane, R. D.; Torgerson, D. F.; *Science* 1976, **191**, 920.

473. Hakansson, P.; Kamensky, I.; Sundqvist, B. *Nucl. Instrum. Meth.* 1982, **198**, 43.

474. Hillenkamp, F. *Int. J. Mass Spectrom. Ion Phys.* 1982, **45**, 305.

475. Krueger, F. R. *Z. Naturforsch.* 1983, **38a**, 385.

476. Grade, H.; Cooks, R. G. *J. Am. Chem. Soc.* 1978, **100**, 5615.

477. Vestal, M. L. *Science* 1984, **226**, 275.

478. Blakley, C. R.; Vestal, M. L. *Anal. Chem.* 1983, **55**, 750.

479. Yamashita, M.; Fenn, J. B. *J. Phys. Chem.* 1984, **88**, 4451.

480. Whitehouse, M.; Dreyer, R. N.; Yamashita, M.; Fenn, J. B. *Anal. Chem.* 1985, **57**, 675.

481. Stimpson, B. P.; Evans, C. A., Jr. *J. Electrostatics* 1978, **5**, 411.

482. Chan, K. W. S.; Cook, K. D. *J. Am. Chem. Soc.* 1982, **104**, 5031.

483. Thomson, B. A.; Iribarne, J. V. *J. Chem. Phys.* 1979, **71**, 4451.

484. Kuwabara, H.; Tsuchiya, M. *Mass Spectroscopy* 1982, **30**, 313.

485. Rhodes, G.; Opsal, R. B.; Meek, J. T.; Reilly, J. P. *Anal. Chem.* 1983, **55**, 280.

486. Tembreull, R. M.; Sin, C. H.; Pang, H. M.; Lubman, D. M. *Anal. Chem.* 1985, **57**, 2911.

487. Beckey, H. D. *Field Ionization Mass Spectrometry*, Pergamon: Oxford, 1970.

488. Anbar, M.; Aberth, W. H. *Anal. Chem.* 1974, **46**, 59A.

489. Schulten, H.-R. In *Soft Ionization Biological Mass Spectrometry*, Morris, H. R. Ed. Heyden and Sons: London, 1981, p. 6.

490. Schulten, H.-R. *Int. J. Mass Spectrom. Ion Phys.* 1979, **32**, 97.

491. Wood, G. W. *Mass Spectrom. Rev.* 1982, **1**, 63.

492. Davis, D. V.; Cooks, R. G. *Org. Mass Spectrom.* 1981, **16**, 176.

493. Houriet, R.; Rufenacht, H.; Stahl, D.; Tichy, M.; Longevialle, P. *Org. Mass Spectrom.* 1985, **20**, 300.

494. Nacson, S.; Harrison, A. G.; Davidson, W. R. *Org. Mass Spectrom.* 1986, **21**, 317.

495. Wood, K. V.; Burinsky, D. J.; Cameron, D.; Cooks, R. G. *J. Org. Chem.* 1983, **48**, 5236.

496. Rollgen, F. W.; Giessmann, U.; Borchers, F.; Levsen, K. *Org. Mass Spectrom.* 1978, **13**, 459.

497. Bursey, M. M.; Marbury, G. D.; Hass, J. R. *Biomed. Mass Spectrom.* 1984, **11**, 522.

498. Wong, C. M.; Crawford, R. W.; Barton, V. C.; Brand, H. R.; Neufeld, K. W.; Bowman, J. E. *Rev. Sci. Instrum.* 1983, **54**, 996.

499. Kirby, H.; Sokolow, S.; Steiner, U. "Data Systems for Tandem Mass Spectrometry." In *Tandem Mass Spectrometry*, McLafferty, F. W., Ed. Wiley, New York, 1983.

500. Youseffi, M.; Cooks, R. G.; McLaughlin, J. L. *J. Am. Chem. Soc.* 1979, **101**, 3400.

501. Pummangura, S.; McLaughlin, J. L.; Davis, D. V.; Cooks, R. G. *J. Nat. Prod.* 1982, **45**, 277.

502. Barofsky, D. F.; Giessmann, U.; Barofsky, E. *Int. J. Mass Spectrom. Ion Phys.* 1983, **53**, 319.

503. Nakamura, T.; Nagaki, H.; Kinoshita, T. *Bull. Chem. Soc. Jpn.* 1985, **58**, 2798.

504. Voyksner, R. D.; Hass, J. R.; Sovocool, G. W.; Bursey, M. M. *Anal. Chem.* 1983, **55**, 744.

505. Slayback, J. R. B.; Taylor, P. A. *Spectra* 1983, **9**, 18.

506. Kondrat, R. W.; Cooks, R. G.; McLaughlin, J. L. *Science* 1978, **199**, 978.

507. Morgan, R. P.; Beynon, J. H.; Bateman, R. H.; Green, B. N. *Int. J. Mass Spectrom. Ion Phys.* 1978, **28**, 171.

508. Stace, A. J.; Shukla, A. K. *Int. J. Mass Spectrom. Ion Phys.* 1981, **37**, 35.

509. la Lau, C. "Mass Discrimination Caused by Electron-Multiplier Detectors." In *Topics in Organic Mass Spectrometry*, *Adv. Anal. Chem. Instrum.*, *Volume* 8, Burlingame, A. L., Ed. Wiley: New York, 1970.

510. Beynon, J. H.; Brothers, D. F.; Cooks, R. G. *Anal. Chem.* 1974, **46**, 1299.

511. Stafford, G. C. *Environ. Health Persp.* 1980, **36**, 85.

512. Beuhler, R. J.; Friedman, L. *Nucl. Instrum. Meth.* 1980, **170**, 309.

513. Rumpf, B. A.; Allison, C. E.; Derrick, P. J. *Org. Mass Spectrom.* 1986, **21**, 295.

514. Proctor, C. J.; McLafferty, F. W. *Org. Mass Spectrom.* 1983, **18**, 193.

515. Beynon, J. H.; Caprioli, R. M.; Baitinger, W. E.; Amy, J. W. *Org. Mass Spectrom.* 1970, **3**, 479.

516. Howells, S.; Brenton, A. G.; Beynon, J. H. *Int. J. Mass Spectrom. Ion Phys.* 1980, **32**, 379.

517. McLafferty, F. W.; Hirota, A.; Barbalas, M. P. *Org. Mass Spectrom.* 1980, **15**, 327.

518. Cheng, M. T.; Kruppa, G. H.; McLafferty, F. W.; Cooper, D. A. *Anal. Chem.* 1982, **54**, 2204.

519. Dawson, P. H. Presented at the 30th Annual Conference on Mass Spectrometry and Allied Topics, Honolulu, HI, 1982.

520. Grotch, S. L. *Anal. Chem.* 1970, **42**, 1214.

521. Dawson, P. H.; Sun, W.-F. Presented at the 31st Annual Conference on Mass Spectrometry and Allied Topics, Boston, MA, 1983.

522. Dawson, P. H.; Sun, W.-F. *Int. J. Mass Spectrom. Ion Phys.* 1983, **55**, 155.

523. Mukhtar, E. S.; Griffiths, I. W.; Harris, F. M.; Beynon, J. H. *Int. J. Mass Spectrom. Ion Phys.* 1981, **37**, 159.

524. Martinez, R. I.; Dheandhanoo, S. *Int. J. Mass Spectrom Ion Proc.* 1986, **74**, 241.

525. Martinez, R. I.; Dheandhanoo, S. *J. Res. Nat. Bur. Stndrds.* 1987, **92**, 229.

526. Martinez, R. I. Presented at the 35th Annual Conference on Mass Spectrometry and Allied Topics, Denver, CO, 1987.

527. Cross, K. P.; Enke, C. G. Presented at the 32nd Annual Conference on Mass Spectrometry and Allied Topics, San Antonio, TX, 1984.

528. Giordani, A. B.; Gregg, H. R.; Hoffman, P. A.; Cross, K. P.; Beckner, C. P.; Enke, C. G. Presented at the 32nd Annual Conference on Mass Spectrometry and Allied Topics, San Antonio, TX, 1984.

529. Cross, K. P.; Enke, C. G. Presented at the 33nd Annual Conference on Mass Spectrometry and Allied Topics, San Diego, CA, 1985.

530. Wade, A. P.; Palmer, P. T.; Hart, K. J.; Enke, C. G. Presented at the 34nd Annual Conference on Mass Spectrometry and Allied Topics, Cincinnati, OH, 1986.

531. Cross, K. P.; Palmer, P. T.; Beckner, C. F.; Giordani, A. B.; Gregg, H. G.; Hoffman, P. A.; Enke, C. G. "Automation of Structure Elucidation from Mass Spectrometry–Mass Spectrometry Data" In *Artificial Intelligence Applications in Chemistry*, Pierce, T. H.; Hohne, B. A. Eds. *ACS Symposium Series* 1986, 306, 321.

532. Giblin, D. E.; Peake, D. A.; Lapp, R. L. Presented at the 32nd Annual Conference on Mass Spectrometry and Allied Topics, San Antonio, TX, 1984.

533. Bass, L. M.; Bowers, M. T. *Org. Mass Spectrom.* 1982, **17**, 229.

534. Van Thuijl, J. *Org. Mass Spectrom.* 1984, **19**, 243.

535. Palmer, P. T.; Enke, C. G. Presented at the 34th Annual Conference on Mass Spectrometry and Allied Topics, Cincinnati, OH, 1986.

536. Bozorgzadeh, M. H.; Morgan, R. P.; Beynon, J. H. *Analyst* 1978, **103**, 613.

537. Palmer, P. T.; Wade, A. P.; Hart, K. J.; Enke, C. G. Presented at the 35th Annual Conference on Mass Spectrometry and Allied Topics, Denver, CO, 1987.

538. Brand, H. R.; Gregg, H. R.; Wong, C. M. Presented at the 34th Annual Conference on Mass Spectrometry and Allied Topics, Cincinnati, OH, 1986.

539. Steiner, U.; Sokolow, S. Presented at the 34th Annual Conference on Mass Spectrometry and Allied Topics, Cincinnati, OH, 1986.

540. Weber, J. J.; van Thuijl, J.; de Jong, H. K. *Anal. Chim. Acta* 1986, **188**, 195.

541. Hunt, D. F.; Shabanowitz, J.; Harvey, T. M.; Coates, M. L. *J. Chrom. Chrom. Rev.* 1983, **271**, 93.

542. Hunt, D. F.; Shabanowitz, J.; Harvey, T. M.; Coates, M. L. *Anal. Chem.* 1985, **57**, 525.

543. Hunt, D. F.; Shabanowitz, J.; Giordani, A. B. *Anal. Chem.* 1980, **52**, 386.

544. Safe, S.; Hutzinger, O.; Cook, M. *J. Chem. Soc. Chem. Commun.* 1971, 446.

545. Safe, S.; Hutzinger, O.; Jamieson, W. D. *Org. Mass Spectrom.* 1973, **7**, 169.

546. Busch, K. L.; Norstrom, A.; Nilsson, C.-A.; Bursey, M. M.; Hass, J. R. *Environ. Health Perspect.* 1980, **36**, 125.

547. Hass, J. R.; Bursey, M. M.; Levy, L. A.; Harvan, D. J. *Org. Mass Spectrom.* 1979, **14**, 319.

548. Chess, E. K.; Gross, M. L. *Anal. Chem.* 1980, **52**, 2057.

549. Harvan, D. J.; Hass, J. R.; Schroeder, J. L.; Corbett, B. J. *Anal. Chem.* 1981, **53**, 1755.

550. Voyksner, R. D.; Hass, J. R.; Sovocool, G. W.; Bursey, M. M. *Anal. Chem.* 1983, **55**, 744.

551. Clement, R. E.; Bobbie, B.; Taguchi, V. *Chemosphere* 1986, **15**, 1147.

552. Bonner, R. F. *Int. J. Mass Spectrom. Ion Phys.* 1983, **48**, 311.

553. Shushan, B. I.; Ngo, A.; Ozvacic, V.; Wong, G.; DeBrou, G.; Bobbie, B.; Clement, R. E. *Abstracts of Papers*, 1986 Pittsburgh Conference and Exposition on Analytical Chemistry and Applied Spectroscopy, March 1986, Atlantic City, NJ, Paper 542.

554. Gale, B. C.; Fulford, J. E.; Thomson, B. A.; Ngo, A.; Tanner, S. D.; Davidson, W. R.; Shushan, B. I. *Adv. Mass Spectrom.* 1986, **10**, 1467.

555. Mahle, N. H.; Cooks, R. G.; Korzeniowski, R. W. *Anal. Chem.* 1983, **55**, 2272.

556. O'Brien, R. J.; Dumdei, B. E.; Hummel, S. V.; Yost, R. A. *Anal. Chem.* 1984, **56**, 1329.

557. Dumdei, B.; Zoldak, J.; Mickunas, D.; Sisak, M. *Abstracts of Papers*, 190th National Meeting of the American Chemical Society, Chicago, IL, Sept. 8–13, 1985. American Chemical Society: Washington, DC, Paper CHAS-31.

558. Boon, J. J.; Genuit, W.; Van der Valk, F.; Dallinga, J.; Koernig, S. A.; Eisma, D. *Abstracts of Papers*, 189th National Meeting of the American Chemical Society, Miami Beach, FL, April 28–May 3 1985. American Chemical Society: Washington, DC, Paper GEOC-44.

559. Lindstrom, K.; Schubert, R. *J. High Res. Chrom. Chrom. Comm.* 1984, **7**, 68.

560. Rivera, J.; Caixach, J.; Ventura, F.; Figueras, A.; Fraisse, D.; Dessalces, G. *Adv. Mass Spectrom.* 1986, **10**, 1453.

561. Schneider, E.; Levsen, K. *Comm. Eur. Communities* [*Rep.*] *EUR* 1986, EUR 10388, 14. CA **106**: 38100c.

562. Rivera, J.; Ventura, F.; Caixach, J.; Figueras, A.; Fraisse, D.; Blondot, V. *Comm. Eur. Communities* [*Rep.*] *EUR* 1986, EUR 10388, 77. CA **106**: 55514h.

563. Rivera, J.; Figueras, A.; Caixach, J.; Ventura, F.; Fraisse, D. Presented at the 35th Annual Conference on Mass Spectrometry and Allied Topics, Denver, CO, 1987.

564. Kurlick, N.; Pritchett, T. Presented at the 35th Annual Conference on Mass Spectrometry and Allied Topics, Denver, CO, 1987.

565. Games, D. E. *Mass Spectrom. Spec. Per. Rep.* 1981, **6**, 241.

566. Games, D. E. *Mass Spectrom. Spec. Per. Rep.* 1984. **7**, 353.

567. Kruger, T. L.; Cooks, R. G.; McLaughlin, J. L.; Raneiri, R. L. *J. Org. Chem.* 1977, **42**, 4161.

568. Pardanani, J. H.; McLaughlin, J. L.; Kondrat, R. W.; Cooks, R. G. *Lloydia* 1977, **40**, 585.

569. Kondrat, R. W.; Cooks, R. G. *Science* 1978, **199**, 978.

570. McClusky, G. A.; Cooks, R. G.; Knevel, A. M. *Tetrahedron Lett.* 1978, **46**, 4471.

571. Kondrat, R. W.; McClusky, G. A.; Cooks, R. G. *Anal. Chem.* 1978, **50**, 2017.

572. Youseffi, M.; Cooks, R. G.; McLaughlin, J. L. *J. Am. Chem. Soc.* 1979, **101**, 3400.

573. Cooks, R. G.; Kondrat, R. W.; Youseffi, M.; McLaughlin, J. L. *J. Ethnopharmacology* 1981, **3**, 299.

574. Unger, S. E.; Cooks, R. G. *Anal. Lett.* 1979, **12**, 1157.

575. Eckers, C.; Games, D. E.; Mallen, D. N. B.; Swann, B. P. *Biomed. Mass Spectrom.* 1982, **9**, 162.

576. Unger, S. E.; Cooks, R. G.; Mata, R.; McLaughlin, J. L. *J. Nat. Prod.* 1980, **43**, 288.

577. Pummangura, S.; McLaughlin, J. L.; Davis, D. V.; Cooks, R. G. *J. Nat. Prod.* 1982, **45**, 277.

578. Meyer, B. N.; Helfrich, J. S.; Nichols, D. E.; McLaughlin, J. L.; Davis, D. V.; Cooks, R. G. *J. Nat. Prod.* 1983, **46**, 688.

579. Davis, D. V.; Cooks, R. G.; Meyer, B. N.; McLaughlin, J. L. *Anal. Chem.* 1983, **55**, 1302.

580. Roush, R. A.; Cooks, R. G. *J. Nat. Prod.* 1984, **47**, 197.

581. Ferrigni, N. R.; Sweetana, S. A.; McLaughlin, J. L.; Singleton, K. E.; Cooks, R. G. *J. Nat. Prod.* 1984, **47**, 839.

582. Roush, R. A.; Cooks, R. G.; Sweetana, S. A.; McLaughlin, J. L. *Anal. Chem.* 1985, **57**, 109.

583. Glish, G. L.; Todd, P. J.; Busch, K. L.; Cooks, R. G. *Int. J. Mass Spectrom. Ion. Phys.* 1984, **56**, 177.

584. Pachuta, R. R.; Cooks, R. G.; Cassady, J. M.; Cong, P. Z.; McCloud, T. M.; Chang, C.-j. *J. Nat. Prod.* 1986, **49**, 412.

585. Maffei Facino, R.; Carini, M.; Traldi, P. *J. Pharm. Biomed. Anal.* 1985, **3**, 201.

586. Maffei Facino, R.; Carini, M.; Da Forno, A.; Belli, B.; Traldi, P. *Adv. Mass Spectrom.* 1986, **10**, 1425.

587. Plattner, R. D.; Grove, M. D.; Powell, R. G. Presented at the 32nd Annual Conference on Mass Spectrometry and Allied Topics, San Antonio, TX, 1982, p. 787.

588. Ohashi, O.; Nagai, Y. *Nippon Kagaku Kaishi* 1986, 1683.

589. Manning, K. S.; Lynn, D. G.; Shabanowitz, J.; Fellows, L. E.; Singh, M.; Schrire, B. D. *J. Chem. Soc. Chem. Commun.* 1985, 127.

590. Masucci, J. A.; Caldwell, G. W.; Wu, W. N.; Isensee, R. K.; Slayback, J. R. B. Presented at the 35th Annual Conference on Mass Spectrometry and Allied Topics, Denver, CO, 1987.

591. Fraisse, D.; Becchi, M. Presented at the 35th Annual Conference on Mass Spectrometry and Allied Topics, Denver, CO, 1987.

592. Greathead, R. J.; Jennings, K. R.; Page, J. A.; Pryce, R. J. *Adv. Mass Spectrom.* 1986, **10**, 1461.
593. Haddon, W. F.; Molyneux, R. J. Presented at the 30th Annual Conference on Mass Spectrometry and Allied Topics, New York, NY, 1980.
594. Dreifuss, P. A.; Brumley, W. C.; Sphon, J. A.; Caress, E. A. *Anal. Chem.* 1983, **55**, 1036.
595. Jones, A. D.; Kelley, R. B.; Winter, C. K. Segall, H. J.; Seiber, J. N. Presented at the 35th Annual Conference on Mass Spectrometry and Allied Topics, Denver, CO, 1987.
596. Jones, A. D.; Meeker, J. E.; Kilgore, W. W. Presented at the 35th Annual Conference on Mass Spectrometry and Allied Topics, Denver, CO, 1987.
597. Plattner, R. D.; Yates, S. G.; Porter, J. K. *J. Agric. Food Chem.* 1983, **31**, 785.
598. Plattner, R. D.; Bennett, G. A. *J. Assoc. Off. Anal. Chem.* 1983, **66**, 1470.
599. Lau, B. P.-Y.; Scott, P. M.; Sakuma, T. Presented at the 32nd Annual Conference on Mass Spectrometry and Allied Topics, San Antonio, TX, 1984.
600. Plattner, R. D.; Bennett, G. A.; Stubblefield, R. D. *J. Assoc. Off. Anal. Chem.* 1984, **67**, 734.
601. Yates, S. G.; Plattner, R. D.; Garner, G. B. *J. Agric. Food Chem.* 1985, **33**, 719.
602. Shotwell, O. D.; Bennett, G. A.; Stubblefield, R. D.; Shannon, G. M.; Kwolek, W. F.; Plattner, R. D. *J. Assoc. Off. Anal. Chem.* 1985, **68**, 954.
603. Plattner, R. D.; Powell, R. G. *J. Nat. Prod.* 1986, **49**, 475.
604. Plattner, R. D. In *Modern Methods in the Analysis and Structural Elucidation of Mycotoxins*, Cole, R. J., Ed. Academic: New York, 1986, Chapter 14.
605. Krishnamurthy, T.; Sarver, E. W. *Anal. Chem.* 1987, **59**, 1272.
606. Betowski, L. D.; Ballard, J. M. *Anal. Chem.* 1984, **55**, 2604.
607. Gale, P. J.; Bentz, B. L.; Chait, B. T.; Field, F. H.; Cotter, R. J. *Anal. Chem.* 1986, **58**, 1070.
608. Burinsky, D. J.; Dilliplane, R. L.; DiDonato, G. C.; Busch, K. L. *Org. Mass Spectrom.* 1988, **23**, 231.
609. Bruins, A. P. *Abstracts of Papers*, Pittsburgh Conference and Exposition on Analytical Chemistry and Applied Spectroscopy, March 1986, Atlantic City, NJ, Paper 546.
610. Weber, R.; Levsen, K.; Louter, G. J.; Boerboom, A. J. H.; Haverkamp, J. *Anal. Chem.* 1982, **54**, 1458.
611. Schneider, E.; Levsen, K.; Dahling, P.; Rollgen, F. W. *Fres. Z. Anal. Chem.* 1983, **316**, 217.
612. Lyon, P. A.; Crow, F. W.; Tomer, K. B.; Gross, M. L. *Anal. Chem.* 1984, **56**, 2278.
613. Lyon, P. A.; Stebbings, W. L.; Crow, F. W.; Tomer, K. B.; Lippstreu, D. L.; Gross, M. L. *Anal. Chem.* 1984, **56**, 8.
614. Lyon, P. A.; Tomer, K. B.; Gross, M. L. *Anal. Chem.* 1985, **57**, 2984.
615. Schneider, E.; Levsen, K.; Boerboom, A. J. H.; Kistemaker, P.; McLuckey, S. A.; Przybylski, M. *Anal. Chem.* 1984, **56**, 1987.
616. Foti, S.; Liquori, A.; Maravigna, P.; Montaudo, G. *Anal. Chem.* 1982, **54**, 674.
617. Craig, A. G.; Derrick, P. J. *J. Chem. Soc. Chem. Commun.* 1985, 891.
618. Craig, A. G.; Derrick, P. J. *J. Am. Chem. Soc.* 1985, **107**, 6707.
619. Craig, A. G.; Derrick, P. J. *Aust. J. Chem.* 1986, **39**, 1421.
620. Kiplinger, J. P.; Bursey, M. M. *Adv. Mass Spectrom.* 1986, **10**, 1475.
621. Shushan, B. I.; Davidson, W. R. *Abstracts of Papers*, Pittsburgh Conference and Exposition on Analytical Chemistry and Applied Spectroscopy, March 1983, Atlantic City, NJ, Paper 197.
622. Shushan, B.; Davidson, B.; Prime, R. B. *Anal. Calorimetry* 1984, **5**, 105.
623. Shushan, B.; Williamson, C.; Parkin, K.; McMorran, D. *Abstracts of Papers*, Pittsburgh Conference and Exposition on Analytical Chemistry and Applied Spectroscopy, March 1984, Atlantic City, NJ, Paper 316.
624. Shushan, B. I.; Prime, R. B.; Neag, C. M.; Carlson, G. M. *Abstracts of Papers*, Pittsburgh Conference and Exposition on Analytical Chemistry and Applied Spectroscopy, March 1985, Atlantic City, NJ, Paper 965.
625. Schulten, H.-R.; Lattimer, R. P. *Mass Spectrom. Rev.* 1984, **3**, 231.

626. Gleria, M.; Audisio, G.; Daolio, S.; Vecchi, E.; Praldi, P.; Krishnamurthy, S. S. *J. Chem. Soc. Dalton Trans.* 1986, 905.

627. Gleria, M.; Audisio, S.; Daolio, S.; Vecchi, E.; Traldi, P. *Org. Mass Spectrom.* 1985, **20**, 498.

628. Campana, J. E.; Doyle, R. J. *J. Chem. Soc. Chem. Commun.* 1985, 45.

629. Doyle, R. J.; Campana, J. E.; Eyler, J. R. *J. Phys. Chem.* 1985 **89**, 5285.

630. Doyle, R. J.; Campana, J. E. *J. Phys. Chem.* 1985, **89**, 4251.

631. Doyle, R. J.; Campana, J. E. *J. Am. Chem. Soc.* 1985, **107**, 7228.

632. Ballistreri, A.; Garozzo, D.; Giuffrida, M.; Maravigna, P.; Montaudo, G. *Macromolecules* 1986, **19**, 2693.

633. Ballistreri, A.; Garozzo, D.; Giuffrida, M.; Montaudo, G. *Polymer Degradation and Stability* 1986, **16**, 337.

634. Ballistreri, A.; Garozzo, D.; Giuffrida, M.; Montaudo, G.; Filippi, A.; Guaita, C.; Manaresi, P.; Pilati, F. *Macromolecules* 1987, **20**, 1029.

635. Lattimer, R. P.; Harris, R. E. *Mass Spectrom. Rev.* 1985, **4**, 369.

636. Lattimer, R. P.; Harris, R. E.; Rhee, C. K.; Schulten, H.-R. *Anal. Chem.* 1986, **58**, 3188.

637. Lattimer, R. P. Presented at the 35th Annual Conference on Mass Spectrometry and Allied Topics, Denver, CO, 1987.

638. Cardaciotto, S. J.; Mowery, P. C.; Thomson, M. L.; Wayne, R. S. *Org. Mass Spectrom.* 1987, **22**, 342.

639. Roach, J. A. G.; Carson, L. J. *J. Assoc. Off. Anal. Chem.* 1987, **70**, 439.

640. Horman, I. *Gazz. Chim. Ital.* 1984, **114**, 297.

641. Horman, I. *Biomed. Mass Spectrom.* 1981, **8**, 384.

642. Horman, I. *Mass Spectrom. Spec. Per. Rep.* 1979, **5**, 211.

643. Games, D. E.; Ramsey, D. E. *J. Chromatogr.* 1985, **323**, 67.

644. Harvey, D. J.; *J. Chromatogr.* 1975, **110**, 91.

645. Davis, D. V.; Cooks, R. G. *J. Agric. Food Chem.* 1982, **30**, 495.

646. Labows, J. N.; Shushan, B. *Amer. Lab.* 1983, **15**(3), 56.

647. Walther, H.; Schlunegger, U. P.; Friedl, F. *Org. Mass Spectrom.* 1983, **18**, 572.

648. Fraisse, D.; Maquin, F.; Stahl, D.; Suon, K.; Tabet, J. C. *Analusis* 1984, **12**, 63.

649. Lau, B. P.-Y.; Michalik, P.; Porter, C. J.; Krolick, S. Presented at the 35th Annual Conference on Mass Spectrometry and Allied Topics, Denver, CO, 1987.

650. Sphon, J. A.; Brumley, W. C.; Andrzejewski, D.; Roach, J. A. G.; White, K. D. *Adv. Mass Spectrom.* 1986, **10**, 1477.

651. Traldi, P.; Daolio, S.; Pelli, B.; Maffei Facino, R; Carini, M. *Biomed. Mass Spectrom.* 1985, **12**, 493.

652. Facino, R. M.; Carini, M.; Traldi, P. In *Chromatography and Mass Spectrometry in Nutrition Science and Food Safety*, Frigerio, A., Ed. Elsevier: Amsterdam, 1984, p. 27.

653. Brumley, W. C.; Min, Z.; Matusik, J. E.; Roach, J. A. G.; Barnes, C. J.; Sphon, J. A.; Fazio, T. *Anal. Chem.* 1983, **55**, 1405.

654. Trehy, M. L.; Yost, R. A.; Dorsey, J. G. *Anal. Chem.* 1986, **58**, 14.

655. Andrzejewski, D.; Havery, D. C.; Sphon, J. A. *Abstracts of Papers*, 192nd National Meeting of the American Chemical Society, Sept. 7–12, 1986, Anaheim, CA; American Chemical Society: Washington, DC, Paper ANYL 62.

656. Farrow, P. E.; Taylor, R. F.; Yelle, L. M. Presented at the 35th Annual Conference on Mass Spectrometry and Allied Topics, Denver, CO, 1987.

657. Fetterolf, D. D. Presented at the 33rd Annual Conference on Mass Spectrometry and Allied Topics, San Diego, CA, 1985.

658. Fetterolf, D. D. *Abstracts of Papers* 189th National Meeting of the American Chemical Society, Miami Beach, FL, April 28–May 3, 1985. American Chemical Society: Washington, DC, Paper ANYL 52.

659. McLuckey, S. A.; Glish, G. L.; Carter, J. A. *J. Forensic Sci.* 1985, **30**, 773.

660. Yinon, J.; Harvan, D. J.; Hass, J. R. *Org. Mass Spectrom.* 1982, **17**, 321.

661. Carper, W. R.; Dorey, R. C.; Tomer, K. B.; Crow. F. W. *Org. Mass Spectrom.* 1984, **19**, 623.
662. Fetterolf, D. D. Presented at the 33rd Annual Conference on Mass Spectrometry and Allied Topics, San Diego, CA, 1985.
663. Pelli, B.; Traldi, P.; Tagliaro, F.; Lubli, G.; Marigo, M. *Biomed. Environ. Mass Spectrom.* 1987, **14**, 63.
664. Kondrat, R. W. In *Tandem Mass Spectrometry*, McLafferty, F. W., Ed. Wiley: New York, 1983, p. 479.
665. Aczel, T. *Mass Spectrometric Characterization of Shale Oils*, ASTM Special Publication 902, American Society for the Testing of Materials, Philadelphia, PA, 1986.
666. Gallegos, E. J. In *High Performance Mass Spectrometry*, Gross, M. L., Ed. American Chemical Society: Washington, DC, 1978, Chapter 15.
667. Warburton, G. A.; Zumberge, J. E. *Anal. Chem.* 1983, **55**, 123.
668. Philip, R. P. *Chem. Eng. News* 1986, **64**(6), 28.
669. Sundararaman, P.; Gallegos, E. J.; Baker, E. W.; Slayback, J. R. B.; Johnston, M. R. *Anal. Chem.* 1984, **56**, 2552.
670. Fukuda, E. K.; Campana, J. E. *Abstracts of Papers*, 1985 Pittsburgh Conference and Exposition on Analytical Chemistry and Applied Spectroscopy, March 1985, New Orleans, LA, Paper 368.
671. Johnson, J. V.; Britton, E. V.; Yost, R. A.; Quirke, J. M. E.; Cuesta, L. L. *Anal. Chem.* 1986, **58**, 1325.
672. Zakett, D.; Shaddock, V. M.; Cooks, R. G. *Anal. Chem.* 1979, **51**, 1849.
673. Ciupek, J. D.; Cooks, R. G.; Wood, K. V.; Ferguson, C. R. *Fuel* 1983, **62**, 829.
674. Hunt, D. F.; Shabanowitz, J. *Anal. Chem.* 1982, **54**, 574.
675. Wood, K. V.; Schmidt, C. E.; Cooks, R. G., Batts, B. D. *Anal. Chem.* 1984, **56**, 1335.
676. Wood, K. V.; Cooks, R. G.; Laugal, J. A.; Benkesser, R. A. *Anal. Chem.* 1985, **57**, 692.
677. Wood, K. V.; Albright, L. F.; Brodbelt, J. S.; Cooks, R. G. *Anal. Chim. Acta* 1985, **173**, 117.
678. Singleton, K. E.; Cooks, R. G.; Wood, K. V.; Tse, K. T.; Stock, L. *Anal. Chim. Acta* 1985, **174**, 211.
679. Brodbelt, J. S.; Cooks, R. G.; Wood, K. V.; Jackson, T. J. *Fuel Sci. Technol.* 1986, **4**, 683.
680. Wong, C. M.; Crawford, R. W.; Yost, R. A. In *Mass Spectrometric Characterization of Shale Oils*, Aczel, T., Ed. ASTM Special Publication 902, American Society for the Testing of Materials, Philadelphia, PA, 1986, p. 106.
681. McLafferty, F. W.; Bockhoff, F. M. *Anal. Chem.* 1978, **50**, 69.
682. Morgan, R. P.; Steele, G.; Harris, J. A.; *Adv. Mass Spectrom.* 1980, **8**, 1725.
683. Wood, K. V.; Ciupek, J. D.; Cooks, R. G.; Ferguson, C. R. *Abstracts of Papers*, SAE Fuels Lubricants Meeting, Toronto, Ontario, Canada, October 18-21, 1982, Technical Paper 821217.
684. Henderson, T. R.; Royer, R. E.; Clark, C. R.; Harvey, T. M.; Hunt, D. F. *Environ. Sci. Technol.* 1983, **17**, 443.
685. Henderson, T. R.; Sun, J. D.; Li, A. P.; Hanson, R. L.; Bechtold, W. E.; Harvey, T. M.; Shabanowitz, J.; Hunt, D. F. *Environ. Sci. Technol.* 1984, **18**, 428.
686. Doretti, L.; Maccioni, A. M.; Traldi, P. *Biomed. Environ. Mass Spectrom.* 1986, **13**, 381.
687. Missler, S. R.; Freas, R. B.; Kelley, P. E. Presented at the 35th Annual Conference on Mass Spectrometry and Allied Topics, Denver, CO, 1987.
688. Niwa, Y.; Ishikawa, K. Presented at the 35th Annual Conference on Mass Spectrometry and Allied Topics, Denver, CO, 1987.
689. Muller, R. A. *Science* 1977, **196**, 489.
690. Litherland, A. E.; Beukens, R. P.; Kilius, L. R.; Rucklidge, J. C.; Gove, H. E.; Elmore, D.; Purser, K. H. *Nucl. Instrum. Meth.* 1981, **186**, 463.
691. Purser, K. H.; Williams, P.; Litherland, A. E.; Stein, J. D.; Storms, H. A.; Gove, H. E.; Stevens, C. M. *Nucl. Instrum. Meth.* 1981, **186**, 487.
692. Elmore, D.; Philips, F. M. *Science* 1987, **236**, 543.

693. Raisbeck, G. M.; Yiou, F.; Bourles, D.; Kent, D. V. *Nature* 1985, **315**, 315.
694. Raisbeck, G. M.; Yiou, F. *Nature* 1979, **277**, 43.
695. Reinhold, V. N.; Carr, S. A. *Mass Spectrom. Rev.* 1983, **2**, 153.
696. Rollgen, F. W.; Schulten, H.-R. *Z. Naturforsch.* 1975, **30a**, 1685.
697. de Jong, E. D.; Heerma, W.; Sicherer, C. A. X. G. F. *Biomed. Mass Spectrom.* 1979, **6**, 242.
698. de Jong, E. D.; Heerma, W.; Dijkstra, G. *Biomed. Mass Spectrom.* 1980, **7**, 127.
699. Aubagnac, J. L.; Devienne, F. M.; Combarieu, R. *Tetrahedron Lett.* 1983, **24**, 2263.
700. Puzo, G.; Maxime, B.; Prome, J. C. *Int. J. Mass Spectrom. Ion Phys.* 1983, **48**, 169.
701. Cerny, R. L.; Tomer, K. B.; Gross, M. L. *Org. Mass Spectrom.* 1986, **21**, 655.
702. Wright, L. G.; Cooks, R. G.; Wood, K. V. *Biomed. Mass Spectrom.* 1985, **12**, 159.
703. Puzo, G.; Fournie, J.-J.; Prome, J.-C. *Anal. Chem.* 1985, **57**, 892.
704. Fournie, J.-J.; Puzo, G. *Anal. Chem.* 1985, **57**, 2287.
705. Puzo, G.; Prome, J. C.; Fournie, J. J. *Carbohydrate Res.* 1985, **140**, 131.
706. Tondeur, Y.; Clifford, A. J.; De Luca, L. M. *Org. Mass Spectrom.* 1985, **20**, 157.
707. Puzo, G.; Prome, J. C. *Org. Mass Spectrom.* 1984, **19**, 448.
708. Puzo, G.; Prome, J. C. *Org. Mass Spectrom.* 1985, **20**, 288.
709. Carr, S. A.; Reinhold, V. N.; Green, B. N.; Hass, J. R. *Biomed. Mass Spectrom.* 1985, **12**, 288.
710. Chen, Y.; Chen, N.; Li, H.; Zhao, F.; Chen, N. *Biomed. Environ. Mass Spectrom.* 1987, **14**, 9.
711. Pamidi, K. M.; Jain, R. K.; Abbas, S. A.; Matta, K. L.; Laine, R. A. Presented at the 35th Annual Conference on Mass Spectrometry and Allied Topics, Denver, CO, 1987.
712. Guevremont, R.; Wright, J. L. C. *Rapid Comm. Mass Spectrom.* 1987, **1**, 12.
713. Levsen, K.; Schulten, H.-R. *Biomed. Mass Spectrom.* 1976, **3**, 137.
714. Schoen, A. E.; Cooks, R. G.; Wiebers, J. L. *Science* 1979, **203**, 1249.
715. Unger, S. E.; Schoen, A. E.; Cooks, R. G.; Ashworth, D. J.; Gomes, J. D.; Chang, C.-j. *J. Org. Chem.* 1981, **46**, 4765.
716. Straub, K. M.; Burlingame, A. L. *Adv. Mass Spectrom.* 1980, **8**, 1127.
717. Thomson, B. A.; Iribarne, J. V.; Dziedzic, P. J. *Anal. Chem.* 1982, **54**, 2219.
718. Sindona, G.; Uccella, N.; Weclawek, K. J. *Chem. Res. (S)* 1982, 184.
719. Neri, N.; Sindona, G.; Uccella, N. *Gazz. Chim. Ital.* 1983, **113**, 197.
720. Panico, M.; Sindona, G.; Uccella, N. *J. Am. Chem. Soc.* 1983, **105**, 5607.
721. Linscheid, H.; Burlingame, A. L. *Org. Mass Spectrom.* 1983, **18**, 245.
722. Crow, F. W.; Tomer, K. B.; Gross, M. L.; McCloskey, J. A.; Bergstrom, D. E. *Anal. Biochem.* 1984, **139**, 243.
723. Cerny, R. L.; Gross, M. L.; Grotjahn, L. *Anal. Biochem.* 1986, **156**, 424.
724. Kingston, E. E.; Beynon, J. H.; Newton, R. P. *Biomed. Mass Spectrom.* 1984, **11**, 367.
725. Kingston, E. E.; Beynon, J. H.; Newton, R. P.; Liehr, J. G. *Biomed. Mass Spectrom.* 1985, **12**, 525.
726. Kralj, B.; Kramer, V.; Susic, R.; Kobe, J. *Biomed. Mass Spectrom.* 1985, **12**, 673.
727. Ashworth, D. J.; Baird, W. M.; Chang, C-j.; Ciupek, J. D.; Busch, K. L.; Cooks, R. G. *Biomed. Mass Spectrom.* 1985, **12**, 309.
728. Isern-Flecha, I.; Jiang, X.-Y.; Cooks, R. G.; Pfleiderer, W.; Chae, W.-G.; Chang, C.-j. *Biomed. Environ. Mass Spectrom.* 1987, **14**, 17.
729. Tomer, K. B.; Gross, M. L.; Deinzer, M. L. *Anal. Chem.* 1986, **58**, 2527.
730. Mallis, L. M.; Raushel, F. M.; Russell, D. H. *Anal. Chem.* 1987, **59**, 980.
731. Clifford, A. J.; Silverman-Jones, C. S.; Creek, K. E.; De Luca, L. M.; Tondeur, Y. *Biomed. Mass Spectrom.* 1985, **12**, 221.
732. Jensen, N. J.; Tomer, K. B.; Gross, M. L. *Anal. Chem.* 1985, **57**, 2018.
733. Tomer, K. B.; Crow, F. W.; Gross, M. L. *J. Am. Chem. Soc.* 1983, **105**, 5487.
734. Jensen, N. J.; Tomer, K. B.; Gross, M. L. *J. Am. Chem. Soc.* 1985, **107**, 1863.
735. Jensen, N. J.; Gross, M. L. *Lipids* 1986, **21**, 362.

736. Jensen, N. J.; Tomer, K. B.; Gross, M. L. *Lipids* 1986, **21**, 580.

737. Cervilla, M.; Puzo, G. *Anal. Chem.* 1983, **55**, 2100.

738. Riviere, M.; Cervilla, M.; Puzo, G. *Anal. Chem.* 1985, **57**, 2444.

739. Peake, D. A.; Gross, M. L. *Anal. Chem.* 1985, **57**, 115.

740. Peake, D. A.; Gross, M. L.; Ridge, D. P. *J. Am. Chem. Soc.* 1984, **106**, 4307.

741. Tomer, K. B.; Jensen, N. J.; Gross, M. L. *Anal. Chem.* 1986, **58**, 2429.

742. Bambagiotti, M.; Coran, S. A.; Vincieri, F. F.; Petrucciani, T.; Traldi, P. *Org. Mass Spectrom.* 1986, **21**, 485.

743. Batrokov, S. G.; Sadovskaya, V. L.; Galyashin, V. N.; Rozynov, B. V.; Bergel'son, L. D. *Biorg. Khim.* 1978, **4**, 1220.

744. Batrokov, S. G.; Sadovskaya, V. L.; Galyashin, V. N.; Rozynov, B. V.; Bergel'son, L. D. *Biorg. Khim.* 1978, **4**, 1390.

745. Batrakov, S. G.; Sadovskaya, V. L.; Rozynov, B. V.; Bergel'son, L. D. *Biorg. Khim.* 1978, **4**, 1398.

746. Sadovskaya, V. L.; Batrakov, S. G.; Rozynov, B. V. *Tezisy Dokl.-Sov.-Indiiskii Simp. Khom. Prir. Soedin.* 5th 1978, 78; Chemical Abstracts **93**: 217409v.

747. Tomer, K. B.; Crow, F. W.; Knoche, H. W.; Gross, M. L. *Anal. Chem.* 1983, **55**, 1033.

748. Sherman, W. R.; Ackermann, K. E.; Bateman, R. H.; Green, B. N.; Lewis, I. *Biomed. Mass Spectrom.* 1985, **12**, 409.

749. Zaretskii, Z. V. I. *Mass Spectrometry of Steroids*. Wiley: New York, 1976.

750. Smith, D. H.; Djerassi, C.; Maurer, K. H.; Rapp, U. *J. Am. Chem. Soc.* 1974, **96**, 3482.

751. Gaskell, S. J.; Millington, D. S. *Biomed. Mass Spectrom.* 1978, **5**, 557.

752. Gaskell, S. J.; Pike, A. W.; Millington, D. S. *Biomed. Mass Spectrom.* 1979, **6**, 78.

753. Gaskell, S. J.; Brownsey, B. G.; Brooks, P. W.; Green, B. N. *Biomed. Mass Spectrom.* 1983, **10**, 215.

754. Maquestiau, A.; van Haverbeke, V.; Flammang, R.; Mispreuve, H.; Kaisin, M.; Braekman, J. C.; Daloze, D.; Tursch, B. *Steroids* 1978, **31**, 2121.

755. Fayret, J.; Lacoste, L.; Alais, J.; Lablache-Combier, A.; Maquestiau, A.; van Harverbeke, Y.; Flammang, R.; Mispreuve, H. *Phytochemistry* 1979, **18**, 431.

756. Dehorter, B.; Brocquet, M. F.; Lacoste, L.; Alais, J.; Lablache-Combier, A.; Maquestiau, A.; van Harverbeke, Y.; Flammang, R.; Mispreuve, H. *Phytochemistry* 1980, **19**, 2311.

757. Bozorgzadeh, M. H.; Brenton, A. G.; Wiebers, J. L.; Beynon, J. H. *Biomed. Mass Spectrom.* 1979, **6**, 340.

758. Larka, E. A.; Howe, I.; Beynon, J. H.; Zaretskii, Z. V. I. *Org. Mass Spectrom.* 1981, **16**, 465.

759. Larka, E. A.; Howe, I.; Beynon, J. H.; Zaretskii, Z. V. I. *Tetrahedron*, 1981, **37**, 2625.

760. Proctor, C. J.; Larka, E. A.; Zaretskii, Z. V. I.; Beynon, J. H. *Org. Mass Spectrom.* 1982, **17**, 131.

761. Zaretskii, Z. V. I.; Dan, P.; Kustanovich, Z.; Larka, E. A.; Herbert, C. G.; Beynon, J. H.; Djerassi, C. *Org. Mass Spectrom.* 1984, **19**, 321.

762. Brown, F. J.; Djerassi, C. *J. Am. Chem. Soc.* 1980, **102**, 807.

763. Prome, D.; Lacave, C.; Roussel, J.; Prome, J. C. *Biomed. Mass Spectrom.* 1982, **9**, 527.

764. Kruger, T. L.; Kondrat, R. W.; Joseph, K. T.; Cooks, R. G. *Anal. Biochem.* 1979, **96**, 104.

765. Cheng, M. T.; Barbalas, M. P.; Pegues, R. F.; McLafferty, F. W. *J. Am. Chem. Soc.* 1983, **105**, 1510.

766. Unger, S. E.; *Int. J. Mass Spectrom. Ion Proc.* 1985, **66**, 195.

767. Liehr, J. G.; Kingston, E. E.; Beynon, J. H. *Biomed. Mass Spectrom.* 1985, **12**, 95.

768. Gaskell, S. J.; Porter, C. J.; Green, B. N. *Biomed. Mass Spectrom.* 1985, **12**, 139.

769. Guenat, C.; Gaskell, S. J. "Secondary Ion Mass Spectrometry SIMS V." Benninghoven, A.; Colton, R. J.; Simons, D. S.; Werner, H. W., Eds., *Springer Series in Chemical Physics*, **44**, Springer-Verlag: New York, 1986.

770. Cole, R. B.; Guenat, C. R.; Gaskell, S. J. *Anal. Chem.* 1987, **59**, 1139.

771. Schweer, H.; Seyberth, H. W.; Schubert, R. *Biomed. Environ. Mass Spectrom.* 1986, **13**, 611.
772. Pelli, B.; Traldi, P.; Gargano, M.; Ravasio, N.; Rossi, M. *Org. Mass Spectrom.* 1987, **22**, 183.
773. Crow, F. W.; Tomer, K. B.; Looker, J. H.; Gross, M. L. *Anal. Biochem.* 1986, **155**, 286.
774. Davis, B. A.; Durden, D. A. In *Analysis of Biogenic Amines*, Baker, G. B.; Coutts, R. T., Eds. Elsevier: New York, 1982, p. 129.
775. Gelpi, E. In *Analysis of Biogenic Amines*, Baker, G. B.; Coutts, R. T., Eds. Elsevier: New York, 1982, p. 151.
776. Addeo, F.; Malorni, A.; Marino, G. *Anal. Biochem.* 1975, **64**, 98.
777. Walther, H.; Schlunegger, U. P.; Friedli, F. *Biomed. Environ. Mass Spectrom.* 1987, **14**, 229.
778. Johnson, J. V.; Yost, R. A.; Faull, K. F. *Anal. Chem.* 1984, **56**, 1655.
779. Biemann, K. *Anal. Chem.* 1986, **58**, 1289A.
780. Biemann, K.; Martin, S. A. *Mass Spectrom. Rev.* 1987, **6**, 1.
781. Johnson, R. S.; Biemann, K. *Biochemistry* 1987, **26**, 1209.
782. Roepstorff, P.; Fohlman, J. *Biomed. Mass Spectrom.* 1984, **11**, 601.
783. Levsen, K.; Wipf, H.-K.; McLafferty, F. W. *Org. Mass Spectrom.*, 1974, **8**, 117.
784. Bradley, C. V.; Howe, I.; Beynon, J. H. *Biomed. Mass Spectrom.* 1981, **8**, 85.
785. Aubagnac, I. L.; Devienne, F. M.; Combarieu, R. *Org. Mass Spectrom.* 1985, **20**, 428.
786. Kinoshita, T.; Nakamura, T.; Nagaki, H. *Nippon Kagaku Kaishi* 1986, 1665.
787. Heerma, W.; Bathelt, E. R. *Biomed. Environ. Mass Spectrom.* 1986, **13**, 205.
788. Schlunegger, U. P.; Hirter, P.; von Felten, H. *Helv. Chim. Acta* 1976, **59**, 406.
789. Steinauer, R.; Walther, H.; Schlunegger, U. P. *Helv. Chim. Acta* 1980, **63**, 610.
790. Steinauer, R.; Schlunegger, U. P. *Biomed. Mass Spectrom.* 1982, **9**, 153.
791. Farncombe, M.; Jennings, K. R.; Mason, R. S.; Schlunegger, U. P. *Org. Mass Spectrom.* 1983, **18**, 612.
792. Farncombe, M.; Mason, R. S.; Jennings, K. R.; Scrivens, J. *Int. J. Mass Spectrom. Ion Phys.* 1982, **44**, 91.
793. Weber, R.; Levsen, K. *Biomed. Mass Spectrom.* 1980, **7**, 314.
794. Matsuo, T.; Matsuda, H.; Katakuse, I.; Shimonishi, Y.; Maruyama, Y.; Higuchi, T.; Kubota, E. *Anal. Chem.* 1981, **53**, 416.
795. Desiderio, D. M.; Sabbatini, J. Z. *Biomed. Mass Spectrom.* 1981, **8**, 565.
796. Katakuse, I.; Desiderio, D. M. *Int. J. Mass Spectrom. Ion Phys.* 1983, **54**, 1.
797. Desiderio, D. M.; Katakuse, I. *Anal. Biochem.* 1983, **129**, 425.
798. Desiderio, D. M. *Int. J. Mass Spectrom. Ion Proc.* 1986, **74**, 217.
799. Seki, S.; Kambara, H.; Naoki, H. *Org. Mass Spectrom.* 1985, **20**, 18.
800. Hunt, D. F.; Bone, W. M.; Shabanowitz, J.; Rhodes, J.; Ballard, J. M. *Anal. Chem.* 1981, **53**, 1704.
801. Hunt, D. F.; Buko, A. M.; Ballard, J. M.; Shabanowitz, J.; Giordani, A. B. *Biomed. Mass Spectrom.* 1981, **8**, 397.
802. Barber, M.; Bordoli, R. S.; Sedgwick, R. D.; Tetler, L. W. *Org. Mass Spectrom.* 1981, **16**, 256.
803. Barber, M.; Bordoli, R. S.; Sedgwick, R. D.; Tyler, A. N. *Biomed. Mass Spectrom.* 1982, **9**, 208.
804. Heerma, W.; Kamerling, J. P.; Slotboom, A. J.; van Scharrenburg, G. J. M.; Green, B. N.; Lewis, I. A. S. *Biomed. Mass Spectrom.* 1983, **10**, 13.
805. Tomer, K. B.; Crow, F. W.; Gross, M. L.; Kopple, K. D. *Anal. Chem.* 1984, **56**, 880.
806. Eckart, K.; Schwarz, H.; Tomer, K. B.; Gross, M. L. *J. Am. Chem. Soc.* 1985, **107**, 6765.
807. Lippstreu-Fisher, D. L.; Gross, M. L. *Anal. Chem.* 1985, **57**, 1174.
808. Eckart, K.; Schwarz, H.; Chorev, M.; Gilon, C. *Eur. J. Biochem.* 1986, **157**, 209.
809. Eckart, K.; Schwarz, H. *Biomed. Environ. Mass Spectrom.* 1986, **13**, 641.
810. Bathelt, E. R.; Heerma, W. *Biomed. Environ. Mass Spectrom.* 1987, **14**, 53.
811. Zwinselman, J. J.; Nibbering, N. M. M.; van der Greef, J.; Ten Noever de Brauw, M. C. *Org. Mass Spectrom.* 1983, **18**, 525.

812. Neumann, G. M.; Derrick, P. J. *Aust. J. Chem.* 1984, **37**, 2261.
813. Mallis, L. M.; Russell, D. H. *Anal. Chem.* 1986, **58**, 1076.
814. Mallis, L. M.; Russell, D. H. Presented at the 35th Annual Conference on Mass Spectrometry and Allied Topics, Denver, CO, 1987.
815. Hunt, D. F.; Shabanowitz, J.; Yates, J. R., III; McIver, R. T., Jr.; Hunter, R. L.; Syka, J. E. P.; Amy, J. *Anal. Chem.* 1985, **57**, 2728.
816. Hunt, D. F.; Shabanowitz, J.; Yates, J. R., III *J. Chem. Soc. Chem. Commun.* 1987, 548.
817. Hunt, D. F.; Shabanowitz, J.; Yates, J. R., III; Zhu, N.-Z.; Russell, D. H.; Castro, M. E. *Proc. Nat. Acad. Sci.* 1987, **84**, 620.
818. Cody, R. B., Jr.; Amster, I. J.; McLafferty, F. W. *Proc. Nat. Acad. Sci.* 1985, **82**, 6367.
819. Yang, C. L.-C. Presented at the 35th Annual Conference on Mass Spectrometry and Allied Topics, Denver, CO, 1987.
820. Welch, M. J.; Sams, R.; White, E., V. *Anal. Chem.* 1986, **58**, 890.
821. Gross, M. L.; McCrery, D.; Crow, F.; Tomer, K. B.; Pope, M. R.; Ciuffetti, L. M.; Knoche, H. W.; Daly, J. M.; Dunkle, L. D. *Tetrahedron Lett.* 1982, **23**, 5381.
822. Kim, S.-D.; Knoche, H. W.; Dunkle, L. D.; McCrery, D. A.; Tomer, K. B. *Tetrahedron Lett.* 1985, **26**, 969.
823. Eckart, K.; Schwarz, H.; Ziegler, R. *Biomed. Mass Spectrom.* 1985, **12**, 623.
824. Ziegler, R.; Eckart, K.; Schwarz, H.; Keller, R. *Biochem. Biophys. Res. Comm.* 1985, **133**, 337.
825. Witten, J. L.; Schaffer, M. H.; O'Shea, M.; Cook, J. C.; Hemling, M. H.; Rinehart, K. L., Jr. *Biochem. Biophys. Res. Comm.* 1984, **124**, 350.
826. Gade, G.; Goldsworthy, G. J.; Schaffer, M. H.; Cook, J. C.; Rinehart, K. L., Jr. *Biochem. Biophys. Res. Comm.* 1986, **134**, 723.
827. Gade, G.; Rinehart, K. L., Jr. *Biol. Chem. Hoppe-Seyler* 1987, **368**, 67.
828. Krishnamurthy, T.; Szafraniec, L.; Sarver, E. W.; Hunt, D. F.; Missler, S.; Carmichael, W. W. "Secondary Ion Mass Spectrometry, SIMS V." Benninghoven, A.; Colton, R. J.; Simons, D. S.; Werner, H. W., Eds.: *Springer Series in Chemical Physics*, **44**. Springer Verlag; New York, 1986.
829. Eckart, K.; Schmidt, U.; Schwarz, H. *Liebigs Ann. Chem.* 1986, 1940.
830. Petit, G. R.; Kamano, Y.; Brown, P.; Gust, D.; Inoue, M.; Herald, C. L. *J. Am. Chem. Soc.* 1982, **104**, 905.
831. Harvan, D. J.; Hass, J. R.; Wilson, W. E.; Hamm, C.; Boyd, R. K.; Yajima, H.; Klapper, D. G. *Biomed. Environ. Mass Spectrom.* 1987, **14**, 281.
832. Hamm, C. W.; Wilson, W. E.; Harvan, D. J. *Comp. Applic. Biosciences* 1986, **2**, 115.
833. Ishikawa, I.; Niwa, Y. *Biomed. Environ. Mass Spectrom.* 1986, **13**, 373.
834. Matsuo, T.; Sakurai, T.; Matsuda, H.; Wollnik, H.; Katakuse, I. *Biomed. Mass Spectrom.* 1983, **10**, 57.
835. Sakurai, T.; Matsuo, T.; Matsuda, H.; Katakuse, I. *Biomed. Mass Spectrom.* 1984, **11**, 396.
836. Scoble, H. A.; Biller, J. E.; Biemann, K. Fres. Z. *Anal. Chem.*, 1987, **327**, 239.
837. Griffin, P. R.; Hunt, D. F.; Shabanowitz, J.; Zhu, N.-Z. Presented at the 35th Annual Conference on Mass Spectrometry and Allied Topics, Denver, CO, 1987.
838. Aubagnac, J. L.; Calmes, M.; Daunis, J.; El Amrani, B.; Jacquier, R.; Devienne, F. M.; Combarieu, R. *Tetrahedron Lett.* 1985, **26**, 1497.
839. Desiderio, D. M.; Dass, C. *Anal. Lett.* 1986, **19**, 1963.
840. Richter, W. J.; Blum, W.; Schlunegger, U. P.; Senn, M. In *Tandem Mass Spectrometry*, McLafferty, F. W., Ed., Wiley: New York, 1983.
841. Duholke, W. K.; Fox. L. E. Presented at the 28th Annual Conference on Mass Spectrometry and Allied Topics, New York, 1980.
842. Catlow, D. A.; Johnson, M.; Monaghan, J. J.; Porter, C.; Scrivens, J. H. *J. Chromatogr.* 1985, **328**, 167.
843. McClusky, G. A.; Kondrat, R. W.; Cooks, R. G. *J. Am. Chem. Soc.* 1978, **100**, 6045.

844. Tondeur, Y.; Shorter, M.; Gustafson, M. E.; Pandey, R. C. *Biomed. Mass Spectrom.* 1984, **11**, 622.

845. Amster, I. J.; McLafferty, F. W. *Anal. Chem.* 1985, **57**, 1208.

846. Soltero-Rigaru, E.; Kruger, T. L.; Cooks, R. G. *Anal. Chem.* 1977, **49**, 435.

847. Unger, S. E. *Anal. Chem.* 1984, **56**, 363.

848. David, L.; Scanzi, E.; Fraisse, D.; Tabet, J. C. *Tetrahedron* 1982, **38**, 1619.

849. Barbalas, M. P.; McLafferty, F. W.; Occolowitz, J. L. *Biomed. Mass Spectrom.* 1983, **10**, 258.

850. Cooper, R.; Unger, S. E. *J. Antibiotics* 1985, **38**, 24.

851. Siegel, M. M.; McGahren, W. J.; Tomer, K. B.; Chang, T. T. *Biomed. Environ. Mass Spectrom.* 1987, **14**, 29.

852. Holzmann, G.; Ostwald, U.; Nickel, P.; Haack, H.-J.; Widjaja, H.; Arduny, U. *Biomed. Mass Spectrom.* 1985, **12**, 659.

853. Roberts, G. D.; Carr, S. A.; Christensen, S. B. Presented at the 35th Annual Conference on Mass Spectrometry and Allied Topics, Denver, CO, 1987

854. Lay, J. D., Jr.; Holder, C. L.; Cooper, W. M.; Korfmacher, W. A. Presented at the 35th Annual Conference on Mass Spectrometry and Allied Topics, Denver, CO, 1987.

855. Finaly, E. M. H.; Games, D. E.; Startin, J. R.; Gilbert, J. *Biomed. Environ. Mass Spectrom.* 1986, **13**, 633.

856. Facino, R. M.; Carini, M.; Da Forno, A.; Traldi, P.; Pompa, G. *Biomed. Environ. Mass Spectrom.* 1986, **13**, 121.

857. Tway, P. C.; Downing, G. V.; Slayback, J. R. B.; Rahn, G. S.; Isensee, R. K. *Biomed. Mass Spectrom.* 1984, **11**, 172.

858. Henion, J. D.; Thomson, B. A.; Dawson, P. H. *Anal. Chem.* 1982, **54**, 451.

859. Henion, J. D.; Maylin, G. A.; Thomson, B. A. *J. Chrom. Chrom. Rev.* 1983, **271**, 107.

860. Brotherton, H. A.; Yost, R. A. *Anal. Chem.* 1983, **55**, 549.

861. Haering, N.; Settlage, J. A.; Sanders, S. W.; Schuberth, R. *Biomed. Mass Spectrom.* 1985, **12**, 197.

862. Gaskell, S. J.; Guenat, C.; Millington, D. S.; Maltby, D. A.; Roe, C. R. *Anal. Chem.* 1986, **58**, 2801.

863. Perchalski, R. J.; Yost, R. A.; Wilder, B. J. *Anal. Chem.* 1982, **54**, 1466.

864. Johanessen, J. N.; Kelner, L.; Hanselman, D.; Shih, M.-C.; Markey, S. P. *Neurochem. Int.* 1985, **7**, 169.

865. Straub, K. M.; Levandoski, P. *Biomed. Mass Spectrom.* 1986, **12**, 338.

866. Lay, J. O., Jr.; Korfmacher, W. A.; Miller, D. W.; Siitonen, P.; Holder, C. L.; Gosnell, A. B. *Biomed. Environ. Mass Spectrom.* 1986, **13**, 627.

867. Korfmacher, W. A.; Betowski, L. D.; Holder, C. L.; Mitchum, R. K. Presented at the 35th Annual Conference on Mass Spectrometry and Allied Topics, Denver, CO, 1987.

868. Rudewicz, P.; Straub, K. M. *Anal. Chem.* 1986, **58**, 2928.

869. Covey, T. R.; Lee, E. D.; Henion, J. D. *Anal. Chem.* 1986, **58**, 2453.

870. Yost, R. A.; Perchalski, R. J.; Brotherton, H. O.; Johnson, J. V.; Budd, M. B. *Talanta* 1984, **31**, 929.

871. Straub, K. M.; Rudewicz, P.; Garvie, C. *Xenobiotica* 1987, **17**, 413.

872. Millington, D. S.; Roe, C. R.; Maltby, D. A. *Biomed. Mass Spectrom.* 1984, **11**, 236.

873. Maltby, D. A.; Millington, D. S.; Roe, C. R. Presented at the 35th Annual Conference on Mass Spectrometry and Allied Topics, Denver, CO, 1987.

874. Rinaldo, P.; Chiandetti, L.; Zacchello, F.; Daolio, S.; Traldi, P. *Biomed. Mass Spectrom.* 1984, **11**, 643.

875. Rinaldo, P.; Miolo, G.; Chiandetti, L.; Zacchello, F.; Daolio, S.; Traldi, P. *Biomed. Mass Spectrom.* 1985, **12**, 570.

876. Hunt, D. F.; Giordani, A. B.; Rhodes, G.; Herold, D. A. *Clin. Chem.* 1982, **28**, 2387.

877. Brodbelt, J. S.; Cooks, R. G. *Anal. Chem.* 1985, **57**, 1153.

878. Maquestiau, A.; Puk, E.; Flammang, R. *Bull. Soc. Chim. Belg.* 1987, **96**, 181.

879. Hutchison, D. W.; Woolfitt, A. R.; Ashcroft, A. E. *Org. Mass Spectrom.* 1987, **22**, 304.
880. Caprioli, R. M. *Mass Spectrom. Rev.* 1987, **6**, 237.
881. Caprioli, R. M.; Smith, L. A.; Beckner, C. F. *Int. J. Mass Spectrom. Ion Phys.* 1983, **46**, 419.
882. Saito, K.; Kato, R. *Proc. Soc. Med. Mass Spectrom.* 1984, **9**, 45.
883. Kostiainen, R.; Hesso, A. *Biomed. Environ. Mass Spectrom.* 1988, **15**, 79.
884. Krishnamurthy, T.; Sarverm E. W. *Biomed. Environ. Mass Spectrom.* 1988, **15**, 185.
885. Lau, B. P.-Y.; Michalik, P.; Porter, C. J.; Krolick, S.; *Biomed. Environ. Mass Spectrom.* 1987, **14**, 723.
886. Maquestiau, A.; Flammang-Barbieux, M.; Flammang, R.; Chen, L.-Z. *Bull. Soc. Chim. Belg.* 1988, **97**, 245.
887. Montaudo, G.; Scamporrino, E.; Puglisi, C.; Vitalini, D. *J. Anal. Appl. Pyrolysis* 1987, **10**, 283.
888. Michel, H.; Hunt, D. F.; Shabanowitz, J.; Bennett, J. *J. Biol. Chem.* 1988, **263**, 1123.
889. Eckart, K.; Schwarz, H. *Helv. Chim. Acta* 1987, **70**, 489.
890. Easton, C.; Johnson, D. W.; Poulos, A. *J. Lipid. Res.* 1988, **29**, 109.
891. Lee, M. S.; Yost, R. A. *Biomed. Environ. Mass Spectrom.* 1988, **15**, 193.
892. Kormacher, W. A.; Holder, C. L.; Betowski, L. D.; Mitchum, R. K. *J. Anal. Toxicology* 1987, **11**, 182.
893. Lay, J. P., Jr.; Potter, D. W.; Hinson, J. A. *Biomed. Environ. Mass Spectrom.* 1987, **14**, 517.
894. Winter, C. K.; Segall, H. J.; Jones, A. D. *Biomed. Environ. Mass Spectrom.* 1988, **15**, 265.
895. Winter, C. K.; Segall, H. J.; Jones, A. D. *Drug Metab. Dispos.* 1987, **15**, 608.
896. Gaskell, S. J. *Biomed. Environ. Mass Spectrom.* 1988, **15**, 99.
897. Weidolf, L. O. G.; Lee, E. D.; Henion, J. D. *Biomed. Environ. Mass Spectrom.* 1988, **15**, 283.
898. Martinez, R. I. *Rapid Comm. Mass Spectrom.* 1988, **2**, 8.
899. Martinez, R. I.; Ganguli, B. *Rapid Comm. Mass Spectrom.* 1988, **2**, 41.

MS/MS Scan Modes on Various Instrument Configurations

The following tables list the operational methods for acquisition of the various types of MS/MS data, daughter ion scans (D), parent ion scans (P), and neutral loss scans (NL), by scanning the various analyzers. Modes in which the acceleration voltage is varied are not included since these are generally inferior to other methods of obtaining the same type of data, due to defocusing effects associated with varying the accelerating voltage. Several reviews on linked scans discuss these methods in more detail for sectors[A1-A4] and for hybrid instruments.[A5] Besides the scan type and necessary parameters to implement that scan type, the kinetic energy range of the dissociating ion is listed (H = high kinetic energies, keV, L = low kinetic energies, eV), as is the reaction region in which the dissociation occurs. For instruments composed entirely of sectors, a deceleration step must be performed to access the low kinetic energy collision regime, followed by reacceleration (this is denoted by a ^ following the scan equation). In hybrid instruments with quadrupoles following the sectors, the deceleration voltage must be varied for any scan involving high kinetic energy ions that dissociate prior to the quadrupoles if the energy of the ions passing through the quadrupole(s) is to be kept constant (this is denoted by a \ following the scan equation). Also, for some scan modes in hybrid instruments, an additional energy filter must be added to implement the scan (these scans are denoted by a $^+$ following the scan equation). For scan modes in which only a magnetic sector is scanned, the mass of the ion being passed is determined by equation (2-8) if the dissociation occurs subsequent to the magnet or at low ion kinetic energy; if the dissociation occurs at high ion kinetic energy, prior to the magnetic sector, equation (2-10) holds. The latter case is denoted by a * following the scan equation. In many hybrid instrument linked scans, two equations are shown, one giving the relationship of the sectors, and the other the relationship between one of the sectors and the quadrupole. The designation of geometry does not include a quadrupole if it is only used as a reaction region. Thus, for example, EBQ is equivalent to EBQQ in which the first quadrupole is operated in an rf-only mode as a reaction region. In the scan equations, B = magnetic field strength, E = electric field strength as a percent of the strength of the electric field strength used to pass the parent ion (i.e., for undissociated ions $E = E_p/E_p = 1$, and for daughter ions $E = E_d/E_p$), $Q =$ quadrupole rf voltage amplitude (the dc voltage is assumed to vary linearly with the rf voltage), and $C =$ constant. For equivalent scans, only the simpler of the scans is listed. For example, daughter ion scans can be implemented on an EBE instrument by linked scans at constant B/E with the dissociations occurring in the first or second reaction regions. In the first reaction region, both electric sectors must be scanned while dissociations in the second reaction region require only scanning of the second electric sector. Only this latter scan is listed. Only two- and three-analyzer instruments are considered and instruments using nonscanning analyzers are not listed in this appendix.

Table A-1 ■ Sector Instruments

Scan type	Energy	RR	Equation(s)
		EB	
D	H	1	$B/E = C$
P	H	1	$B_2/E = C$
NL	H	1	$(B_2/E_2)(1 - E) = C$
		BE	
D	H	1	$B/E = C$
P	H	1	$B^2/E = C$
NL	H		$(B^2/E^2)(1 - E) = C$
D	H	2	SCAN E
P	H	2	$B^2E = C$
NL	H	2	$B^2(1 - E) = C$
		BB	
D	H	2	SCAN B_2*
P	H	2	$B_1B_2 = C$
NL	H	2	$B_1^2 - B_2^2 = C$
D	L	2	SCAN $B_2\hat{\,}$
P	L	2	SCAN $B_1\hat{\,}$
NL	L	2	$B_1^2 - B_2^2 = C\hat{\,}$
		EBE	
D	H	2	$B/E_2 = C$
P	H	2	$B^2/E_2 = C$
NL	H	2	$(B^2/E_2^2)(1 - E_2) = C$
D	H	3	SCAN E_2
P	H	3	$B^2E_2 = C$
NL	H	3	$B^2(1 - E) = C$

Table A-1   Continued

Scan Type	Energy	RR	Equation(s)
		BEE	
D	H	1	$B/E_1 = C,\ E_1 = E_2$
P	H	1	$B^2/E = C,\ E_1 = E_2$
NL	H	1	$(B^2/E^2)(1 - E) = C,\ E_1 = E_2$
D	H	3	SCAN E_2
P	H	3	$B^2 E_2 = C$
NL	H	3	$B^2(1 - E_2) = C$
		BEB	
D	H	2	$B_2/E = C$
P	H	2	$B_2^2/E = C$
NL	H	2	$(B_2^2/E^2)(1 - E) = C$
D	L	2	SCAN $B_2\,\hat{}$
P	L	2	SCAN $B_1\,\hat{}$
NL	L	2	$B_1^2 - B_2^2 = C\hat{}$
D	H	3	SCAN B_2*
P	H	3	$B_1 B_2 = C$
NL	H	3	$B_1^2 - B_1 B_2 = C$
D	L	3	SCAN $B_2\,\hat{}$
P	L	3	SCAN $B_1\,\hat{}$
NL	L	3	$B_1^2 - B_2^2 = C\hat{}$

Table A-2 ■ Quadrupole Instruments

Scan Type	Energy	RR	Equation(s)
		QQ	
D	L	1	SCAN Q_2
P	L	1	SCAN Q_1
NL	L	1	$Q_1 - Q_2 = C$

Table A-3 ■ Hybrid Instruments

Scan Type	Energy	RR	Equation(s)
		EQ	
D	H	1	$E/Q = C\backslash$
P	H	1	SCAN $E\backslash$
NL	H	1	$Q/E - Q = C\backslash$
		BQ	
D	H	1	$B^2/Q^2 = C\backslash$
P	H	1	SCAN $B\backslash$ *
NL	H	1	$Q^2/B^2 - Q = C\backslash$
D	H	2	SCAN $Q\backslash^{+}$
P	H	2	SCAN $B\backslash^{+}$
NL	H	2	$B^2 - Q = C\backslash^{+}$
D	L	3	SCAN Q
P	L	3	SCAN B
NL	L	3	$B^2 - Q = C$
		EBQ	
D	H	1	$B/E = C_1, Q/E = C_2\backslash$
P	H	1	$B^2/E = C\backslash$
NL	H	1	$B^2/E^2(1 - E) = C_1, (Q/E) - Q = C_2\backslash$
D	H	3	SCAN $Q\backslash^{+}$
P	H	3	SCAN $B\backslash^{+}$
NL	H	3	$B^2 - Q = C\backslash^{+}$
D	L	4	SCAN Q
P	L	4	SCAN B
NL	L	4	$B^2 - Q = C$
		BEQ	
D	H	1	$B/E = C_1, Q/E = C_2\backslash$
P	H	1	$B^2/E = C\backslash$
NL	H	1	$B^2/E^2(1 - E) = C_1, (Q/E) - Q = C_2\backslash$
D	H	2	$E/Q = C\backslash$
P	H	2	$B^2E = C\backslash$
NL	H	2	$B^2(1 - E) = C_1, (Q/E) - Q = C_2\backslash$
D	H	3	SCAN $Q\backslash^{+}$
P	H	3	SCAN $B\backslash^{+}$
NL	H	3	$B^2 - Q = C\backslash^{+}$
D	L	4	SCAN Q
P	L	4	SCAN B
NL	L	4	$B^2 - Q = C$
		QE	
D	H	1	SCAN E
P	H	1	$EQ = C$
NL	H	1	$Q - EQ = C$

Table A-3   Continued

Scan Type	Energy	RR	Equation(s)
		QB	
D	L	1	SCAN B
P	L	1	SCAN Q
NL	L	1	$Q - B^2 = C$
D	H	2	SCAN B^*
P	II	2	$B^2Q = C$
NL	H	2	$Q - BQ^{0.5} = C$
		QEB	
D	L	1	SCAN B
P	L	1	SCAN Q
NL	L	1	$Q - B^2 = C$
D	H	2	$B/E = C$
P	H	2	$B^2/E = C_1, QE = C_2$
NL	H	2	$(B_2^2/E^2)(1 - E) = C_1, Q - EQ = C_2$
D	H	3	SCAN B^*
P	H	3	$QB^2 = C$
NL	H	3	$Q - BQ^{0.5} = C$
D	L	3	SCAN $B\hat{}$
P	L	3	SCAN $Q\hat{}$
NL	L	3	$Q - B^2 = C\hat{}$
		QBE	
D	L	1	SCAN B
P	L	1	SCAN Q
NL	L	1	$Q - B^2 = C$
D	H	2	$B/E = C$
P	H	2	$B^2/E = C_1, QE = C_2$
NL	H	2	$(B_2^2/E^2)(1 - E) = C_1, Q - EQ = C_2$
D	H	3	SCAN E
P	H	3	$Q/B^2 = C_1, B^2E = C_2$
NL	H	3	$Q/B^2 = C_1, B^2(1 - E) = C_2$

References

A1. Boyd, R. K.; Shushan, B. *Int. J. Mass Spectrom. Ion Phys.* 1981, **37**, 355.

A2. Jennings, K. R.; Mason, R. S. In *Tandem Mass Spectrometry*, McLafferty, F. W. Ed. Wiley: New York, 1983.

A3. Macdonald, C. G.; Lacey, M. J. *Org. Mass Spectrom.*, 1984, **19**, 55.

A4. Fraefel, A.; Seibl, J. *Mass Spectrom. Rev.* 1985, **4**, 151.

A5. Glish, G. L.; McLuckey, S. A. Submitted for publication.

Appendix B

Frequently Used Symbols and Acronyms

a	acceleration
AE	appearance energy
α	empirical parameter usually taken as 0.44
B	magnetic sector
b	impact parameter
CA	collisional activation
CI	chemical ionization
CID	collision induced dissociation
CM	center-of-mass
D	disproportionation factor
d	9.64×10^7
DC	DC voltage
DI	desorption ionization
ΔE_{LAB}	laboratory kinetic energy change of the projectile
ΔE_N	laboratory kinetic energy change of the target
ΔG_{rxn}	free energy change of reaction
ΔH_{rxn}	enthalpy change of reaction
E	electric sector
e	fundamental unit of charge, 1.6×10^{-19} coulombs
EA	electron affinity
ECID	electron capture induced dissociation
EI	electron ionization
E_{LAB}	kinetic energy in the laboratory reference system
E_{REL}	relative kinetic energy
ϵ	internal energy
ϵ_0	critical energy
ϵ^r	reverse critical energy
$\epsilon^{\neq}$	non-fixed energy
F	force
FAB	fast atom bombardment

FD	field desorption
FT–ICR	Fourier transform ion cyclotron resonance
f_c	cyclotron frequency
GA	gas phase acidity
GB	gas phase basicity
GC/MS	gas chromatography/mass spectrometry
γ	characteristic period of motion
I	main beam current
I_0	unattenuated main beam current
I_{ab}	flux of absorbed photons
ICR	ion cyclotron resonance
IKES	ion kinetic energy spectrometry
IP	ionization potential
ITMS	ion trap mass spectrometer
K	equilibrium constant
$k(\epsilon)$	internal energy dependent rate constant
L	distance
l	path length
LC/MS	liquid chromatography/mass spectrometry
LD	laser desorption
m_d^+	daughter ion
m_{gd}^+	granddaughter ion
m_n	neutral fragment
m_p^+	parent ion, m_p represents the mass of the parent ion or molecule
MIKES	mass-analyzed ion kinetic energy spectrometry
MS/MS	mass spectrometry/mass spectrometry
N	target species, may also represent the mass of the target species
n	target gas number density
NI	nebulization ionization
ν	ratio of the product of the vibrational frequencies of the activated complex to that of the parent ion
PA	proton affinity
PD	photodissociation
Φ	number of degrees of freedom
Q	quadrupole mass filter
q	endothermicity of collision
R	gas constant
r	radius of deflection
RE	recombination energy
rf	radio frequency voltage
RR	reaction region
s	effective number of oscillators
SID	surface induced dissociation
SIM	single ion monitoring
SIMS	secondary ion mass spectrometry
SRM	single reaction monitoring
σ	total loss cross-section

σ_{CID}	CID cross-section
σ_{PD}	photodissociation cross-section
T	kinetic energy release
t	time
t_c	collision duration
T°	temperature in degrees Kelvin
T^r	kinetic energy release from ϵ^r
$T^{\neq}$	kinetic energy release from $\epsilon^{\neq}$
TOF	time-of-flight
θ_{CM}	scattering angle in the CM reference system
θ_{LAB}	scattering angle in the laboratory reference system
U_{mp}	velocity of the parent ion wrt the CM
U_N	velocity of the target wrt the CM
V	potential drop in volts
v	velocity
v_i	initial relative velocity
v_f	final relative velocity
$V(r)$	intermolecular potential where r is the distance between particles
W	Wien filter
W_0	total spread in E_{REL}
W_1	spread in E_{LAB}
W_2	spread in E_N
z	number of unit charges

Index